Foundations of Synergetics II

Springer
Berlin
Heidelberg
New York
Barcelona
Budapest
Hong Kong
London
Milan
Paris
Santa Clara
Singapore
Tokyo

Springer Series in Synergetics

Editor: Hermann Haken

An ever increasing number of scientific disciplines deal with complex systems. These are systems that are composed of many parts which interact with one another in a more or less complicated manner. One of the most striking features of many such systems is their ability to spontaneously form spatial or temporal structures. A great variety of these structures are found, in both the inanimate and the living world. In the inanimate world of physics and chemistry, examples include the growth of crystals, coherent oscillations of laser light, and the spiral structures formed in fluids and chemical reactions. In biology we encounter the growth of plants and animals (morphogenesis) and the evolution of species. In medicine we observe, for instance, the electromagnetic activity of the brain with its pronounced spatio-temporal structures. Psychology deals with characteristic features of human behavior ranging from simple pattern recognition tasks to complex patterns of social behavior. Examples from sociology include the formation of public opinion and cooperation or competition between social groups.

In recent decades, it has become increasingly evident that all these seemingly quite different kinds of structure formation have a number of important features in common. The task of studying analogies as well as differences between structure formation in these different fields has proved to be an ambitious but highly rewarding endeavor. The Springer Series in Synergetics provides a forum for interdisciplinary research and discussions on this fascinating new scientific challenge. It deals with both experimental and theoretical aspects. The scientific community and the interested layman are becoming ever more conscious of concepts such as self-organization, instabilities, deterministic chaos, nonlinearity, dynamical systems, stochastic processes, and complexity. All of these concepts are facets of a field that tackles complex systems, namely synergetics. Students, research workers, university teachers, and interested laymen can find the details and latest developments in the Springer Series in Synergetics, which publishes textbooks, monographs and, occasionally, proceedings. As witnessed by the previously published volumes, this series has always been at the forefront of modern research in the above mentioned fields. It includes textbooks on all aspects of this rapidly growing field, books which provide a sound basis for the study of complex systems.

A selection of volumes in the Springer Series in Synergetics:

A.S. Mikhailov   A.Yu. Loskutov

Foundations of Synergetics II

Chaos and Noise

Second Revised and Enlarged Edition
With 120 Figures

 Springer

Professor Alexander S. Mikhailov

Fritz-Haber-Institut, Max-Planck-Gesellschaft, Faradayweg 4-6
D-14195 Berlin, Germany

Professor Alexander Yu. Loskutov

Department of Physics, Lomonosov Moscow State University
117234 Moscow, Russia

Series Editor:

Professor Dr. Dr. h.c.mult. Hermann Haken

Institut für Theoretische Physik und Synergetik der Universität Stuttgart
D-70550 Stuttgart, Germany
and
Center for Complex Systems, Florida Atlantic University
Boca Raton, FL 33431, USA

Cip-data applied for.
Die Deutsche Bibliothek – CIP-Einheitsaufnahme.
Foundations in synergetics. – Berlin; Heidelberg; New York; Barcelona; Budapest; Hong Kong; London; Milano; Paris; Santa Clara; Singapur; Tokyo: Springer. 2. Complex patterns/A.S. Mikhailov; A.Yu. Loskutov. – 2., rev. and enl. ed. – 1996. (Springer series in synergetics; Vol. 52). ISBN-13:978-3-642-80198-3
NE: Michajlov, Aleksandr S.; GT

ISSN 0172-7389

ISBN-13:978-3-642-80198-3   e-ISBN-13:978-3-642-80196-9
DOI: 10.1007/978-3-642-80196-9

© Springer-Verlag Berlin Heidelberg 1991, 1996
Softcover reprint of the hardcover 2nd edition 1996

Typesetting: Data conversion by Springer-Verlag
SPIN 10125137   55/3144 - 5 4 3 2 1 0 - Printed on acid-free paper

Preface to the Second Edition

The second edition of this volume has been extensively revised. A different version of Chap. 7, reflecting recent significant progress in understanding of spatiotemporal chaos, is now provided. Much new material has been included in the sections dealing with intermittency in birth–death models and noise-induced phase transitions. A new section on control of chaotic behavior has been added to Chap. 6. The subtitle of the volume has been changed to better reflect its contents.

We acknowledge stimulating discussions with H. Haken and E. Schöll and are grateful to our colleagues M. Bär, D. Battogtokh, M. Eiswirth, M. Hildebrand, K. Krischer, and V. Tereshko for their comments and assistance. We thank M. Lübke for her help in producing new figures for this volume.

Berlin and Moscow
April 1996

A. S. Mikhailov
A. Yu. Loskutov

Preface to the First Edition

This textbook is based on a lecture course in synergetics given at the University of Moscow. In this second of two volumes, we discuss the emergence and properties of complex chaotic patterns in distributed active systems. Such patterns can be produced autonomously by a system, or can result from selective amplification of fluctuations caused by external weak noise.

Although the material in this book is often described by refined mathematical theories, we have tried to avoid a formal mathematical style. Instead of rigorous proofs, the reader will usually be offered only "demonstrations" (the term used by V. I. Arnold) to encourage intuitive understanding of a problem and to explain why a particular statement seems plausible. We also refrained from detailing concrete applications in physics or in other scientific fields, so that the book can be used by students of different disciplines.

While preparing the lecture course and producing this book, we had intensive discussions with and asked the advice of V. I. Arnold, S. Grossmann, H. Haken, Yu. L. Klimontovich, R. L. Stratonovich and Ya. B. Zeldovich. We wish to express our gratitude to all of them. We are especially grateful to Ya. G. Sinai who read the chapters on chaotic dynamics and made important comments. We also wish to thank our students who posed naive questions and made sharp criticisms, and thus contributed very much to the improvement of this text. Most of the graphic illustrations in this volume were produced by one of the authors (A. Yu. L.).

Moscow and Stuttgart
March 1991

A. S. Mikhailov
A. Yu. Loskutov

Contents

1. Introduction

Interactions between elements in distributed active systems can produce a great variety of different patterns, either stationary or time-dependent. The process of their spontaneous formation is known as self-organization. In the first volume of this book we described typical *coherent* structures in active media, such as self-propagating fronts and pulses, spiral waves, localized stationary dissipative patterns, etc. We also discussed the coherent behavior of neural and reproductive networks. The second volume deals mainly with much more complex *chaotic* patterns that are also found in distributed active systems.

The distinctive feature of a coherent pattern is the presence of long-range order. When such a pattern is fixed at one moment in time at one point in the medium, this uniquely determines it at all other points for all later moments. In contrast, chaotic patterns are characterized by short-range order with respect to time and/or spatial coordinates. When we fix the pattern at a given point at some moment, its behavior will thus be determined only within a short interval of time and/or in a small neighborhood of this point. For later moments and larger distances, the correlation is lost. Hence, chaotic patterns can be viewed as being composed of many uncorrelated small pieces.

The difference between coherent and chaotic patterns is well illustrated by an example from the physics of condensed matter. At thermal equilibrium condensed matter can exist in two basic forms – as a crystal or as a liquid. In crystals, all atoms occupy the sites of a perfect periodic lattice and the system possesses long-range order. In liquids, which are the systems with short-range order, regularity in the positions of atoms is maintained only within small domains; at larger separations the motion of atoms is not mutually correlated. Thus, broadly speaking, coherent patterns correspond to a "solid" state of a distributed active system, while chaotic patterns are "liquid" forms of its self-organization.

When long-range correlation in a distributed active system is broken in both time and space, this is the situation of *developed turbulence*. A further possibility is that a pattern remains coherent in space but varies chaotically with time (so that its dynamics is predictable only within a certain time interval). In this case, called *early* or *few-mode turbulence*, spatial long-range order is maintained while temporal long-range order is missing[1].

---

[1] A situation with long-range temporal order and short-range spatial order is also possible. The system then has a frozen chaotic spatial structure (usually produced by application of random initial conditions). This case corresponds to *amorphous solids*.

Chaotic time evolution of spatially coherent patterns in the regime of few-mode turbulence is usually described by simple sets of ordinary differential equations with only small numbers of dynamical variables. It turns out that the mathematical solutions to such equations can be so utterly unstable and unpredictable that they must be treated as random processes. The chaotic behavior (in time alone, or both in time and space) might thus be a perfectly autonomous property of an active system, and one that does not require any external random perturbations.

Chaotic patterns can also be produced by the process of *selective amplification* of fluctuations induced by some weak external noise. This mechanism operates when an active system, although it may have a perfectly regular dynamics, is extremely sensitive to small variations in the values of its parameters (this is usually oberserved near the points of transitions between two distinct steady regimes). Because of this sensitivity, the system responds strongly to noise provided by the environment. It enhances and nonlinearly transforms some of the spatial modes present in the noise, and suppresses the others. The result is the production of chaotically varying spatial patterns. Since the properties of such patterns are again determined by the active system itself, and depend only to a small extent on the properties of the applied noise, this process must also be considered as a self-organization phenomenon.

1.1 Chaotic Dynamics

It seems quite natural that large complex systems may demonstrate very complicated and unpredictable behavior. If we take a volume filled by a gas, motion of the molecules in this volume is obviously chaotic. Another standard example of chaotic dynamics is provided by turbulent hydrodynamical flows. In both cases we have extremely complicated systems formed by huge numbers of interacting elements.

For a long time, it was thought that system complexity was a necessary prerequisite for chaotic dynamics. However, in 1963 *Lorenz* [1.1] found a simple set of three first-order differential equations which demonstrated unpredictable behavior. These equations were derived for a distributed system in the study of thermal convection in a horizontal layer of liquid. To obtain them, an infinite chain of coupled equations for the amplitudes of different modes was truncated, so that finally the description used only three dynamical variables. In the same year *Sinai* [1.2] proved that even the motion of two balls in a box is chaotic, i.e. that the trajectories of colliding balls are absolutely unstable with respect to small perturbations.

These important results stimulated an intense search for small dynamical systems with chaotic dynamics. Within two decades such dynamical systems had been found in a great number of applications, even in classical fields which had seemed to be thoroughly understood (for instance, *Huberman* et al. [1.3] noted that the motion of a periodically forced pendulum can be chaotic). It turned out that chaotic systems are "typical": they represent a finite fraction in the set of all conceivable dynamical systems. Such developments were accompanied by rapid progress in the mathematical theory of chaotic dynamics, particularly the concept of a *strange attractor* introduced by *Ruelle* and *Takens* [1.4] in 1971.

Although there are several good textbooks on chaotic dynamics, e.g. [1.5–7], they are not quite suitable for certain categories of students. These books have a special emphasis on physical applications and thus require much preliminary knowledge in the field of physics. They might also be too detailed for a first acquaintance with the subject.

In Chaps. 1–7 we give a concise elementary introduction to chaotic dynamics in small systems (and, hence, to the theory of few-mode turbulence). We begin with a brief discussion of the Hamiltonian systems studied in classical mechanics. Although they constitute a very special class of models, one should know some fundamental aspects of their behavior. Indeed, it was in the context of such systems that many important notions of the modern mathematical theory of chaotic dynamics were developed.

The nonspecialists often state that the chaotic dynamics of a system is equivalent to ergodicity of its behavior. We show in Chap. 2 that this claim is erroneous. As a matter of fact, short-range temporal order (i.e. fading of temporal correlations) is ensured only by the property of *mixing*, which also implies divergence of trajectories and instability of dynamics.

The principal definitions relevant for general active dynamical systems are formulated in Chap. 3. Here the concept of a strange attractor is first explained and explicit examples are given. Furthermore, we indicate practical criteria that can be used to identify the chaotic dynamics of a particular model.

The precise mathematical description of strange attractors involves the geometric concept of *fractal sets*. This general concept is introduced in Chap. 4, where we give several typical examples of fractal geometric patterns and define the family of fractional *dimensions* which provide a qualitative measure for such complex patterns. At the end of Chap. 4 we show how the fractional dimension of a strange attractor can be estimated.

Dynamical systems with continuous time are intimately related to a simpler class of models where time is discrete. Such models are known as *iterative maps*; they are discussed in Chap. 5. We formulate the conditions under which discrete dynamics can be chaotic and provide a few concrete examples. We also approach, for the first time, the problem of transition from regular to chaotic dynamics. It turns out that in many cases, such transition has universal properties which depend only weakly on a particular map.

Basic scenarios of transitions to temporal chaos in dynamical systems with continuous time are considered in Chap. 6. We further discuss in this chapter how chaotic dynamics can be controlled, i.e., suppressed, by applying corrective perturbations or introducing additional feedbacks.

As already noted, models involving a small number of dynamical variables can also be used to describe the early stage of turbulence in distributed active systems, where the behavior of such systems remains coherent in space but has already lost long-range temporal order. The chaotic temporal dynamics of a spatially coherent pattern can often be effectively described by a model involving only a few variables which correspond to different "degrees of freedom" of the pattern. Such variables can also be considered as the "order parameters" of such a pattern.

Remarkably, there are some numerical methods which allow us to determine, proceeding from the experimental time series of data, the minimal number of independent variables that should be employed in the construction of an effective dynamical model. In other words, these methods allow us to find the number of "degrees of freedom" in the observed behavior. We discuss some such methods in the first section of Chap. 7.

The transition from few-mode early turbulence to developed turbulence in distributed active systems is a phenomenon that is now being intensively investigated. A detailed numerical study was performed by *Kaneko* [1.8] who considered systems of coupled maps with chaotic dynamics. The classical example of turbulence in a distributed active system with continuous variables is provided by the complex Ginzburg–Landau equation and the closely related Kuramoto–Sivashinsky equation. Essential contributions to studies of spatio-temporal chaos in these systems have been made by *Kuramoto* [1.9] and by *Coullet* et al. [1.10].

In many respects, the transition to developed turbulence looks like the melting of a crystal. It is accompanied by rapid accumulation of defects and dislocations that destroy long-range spatial order. These defects behave as individual particles. They can die or reproduce in the medium. These processes are well described by stochastic reaction models.

After a transition to developed turbulence, the medium breaks down into an ensemble of subsystems (i.e., small spatial domains with the size of the correlation radius). At this stage one can treat the action of all other elements of the medium on a given subsystem as the action of a stochastic *environment*. Such an effective environment is only weekly coupled to the subsystem and its influence can often be modeled by introducing an external *noise*.

1.2 Noise-Induced Complex Patterns

As we have pointed out, there are many situations when the behavior of a dynamical system is intrinsically chaotic, so that no external noise is required to produce unpredictable (or partially predictable) dynamics. However, the development of large-scale complex patterns may be observed also as a result of selective amplification of some fluctuation modes generated by weak noise. These phenomena are discussed in Chaps. 8–12.

Our analysis is preceded in Chaps. 8 and 9 by a brief outline of the general mathematical theory of random processes, which provides the framework for studies of noise-induced complex patterns. This part of the text is not intended for use as an introduction to this advanced mathematical theory (which is fully explained in e.g. [1.11–18]). Here we wish merely to bring together some practical notions and definitions.

The simplest examples of noise-induced complex patterns are found in "explosive" systems that evolve from an unstable state. In the process of such an evolution, amplitudes of many fluctuation modes increase. Since the rates of growth are generally different, there will be a relative enhancement of some fluctuations at the expense of others. If the most rapidly growing fluctuations have large characteristic length scales, this process will lead to the creation of macroscopic random patterns in the unstable medium.

Such a situation is typical in cosmology, and occurred during the very first moments of the existence of the Universe after the Big Bang. Similar effects are found in chain explosions that can be described in their initial stage by simple linear models, which involve breeding (reproduction) of some active particles, occurring in the otherwise inert medium. We analyze such processes in Chap. 10.

Surprisingly, the relative enhancement of fluctuations is observed not only in systems with reproduction of active particles, but also in systems where the total number of particles *decreases* with time. An example of this is provided by the binary annihilation reaction. Let any two active particles, A and B, annihilate when they meet in the process of a diffusional random walk. The total number of particles A and B in the medium will then decrease with time, but the *difference* between these numbers will remain constant. If there is initially an excess of particles A in the system, then in the final state, after all particles B have been annihilated in collisions, only particles A will persist.

Even if at the outset the total numbers of particles A and B in the medium are the same, their balance can be locally shifted due to random statistical variations in the concentrations of both. In other words, there can be regions with an initial excess of particles A, and other regions with an excess of particles B. After a rapid annihilation, these regions will be occupied by the particles of a single type (A or B). Therefore, the relative fluctuations in the concentrations of particles will be enhanced, leading to the formation of a complicated spatial pattern. If the particles wander slowly in the medium, they can penetrate into neighboring regions that are occupied by particles of an opposite type. As a result, annihilation will be maintained at the boundaries. In the evolution of the Universe, this mechanism may have been responsible for the creation of regions filled exclusively by matter and antimatter. We discuss phenomena related to the formation of large-scale complex patterns in the distribution of reagents in binary annihilation or recombination reactions in Chap. 11.

In general, distributed active systems become extremely sensitive to the action of noise near the points of *catastrophes*[2] and nonequilibrium *phase transitions*, where the system organization undergoes a qualitative change. We consider these phenomena in Chap. 12.

---

[2] This term is not meant to imply any negative or destructive consequences. Catastrophes might actually be quite beneficial to a developing system.

Catastrophes represent crises of attractors. When the control parameter reaches the critical point of a catastrophe, one of the system's attractors completely disappears. When this occurs, the system can no longer reside on the previous attractive set. It starts to move, exploring the phase space until a new attractor is found. As a consequence, a sudden qualitative change in the established behavior of the system is produced. Usually, a catastrophe can occur earlier if a sufficiently strong perturbation is applied. Hence, the principal effect of external noise consists here in the early triggering of a catastrophe. *Thom* [1.19] and *Arnold* [1.20] have proposed a classification of different catastrophes for special case of potential dynamical systems.

In contrast to catastrophes, soft transitions are not accompanied by sudden and drastic changes in the state of a system. Rather, the system arrives at a crossroads where new vistas become accessible. At the point of such a transition, the path splits into several distinct branches. They correspond initially to states which differ only slightly. However, the separation between the branches gradually increases with distance from the transition point, so that eventually they give rise to entirely different qualitative regimes. Since the system itself does not 'know' which further path to choose at the point of a soft transition, any external noise or a perturbation can play an important role in selecting a particular branch of development.

The classical example of this phenomenon is a *second-order phase transition* which represents the branching of a steady uniform state. The fluctuation phenomena occurring in the vicinity of such a transition are analyzed in the first section of Chap. 12. In the next two sections, we consider the process of sweeping through the critical region and the effects of a steady external bias.

A special kind of a soft transition is found when a fluctuating medium becomes populated by reproducing and decaying particles (Sect. 12.4). We conclude Chap. 12 with the analysis of an ecological model with two competing populations where a noise-induced phase transition takes place.

1.3 Chaos, Noise, and Self-Organization

It is sometimes believed that chaos is the exact opposite of organization. For many years the most favorable mathematical example of chaotic behavior has been noise produced by the superposition of weak signals from many independent random sources. Such structureless chaos is indeed found in certain applications. However, the chaotic patterns produced by *synergetic* systems bear an intricate inner organization determined by interactions among the elements of the system.

As already noted, complex chaotic patterns in distributed active systems maintain partial coherence within correlated spatio-temporal domains. Inside such regions, the behavior of a system remains fairly regular and predictable. On the other hand, very interesting self-organization phenomena in chaotic distributed active systems can be observed on large spatial scales. In the fully developed turbulence regime, where only short-range order in space and time is preserved, a distributed active medium breaks down into a collection of weakly interacting units, each corresponding to a correlated domain. These new units can play the role of elements of a next level in the structural hierarchy. Interactions between them can produce large-scale coherent patterns which belong to a higher hierarchical level.

The building up of a hierarchical structure is most evident in fluid dynamics. We know that any fluid consists of individual molecules which move along unpredictable irregular trajectories. Nevertheless, the chaotic motion of single molecules is not seen in the properties of a "physically small" element of the fluid that includes a large number of molecules. Effective self-averaging takes place, and we can specify this element by a few principal variables such as density, pressure, and temperature. In the interactions between separate fluid elements, only these macroscopic variables are significant.

Thus, despite the inherent microscopic chaoticity of fluid motion, coherent patterns of flow (such as stable vortices) can be observed in the laminar regime corresponding to the low Reynolds numbers. These deterministic coherent patterns belong to the "macroscopic" level of the structural hierarchy. At high Reynolds numbers, the patterns of flow become chaotic with respect to both space and time. The resulting turbulent state can be viewed as a system of eddies of different sizes. The smaller eddies are captured by the larger ones and are thus incorporated into their flow.

At the largest spatial scales one can sometimes observe very regular and predictable patterns. To describe such "supermacroscopic" phenomena, one can again use simple models of continuous media where small-scale turbulence is taken into account by using the renormalized constants of diffusion, viscosity, etc.

Hence, developed hydrodynamical turbulence usually has a very complicated hierarchical nature and consists of many interacting and partially coherent patterns of various sizes. Similar behavior should be observed in other distributed active systems. Thus, self-organization in media with locally chaotic dynamics may be operating not only within small correlated domains but also at much larger scales. The patterns produced at this stage may be coherent and predictable at larger scales, while at smaller characteristic lengths their properties will be chaotic. It might further happen that, at still larger scales, the coherence is again missing and chaotic behavior is observed, giving way to a new twist in the structural hierarchy. We see that, when perfect long-range order is lost, the medium effectively transforms from a rigid to a more flexible state, which opens up a variety of new modes of its organization.

When dealing with complex patterns, it is useful to have some qualitative criterion (or quantitative measure) of their *complexity*. Obviously, a pattern would be simple if it was constructed by regular repetition of identical elements (as, for example, the crystalline lattice). This implies that complexity should somehow be related to the diversity of a pattern. At this point, however, an important remark should be made. If we consider a completely random pattern, produced by the superposition of many independent weak noises, it appears complicated but actually is still very simply organized. Indeed, if we look at this pattern with a lower resolution, so that the "microscopic" details are washed out, we would see that it is nearly uniform.

This remark indicates that the true criterion of complexity cannot be obtained without analysis of the hierarchical nature of a complex pattern. It seems that we should examine the pattern many times at different increasing scales and determine the degree of diversity at each of the hierarchical levels. The simplest patterns are uniform even at the lowest level. Slightly more complex are patterns that only become uniform a few steps higher in the structural hierarchy.

An important class of self-similar *fractal* patterns (which we discuss briefly in Chap. 4) has the property of reproducing the same unique structure at each magnification. Since the structure is never washed out and persists at all scales, such patterns are characterized by higher complexity. Much more complex patterns are also possible; at each level of the hierarchy they would have a large diversity of observed elementary structures.

Using the above arguments, one can qualitatively compare the complexity of different patterns. However, it is not easy to cast them into a quantitative form, thus providing a rigorous *measure* of complexity for an arbitrary pattern. An important step in this direction was made by *Ceccatto* and *Huberman* [1.21] who constructed the complexity measure for discrete trees.

The emergence of a hierarchy of *coherent* patterns in the process of complex self-organization can play an important role in the advanced functional behavior of natural and artificial living systems. However, an essential requirement for the successful operation of living systems might also be the presence of some internal *noise sources*. Because of noise generation, a system might gain the property of *creativity* which allows its effective adjustment to environmental changes.

We know that random mutations are crucial for biological evolution. Because of them, the seeds of new biological species that enter into the competition are continually being produced. The progressive development of life and the emergence of new genetic information are thus intimately related to stochastic processes.

Molecular genetic mutations are induced by external factors (such as radiation) or represent errors of replication. However, at higher structural levels, similar effects may well result from the intrinsic dynamical chaoticity of certain subsystems.

Similar effects might be involved in the operation of human societies. A limited degree of instability is beneficial for a society; it should not suppress too severely the fluctuations that result in the emergence of new public structures. Instead, conditions must be provided for the lawful competition of arising patterns. It is this process that can guarantee the continual adjustment of a society organization to the challenges of new technology and environmental variations. Thus, in flexible self-organized social systems, fluctuations represent an important factor in their progressive development.

The creative ability of humans, which is considered a central feature of human intelligence, might ultimately be a consequence of chaotic behavior in certain regions of the brain. The successful operation of the brain may require the presence of a random source to generate a flow of "action plans" or "ideas". Once produced, these "ideas" become contestants in the mental competition which reproduces the conditions of the outside world. The surviving "idea" is eventually accepted as the valid action plan. It seems that in most cases, the entire process remains unconscious. Only if the best "idea" cannot be worked out at the unconscious level and more than one potential candidates persist, do these ideas penetrate the threshold of consciousness and the final decision is made using the mechanisms of rational thought.

In psychological terms, sensitivity to small variations in the outside world and the spontaneous creative activity of an individual might be viewed as aspects of its "artistic" behavior.

Hence we see that complex patterns produced by distributed active systems can play an important role in the self-organization of purposeful behavior. To analyze such processes in realistic systems and to engineer them in artificial devices, one needs, however, a much more complete knowledge of the basic mathematical models that describe the fundamental features of these phenomena. The discussion of some such basic models is the aim of this book.

2. Unpredictable Dynamics

At first glance it would appear that, when the dynamical equations of a system are known, we are able to accurately predict its state at any future moment in time. A closer examination, however, reveals the counter-example of molecular motion in gases. Although all the equations of motion of individual molecules and the laws of their collisions are known in this case, it is useless to solve these equations in an attempt to predict the precise positions and velocities of molecules at some future moment. The deterministic prediction fails in this case because of the extreme sensitivity of such a system to small variations in its initial conditions. The slightest perturbation in the coordinates and velocities of the molecules is sufficient to completely change their motion.

Hence, even within the realm of classical physics there are examples of systems with unpredictable dynamics. However, the existence of such behavior was traditionally attributed to the fact that these systems were very large (e.g., each cubic centimeter of gas contains about 10^{23} molecules). Therefore, about two decades ago, it came as a great surprise when numerical simulations revealed that even *small* dynamical systems are able to produce unpredictable behavior, i.e. that their dynamics might be very sensitive to slight perturbations. Today we have an elaborate theory of such phenomena. It turns out that, under certain conditions, a solution of a purely deterministic system of ordinary differential equations may represent a random process.

We begin our discussion of dynamical chaos with a study of mechanical Hamiltonian systems. Although they are described by a special class of mathematical models, many general properties and concepts were originally introduced in relation to these systems. A minimum knowledge of these models is essential for any student.

The main aim of this chapter is to formulate the concepts of ergodicity and mixing. We show that the presence of mixing destroys long-term correlations and leads to unpredictable dynamics.

2.1 Hamiltonian Systems

Mechanical motion is described by differential equations of a special form, namely

$$\dot{q}_i = \frac{\partial H}{\partial p_i}\,, \quad \dot{p}_i = -\frac{\partial H}{\partial q_i}\,. \tag{2.1.1}$$

Here the variables $q = \{q_i\}$ and $p = \{p_i\}$ are the (generalized) coordinates and (generalized) momenta ($i = 1, 2, \cdots, n$). The total number n of generalized coordinates is called the number of degrees of freedom. A mechanical system is specified by its Hamiltonian function $H = H(q, p)$ which constitutes the energy of the system.

The analysis of various Hamiltonian systems represents a separate branch of physics. There are many books, e.g. [2.1–4], where all significant details of this analysis are given. In the present chapter, we use these systems only as illustrations of some general concepts and ideas. However, to make our discussion self-consistent, we cannot avoid several special definitions and remarks concerning the properties of Hamiltonian motion.

The solution to (2.1.1) determines the dependence of q and p on time for a given choice of the initial conditions q_0 and p_0. It can be visualized as the motion of a point along some phase trajectory in the $2n$-dimensional phase space with the coordinates q and p. Note that no two phase trajectories can intersect (because the initial conditions uniquely determine the entire solution).

From a general point of view, the Hamiltonian systems (2.1.1) belong to the class of conservative systems: they neither compress, nor inflate the phase-space volume. This statement, known as the Liouville theorem, follows immediately from a general analysis to be presented in Sect. 3.1. It can also be shown that any initial closed region D_0 in the phase space will continuously (i.e. without breaking) transform in the course of time t into some other closed region D_t (Fig. 2.1).

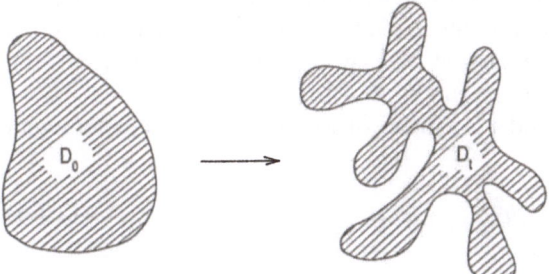

Fig. 2.1. Conservation of phase-space volume in the evolution of a Hamiltonian system

The integrals of motion are the functions $f(q, p)$ which remain constant along phase-space trajectories of a Hamiltonian system. When the system is autonomous, i.e. $H(q, p)$ does not depend explicitly on time t, it should possess at least one integral of motion, i.e. its energy $E = H(q, p)$. All trajectories with a given value of energy E belong to the same energetic hypersurface of dimensionality $2n - 1$ in phase space. If the phase-space trajectories do not go to infinity (at a given energy), the motion is called finite. Only finite motion is considered below.

Using so-called canonical transformations, it is possible to go from one set (q, p) of generalized coordinates and momenta to another set (q', p'), in such a way that the form of (2.1.1) remains unchanged. There is a class of Hamiltonian systems that permit the construction of a special canonical transformation $(q, p) \rightarrow (\alpha, J)$, with the function H independent of new generalized coordinates α. Then (2.1.1) will simply read

$$\dot{\alpha}_i = \frac{\partial H}{\partial J_i}, \quad \dot{J}_i = 0, \quad i = 1, 2, \cdots, n . \tag{2.1.2}$$

These equations can be easily integrated, yielding

$$\alpha_i = \alpha_i^0 + \omega_i(J)t, \quad J_i = \text{const}, \tag{2.1.3}$$

where $\omega_i = \partial H/\partial J_i$. Generalized coordinates α_i are *angles*[1] and generalized momenta J_i are *actions*. If a Hamiltonian system permits a canonical transformation to the action–angle variables, such that the transformation yields $H = H(J)$, this system is called (completely) integrable.

A Hamiltonian system with a single degree of freedom ($n = 1$) is always integrable since we can choose its energy $E = H(q, p)$ as an action variable. The dynamics of such a system in the space (α, J) is most conveniently described using polar coordinates. Then J is measured as the distance from the origin of the coordinates and α as the polar angle (Fig. 2.2). For a fixed value of J, the point portraying the system moves along a curve representing a circle of radius J. Varying the radius, we obtain a set of concentric circles that fill the entire "phase space" of this dynamical system. Note that generally (for a nonlinear oscillator), the velocity of motion along any particular circle depends on its radius, i.e. $\omega = \omega(J)$.

The phase space of a system with two degrees of freedom is four-dimensional. After transformation to the action–angle variables (when such a transformation exists!), the dynamics at fixed action variables J_1 and J_2 can be viewed as the motion of the phase-space point over the surface of a two-dimensional torus (Fig. 2.3), with the trajectory winding around the torus.

As a simple illustration, consider a system of two harmonic oscillators with the Hamiltonian function

$$H = \tfrac{1}{2} \left(p_1^2 + p_2^2 \right) + \tfrac{1}{2} \left(\omega_1^2 q_1^2 + \omega_2^2 q_2^2 \right) . \tag{2.1.4}$$

Its dynamical equations are

$$\dot{q}_i = p_i, \quad \dot{p}_i = -\omega_i^2 q_i, \quad i = 1, 2 . \tag{2.1.5}$$

The canonical transformation to the action–angle variables has the form

---

[1] When the motion is finite, the coordinates q_i and momenta p_i are 2π-periodic in terms of angles α_i.

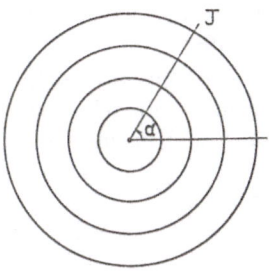

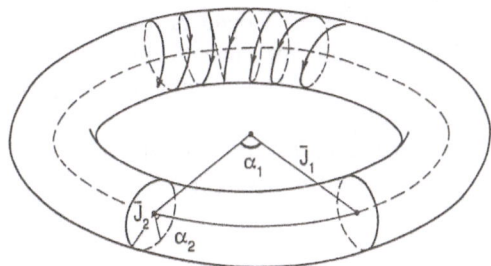

Fig. 2.2. Phase space of a Hamiltonian system with one degree of freedom in action–angle variables

Fig. 2.3. A toroidal surface in the phase space of an autonomous Hamiltonian system with two degrees of freedom, obtained by fixing of two action variables

$$p_i = (2\omega_i J_i)^{1/2} \cos \alpha_i \, ,$$

$$q_i = -\left(\frac{2J_i}{\omega_i}\right)^{1/2} \sin \alpha_i \, , \quad i = 1, 2 \tag{2.1.6}$$

It yields

$$H = \omega_1 J_1 + \omega_2 J_2 \, . \tag{2.1.7}$$

A general solution of the dynamical equation in terms of the action–angle variables can be written as

$$J_1 = J_1^0 \, , \quad \alpha_1 = \omega_1 t + \alpha_1^0 \, , $$
$$J_2 = J_2^0 \, , \quad \alpha_2 = \omega_2 t + \alpha_2^0 \, , \tag{2.1.8}$$

By convention, we measure the actions J_1 and J_2 as the inner and cross-section radii of the torus, and α_1 and α_2 as the angles on its surface (Fig. 2.3). Then the entire four-dimensional phase space of such a system will be split into a set of enclosed two-dimensional tori (the three-dimensional projection of this set is shown in Fig. 2.4).

In effect, a similar picture holds for an arbitrary integrable Hamiltonian system with two degrees of freedom. Its phase-space trajectories wind over a set of two-dimensional concentric tori. The only difference is that, generally, the rota-

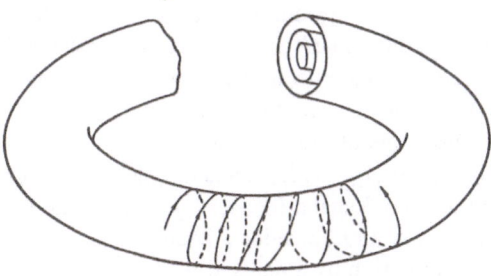

Fig. 2.4. The phase space structure of an integrable Hamiltonian system with two degrees of freedom in action–angle variables

tion frequencies depend on the actions, i.e. $\omega_1 = \omega_1(J_1, J_2)$ and $\omega_2 = \omega_2(J_1, J_2)$. Therefore, the angular velocities change from one torus to another.

The behavior of trajectories on a given torus is essentially different depending on whether the ratio $\omega_1/\omega_2 = \omega_1(J_1, J_2)/\omega_2(J_1, J_2)$ represents a rational or an irrational number. If it is rational, i.e. $\omega_1/\omega_2 = k/m$ where k and m are some integers, we find from (2.1.8) that

$$\alpha_1(t + T) = \alpha_1(t) + 4\pi k \ ,$$
$$\alpha_2(t + T) = \alpha_2(t) + 4\pi m \ , \qquad\qquad (2.1.9)$$

for $T = 2\pi(k/\omega_1 + m/\omega_2)$. Since values of the angle variables which differ by an integer multiple of 2π are identified, this implies that after time T the trajectory returns to the same point on the torus surface where it was at time t.

Consequently, for such commensurable frequencies ω_1 and ω_2, the trajectory represents a closed curve on the torus. On the other hand, when ω_1/ω_2 is irrational (i.e. frequencies are *incommensurable*), the phase trajectory would never close. Instead it winds over the torus, passing arbitrarily close to every point if we wait sufficiently long. Such motion is called *quasiperiodic*.

Finally, we can discuss the general case of an integrable Hamiltonian system with n degrees of freedom. In this case the phase space is $2n$-dimensional; in the action–angle variables it splits into a set of n-dimensional tori. Every possible trajectory belongs to one of these tori. Some of the trajectories are closed, while others densely cover the entire surface of their corresponding torus.

An n-dimensional torus with given actions $J_1, \cdots, J_n$ is *resonant* if its frequencies $\omega_1(J_1, \cdots, J_n), \cdots, \omega_n(J_1, \cdots, J_n)$ satisfy the condition

$$\sum_{i=1}^{n} k_i \omega_i(J_1, \cdots, J_n) = 0 \qquad\qquad (2.1.10)$$

for some choice of nonvanishing integers k_i.

When a torus is not resonant, it corresponds to a quasiperiodic motion, with the phase trajectory densely covering its entire surface. Since the rotation frequencies $\omega_1, \cdots, \omega_n$ depend continuously on the actions $J_1, \cdots, J_n$, by varying the latter we can go from a resonant torus to a nonresonant one, then again to a resonant torus, and so on.

An important concept often used in studies of dynamical systems is the Poincaré map. To define it, we first consider a system with only two degrees of freedom. In this case, if the Hamiltonian function H does not explicitly depend on time, each trajectory belongs to a certain energetic hypersurface in the four-dimensional phase space (q_1, q_2, p_1, p_2). Using the equation $H(\boldsymbol{q}, \boldsymbol{p}) = E$, which determines this hypersurface, we can express one of the variables (for instance, p_2) as a function of the other three, i.e. $p_2 = p_2(q_1, q_2, p_1, E)$. Hence, it is enough to consider the trajectories in the three-dimensional space (q_1, q_2, p_1). Let us choose in this space some two-dimensional surface S and consider a sequence of points where this surface is intersected in certain direction by a phase trajectory (Fig. 2.5).

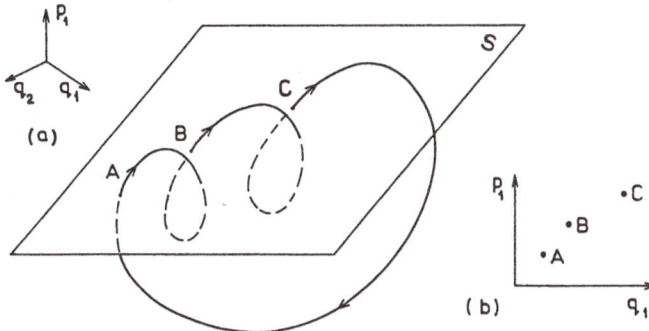

Fig. 2.5a,b. Construction of the Poincaré map for a Hamiltonian system with two degrees of freedom

In the process of motion, any point A of intersection of the surface S with the phase trajectory is mapped into some other point $B = \Phi(A)$ on the same surface. This function Φ, which generates the sequence of intersection points $B = \Phi(A)$, $C = \Phi(B)$, etc., is called the *Poincaré map*. Sometimes it is convenient to choose S as a piece of a plane; then Φ maps this piece of the plane into itself.

The concept of the Poincaré map can be generalized to systems with a higher number of degrees of freedom $(n > 2)$. In this case, the energetic hyperspace has $2n - 1$ dimensions. After one of the variables is excluded, as was done above, we can consider the subsequent points of intersection of a phase trajectory with some $(2n - 2)$-dimensional hypersurface S. If we exclude the case in which a loop of the trajectory meets S in a single point, i.e. where S is locally tangential to a point on the trajectory, then we can again define a map which relates each point M to the next point M' at which the phase trajectory passes through the same hypersurface.

The use of Poincaré maps significantly simplifies the investigation of Hamiltonian systems. If the map Φ is known (or we know the pattern of traces left by phase trajectories on the surface S), this allows us to draw important conclusions about the behavior of a system. Indeed, when the Poincaré map generates a finite sequence of points (for example, $A \to B \to C \to A$, as shown in Fig. 2.5), this indicates a closed trajectory in phase space and, hence, the presence of periodic motion. If the set of points generated by a Poincaré map fills some closed curve on the surface S, this manifests quasiperiodic motion (Fig. 2.6). Finally, there are also systems where a trajectory intersects the surface S in a "random" set of points that seems apparently irregular. Then the trajectory wanders chaotically through phase space. Such behavior differs from both periodic and quasiperiodic motion.

We illustrate the latter possibility by an example proposed in 1964 by *Henon* and *Heiles* [2.5]. They studied the problem of the motion of a star in the mean gravitational field of a galaxy. After a number of simplifications taking into account the integrals of motion, this problem was reduced to the dynamics of a system with two degrees of freedom described by a Hamiltonian function

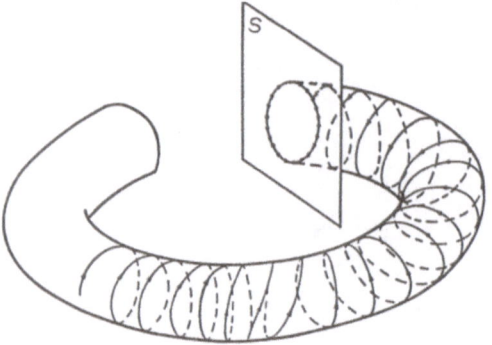

Fig. 2.6. The Poincaré map of the quasiperiodic motion

$$H = \tfrac{1}{2}\left(p_1^2 + p_2^2\right) + \tfrac{1}{2}\left(q_1^2 + q_2^2\right) + q_1^2 q_2 - \tfrac{1}{3}q_2^3 . \qquad (2.1.11)$$

If this system were completely integrable, it would have allowed a canonical transformation to the action–angle variables which would yield $H = H(J_1, J_2)$. Hence at fixed J_1 and J_2 (or E and J_2), its trajectory in phase space (q_1, q_2, p_1, p_2) would have belonged to a surface topologically equivalent to a torus. Generally, such a surface would have been densely covered by a trajectory and the points of the Poincaré map would have formed some closed curve in the cross-section.

Since this system conserves energy $E = H(q_1, q_2, p_1, p_2)$, we can exclude one of the variables, $p_1 = p_1(q_1, q_2, p_2, E)$.

Henon and *Heiles* carried out numerical simulations of the system (2.1.11). They marked the subsequent points where a phase trajectory crossed the coordinate plane (q_2, p_2), repeating this procedure for different values of energy E. The results were quite surprising. At sufficiently small energies, the sets of points generated by the Poincaré map looked like a system of closed curves, contained one within another, as would have been expected for quasiperiodic motion on the surface of a torus (Fig. 2.7a).

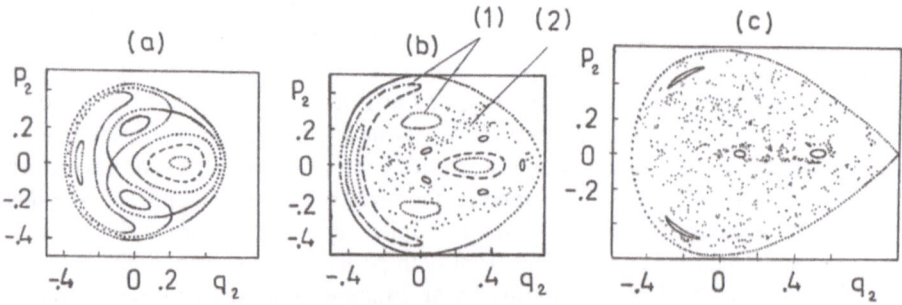

Fig. 2.7. Poincaré maps for the Henon–Heiles model: (a) $E = 0.04$, (b) $E = 0.125$, (c) $E = 0.167$. From [2.5]

When the energy E was increased, this picture underwent significant changes. Some invariant curves started to disintegrate and formed a dense set of points that were apparently irregularly scattered over the plane (q_2, p_2). For instance, at $E = 0.125$ in this cross-section one observes both invariant closed curves (1) and the set consisting of points (2) chaotically scattered between these curves (Fig. 2.7b). In other words, at this energy value, the system had the domains of regular (1) and irregular (2) behavior in its phase space. For still larger energy values, the area occupied by tori shrank to the tiny regions (Fig. 2.7c). Outside these very small regions, a trajectory randomly visited all parts of the energetic hypersurface. At very high energies the motion in the Henon-Heiles system ceased to be finite.

Further studies of the Henon-Heiles system showed (see [2.6]) that even at small energies (Fig. 2.7a) it has, besides regular trajectories, some very narrow (exponentially small in E^{-1}) layers where the phase trajectories are chaotic. However, since these layers are extremely narrow, their presence cannot be detected by numerical simulations. *Holmes* [2.7] proved analytically that the Henon-Heiles system is never strictly integrable; it possesses no additional integral of motion.

Completely integrable systems actually represent something of an exception. In many cases, however, a system is close to some completely integrable one, so that the difference of their Hamiltonian functions may be treated as a small perturbation. This situation is discussed in the following section.

2.2 Destruction of Tori

Suppose that, by finding the appropriate variables (α, J), we can represent the Hamiltonian function H of a system as

$$H(\alpha, J) = H_0(J) + \varepsilon H_1(\alpha, J) , \qquad (2.2.1)$$

where ε is a small parameter ($\varepsilon \ll 1$). If a system allows such representation it is *nearly integrable*.

For example, the Hamiltonian function (2.1.11) of the Henon-Heiles system can be represented in the form (2.2.1) with

$$H_0 = \tfrac{1}{2} \left(p_1^2 + p_2^2 \right) + \tfrac{1}{2} \left(q_1^2 + q_2^2 \right) , \qquad (2.2.2)$$

$$\varepsilon H_1 = q_1^2 q_2 - \tfrac{1}{3} q_2^3 . \qquad (2.2.3)$$

Although the right-hand of (2.2.3) includes no explicit small parameter, at small energies it is proportional to $E^{1/2}$ and can indeed be treated as a perturbation [2.6]. Hence, at small energies the Henon-Heiles system is nearly integrable.

The equations of motion of an nearly integrable system in the variables (α, J) are

$$\dot{J}_i = -\varepsilon \frac{\partial H_1}{\partial \alpha_i} \,,$$

$$\dot{\alpha}_i = \frac{\partial H_0}{\partial J_i} + \varepsilon \frac{\partial H_1}{\partial J_i} \,, \quad i = 1, 2, \cdots, n \,. \tag{2.2.4}$$

When $\varepsilon = 0$ the system is completely integrable and its trajectories cover n-dimensional tori. How is this motion modified by a small perturbation that makes the system nonintegrable?

The discussion of integrable systems in Sect. 2.1 showed that at given values of actions $J_1, \cdots, J_n$ they behave as a set of oscillators with frequencies $\omega_1(J_1, \cdots, J_n), \cdots, \omega_n(J_1, \cdots, J_n)$. The perturbation H_1 introduces a weak interaction between these oscillators. Because of resonance effects, the result of such an interaction is very sensitive to the relationships between the oscillation frequencies.

Let us begin the analysis with the simplest problem of the resonance of a single nonlinear oscillator under the action of a weak periodic *external* perturbation:

$$H = H_0(J) + \varepsilon h_{km} \cos(k\alpha - m\nu t) \,, \tag{2.2.5}$$

where h_{km} is independent of J, k and m are some integers; $\varepsilon \ll 1$. It is convenient to introduce the phase $\phi = k\alpha - m\nu t$. Then the equations of motion in the variables J and ϕ are

$$\dot{J} = \varepsilon h_{km} k \sin \phi \,, \quad \dot{\phi} = k\omega(J) - \nu m \,. \tag{2.2.6}$$

The resonance occurs when $k\omega(J) = m\nu$. Since the oscillation frequency $\omega(J)$ depends on the action J, this condition determines a certain J_0, such that $\omega(J_0) = m\nu/k$. Let us suppose that the deviation of J from J_0 is sufficiently small and expand $H_0(J)$ and $\omega(J)$ in terms of $\Delta J = J - J_0$:

$$\omega(J) = \omega(J_0) + \frac{\partial \omega}{\partial J} \Delta J + \frac{1}{2} \frac{\partial^2 \omega}{\partial J^2} (\Delta J)^2 + \cdots \,,$$

$$H_0(J) = H_0(J_0) + \frac{\partial H_0}{\partial J} \Delta J + \frac{1}{2} \frac{\partial^2 H_0}{\partial J^2} (\Delta J)^2 + \cdots \,. \tag{2.2.7}$$

Then the equations of motion (2.2.6) become

$$\Delta \dot{J} = \varepsilon h_{km} k \sin \phi \,, \quad \dot{\phi} = k \left(\frac{d\omega}{dJ} \right)_0 \Delta J \,, \tag{2.2.8}$$

where the derivative $(d\omega/dJ)_0$ is taken at $J = J_0$.

The effective dynamical system (2.2.8) can be described by a *universal Hamiltonian function* of the nonlinear resonance

$$\bar{H} = \frac{1}{2} k \left(\frac{d\omega}{dJ} \right)_0 (\Delta J)^2 + \varepsilon h_{km} k \cos \phi \,. \tag{2.2.9}$$

Actually, it coincides with the Hamiltonian function of a pendulum. Indeed, if we identify ΔJ with the momentum p of the pendulum and $\Phi = \phi - \pi$ with its angle, the dynamical equations (2.2.8) are equivalent to

$$\frac{d^2\Phi}{dt^2} = -\Omega^2 \sin \Phi \,, \tag{2.2.10}$$

where $\Omega^2 = \varepsilon h_{km} k^2 (d\omega/dJ)_0$. When the energy E is small, the pendulum performs small harmonic oscillations with period $2\pi/\Omega$ near its rest state $\Phi = 0$. If we increase E, deviations from the harmonic form appear and the oscillation period begins to grow. The oscillation period diverges when the energy approaches the value $E = \Omega^2$, which corresponds to a separatrix trajectory (this trajectory includes the unstable upper stationary position $\Phi = \pi$ of the pendulum). For higher energies, the oscillations are replaced by pendulum rotations. The family of phase trajectories of the pendulum in the coordinates p and Φ is shown in Fig. 2.8.

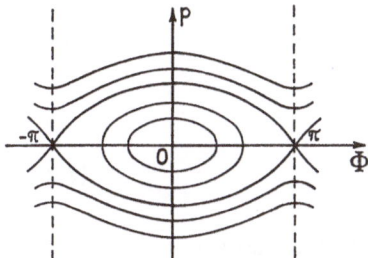

Fig. 2.8. Phase trajectories of a pendulum

Proceeding from this analogy, we can discuss the properties of solutions of (2.2.6) in the vicinity of a resonance. Note that, to obtain the universal Hamiltonian function (2.2.9), we went to the rotating coordinate system (since $\phi = k\alpha - m\nu t$) and measured the "momentum" ΔJ from the resonance value J_0, so that $J = J_0 + \Delta J$. The family of phase trajectories for the first-order resonance ($k = 1$) in the coordinate system (J, α), which rotates with the angular velocity $m\nu = \omega(J_0)$, can be readily obtained from the corresponding family for the pendulum (Fig. 2.8); this is shown in Fig. 2.9b.

The motion undergoes significant changes only within a narrow layer near the resonance trajectory of the unperturbed oscillator. Within this layer, a pattern of crescents is formed. The characteristic width of this pattern (or the *width of a nonlinear resonance*) can be estimated as

$$\Delta J_{\max} \sim \varepsilon^{1/2} h_{km}^{1/2} \left| \left(\frac{d\omega}{dJ} \right)_0 \right|^{-1/2} \tag{2.2.11}$$

Note that $\Delta J_{\max}$ should be small in comparison to J_0, as was assumed in (2.2.7). This implies that $\varepsilon \ll \gamma$, where

$$\gamma = \left| \left(\frac{d\omega}{dJ} \right)_0 \right| \frac{J_0^2}{h_{km}} \tag{2.2.12}$$

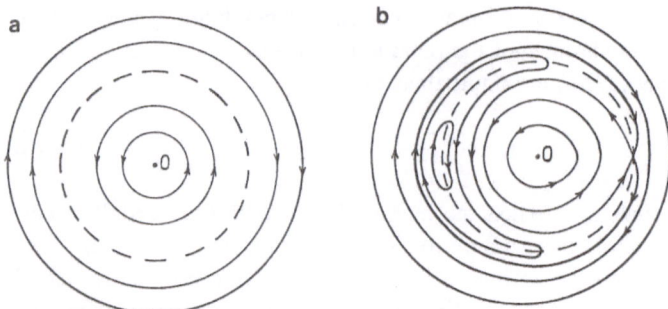

Fig. 2.9. Phase trajectories of the unperturbed oscillator (**a**) and of the same oscillator under a periodic external perturbation (**b**) in the case of a first-order resonance. The dashed curve is the resonance circle $J = J_0$ of the unperturbed motion. The polar coordinate system (J, α) rotates with the angular velocity $\omega(J_0)$. Thus phase points move in the opposite direction (i.e. counterclockwise) inside the resonance circle

is the *nonlinearity parameter*. The width of a nonlinear resonance can also be estimated in terms of the oscillation frequencies. Since $\Delta\omega = (d\omega/dJ)_0\Delta J$, we find

$$\Delta\omega_{\max} \sim \left(\varepsilon h_{km}\left|\left(\frac{d\omega}{dJ}\right)_0\right|\right)^{1/2} . \tag{2.2.13}$$

The same arguments can be applied to higher-order resonances (with $k > i$). Since $\phi = k\alpha - m\nu t$, the resonance layer in the phase plane now consists of the garland of k crescents, replacing the resonance trajectory of the unperturbed system (Fig. 2.10). The width of such a layer can again be estimated as (2.2.11).

Note that an arbitrary periodic perturbation εH_1 can be expanded into the Fourier series

$$\varepsilon H_1 = \sum_{k,m} \varepsilon h_{km}(J)\cos(k\alpha - m\nu t) . \tag{2.2.14}$$

Each term in this series can be further treated as an independent resonance perturbation, as long as the widths (2.2.13) of different resonances remain smaller than the intervals between different frequencies $\omega = m\nu/k$. Convergence of the Fourier series (2.2.14) implies that the coefficients h_{km} tend to zero for large k and m. Therefore, the resonance layers of the higher resonances should be increasingly narrow.

The situation for autonomous, nearly integrable Hamiltonian systems is very similar to the effects produced by an external periodic perturbation. As already mentioned, we have in this case a set of nonlinear oscillators that interact weakly with one another. These interactions can be effectively described (e.g. [2.8–9]) as an external periodic perturbation. Therefore, we can expect that the vicinity of each resonance torus of the integrable system will be transformed by the interactions

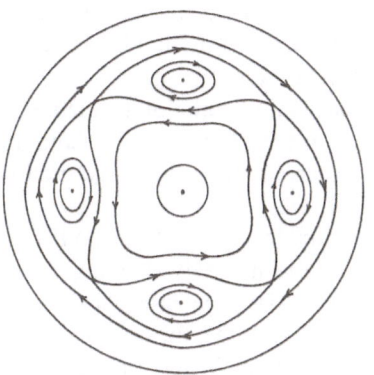

Fig. 2.10. Family of phase-space trajectories for nonlinear resonance with $k = 4$

into a layer of crescents, as shown in Fig. 2.10. The higher the resonance, the narrower such a layer.

The approximate analysis given above cannot be used for predictions of long-term behavior. It neglects the influence of small higher-order perturbations, which might bring about significant long-term effects. The system is especially sensitive to such perturbations near the separatrices which set the borders of the crescents. Therefore the question still remains: what are the long-term consequences of adding a small perturbation to a completely integrable system?

The rigorous answer to this question is provided by the *Kolmogorov-Arnold-Moser* (**KAM**) *theorem* [2.4,10–14]. Its correct formulation and proof are very complicated; below we give only a summary of its main results.

Suppose that the Hamiltonian function of a system can be written in the form (2.2.1) and the unperturbed function H_0 satisfies the condition

$$\det \frac{\partial \omega_i}{\partial J_k} = \det \frac{\partial^2 H_0}{\partial J_i \partial J_k} \neq 0 \ . \tag{2.2.15}$$

The theorem applies to any region of the phase space which lies far from all resonance tori, so that the condition[2]

$$\left| \sum_i \omega_i k_i \right| > C|\boldsymbol{k}|^{-\kappa} \ , \quad \boldsymbol{k} = (k_1, \cdots, k_n) \tag{2.2.16}$$

holds (here C and κ depend on the properties of the functions H_0 and H_1, and the coefficient C vanishes when $\varepsilon \to 0$).

If the above conditions are satisfied and ε is sufficiently small, the considered region of the phase space (of the energetic hypersurface $H = E$) can be divided into two regions of a nonvanishing volume. The larger region contains only

---

[2] Note that this condition is less stringent for resonances of higher orders which correspond to larger values of $|k|$. The tori with very high resonances may lie in increasingly close proximity to the region considered.

slightly perturbed nonresonant tori of the unperturbed problem. In the smaller region (whose volume tends to zero as $\varepsilon \to 0$), the character of motion is extremely complicated; it differs both from periodic and quasiperiodic.

Hence, for the majority of initial conditions, quasiperiodic motion is retained if the system is close to an integrable one. There are, however, some initial conditions for which the nonresonant tori that existed at $\varepsilon = 0$ are destroyed and the motion becomes irregular.

The KAM theorem has an important implication for the global stability of the behavior of Hamiltonian systems. When the system has only two degrees of freedom ($n = 2$), its energetic hypersurface $H = E$ is three dimensional and the invariant tori are two dimensional. These tori divide the three-dimensional energetic hypersurface into disconnected regions (Fig. 2.11). When a perturbation is added, the preserved tori would isolate the regions where the initial invariant tori have been destroyed (these regions constitute the gaps between the preserved tori, see Fig. 2.11). Consequently, a phase trajectory that was initially in one of the narrow gap would never leave it. This means that very weak perturbations never lead to a global instability of a system with two degrees of freedom.

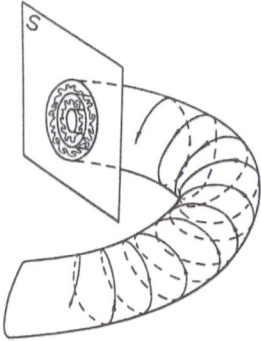

Fig. 2.11. The remaining invariant tori separate the regions where the tori are destroyed

When the number of degrees of freedom is larger than two ($n > 2$), the invariant tori no longer divide the $(2n-1)$-dimensional energetic hypersurface into disconnected parts. As a consequence, different regions with destroyed tori merge and form a single complicated cobweb. Moving along the threads of this cobweb, the phase-space point can travel a large distance from its original position at one of the invariant tori. This phenomenon is known as *Arnold diffusion* [2.4,6,15–18]. The characteristic feature of Arnold diffusion is that it has no threshold, i.e. it persists at all values of ε, however small. The existence of such diffusion proves that all nearly integrable Hamiltonian systems with $n > 2$ are globally unstable.

The KAM theorem describes the situation for small perturbations. When ε increases, the fraction of the phase space volume occupied by a stochastic cobweb grows. If ε is sufficiently large, the stochastic region might expand over a significant part of the phase space, as is evident from the Henon-Heiles example. Note that at sufficiently large perturbations, another mechanism leading to stochastic dynamics can appear. It is related to overlap between different resonances of lowest orders [2.8–9,16].

Our discussion shows that in some cases the dynamics of a Hamiltonian system may be extremely irregular and apparently chaotic. To further analyze these complex dynamic regimes, we should introduce some new concepts.

2.3 Ergodicity and Mixing

In this section we consider general (conservative) dynamical systems given by the differential equations

$$\dot{x} = v(x) \ . \tag{2.3.1}$$

Here $x = \{x_1, \cdots, x_n\}$ is an n-dimensional vector which specifies the state of the system and $v = \{v_1, \cdots, v_n\}$ is a vector function. Note that the dynamical equations (2.1.1) of a Hamiltonian system with k degrees of freedom can be cast into the form (2.3.1) if we assume that $x_i = q_i$ and $x_{i+k} = p_i$ for $i = 1, \cdots, k$.

When the initial conditions $x(0) = x_0$ are fixed, the equations (2.3.1) have a unique solution

$$x(t) = F(t, x_0) \ . \tag{2.3.2}$$

Any solution $x(t)$ determines a trajectory in the phase space M formed by all vectors x.

Since the solution (2.3.2) relates each initial point x_0 to some other point $x(t)$ at time t, we can say that it defines a certain mapping $F^t x_0 = x(t)$ in the phase space. Any initial region Ω_0 of the phase space is thus mapped by F^t into another region $\Omega_t = F^t \Omega_0$ (Fig. 2.12). The mapping F^t is called the *phase flow* of a dynamical system (2.3.1).

For example, in the angle–action variables, the equations of motion of a harmonic oscillator are

$$\dot{J} = 0 \ , \quad \dot{\alpha} = \omega \ . \tag{2.3.3}$$

Their solution is $J(t) = \text{const}$, $\alpha(t) = \alpha_0 + \omega t$. Therefore, in this case the mapping F^t consists of the rotation of phase-space regions (Fig. 2.13).

Another dynamical system

$$\dot{x} = ax \ , \quad \dot{y} = ay \ , \quad a > 0 \ , \tag{2.3.4}$$

does not conserve the phase volume. The action of its phase flow is shown in Fig. 2.14.

In this remainder of this section we consider only conservative dynamical systems that keep the phase-space volume invariant.

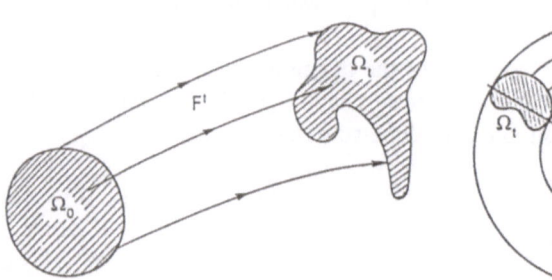

Fig. 2.12. Action of the mapping $F^t : M \rightarrow M$ for a Hamiltonian system

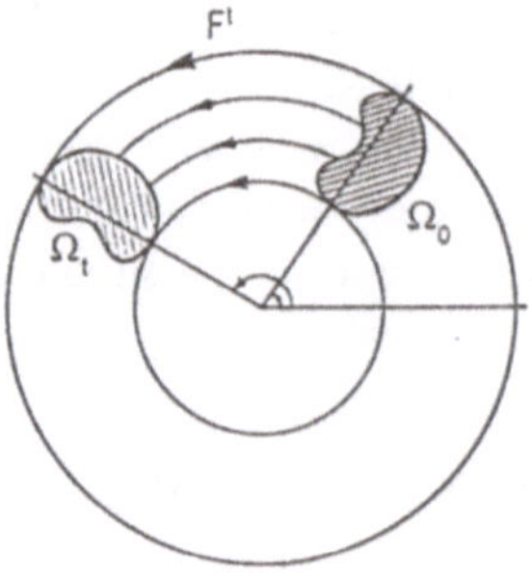

Fig. 2.13. The phase flow of a harmonic oscillator

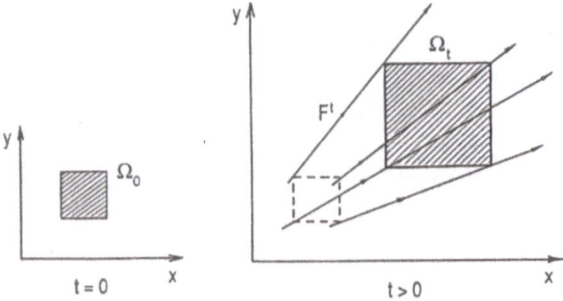

Fig. 2.14. The phase flow for the system (2.3.4)

To introduce the concept of ergodicity, some preliminary definitions are needed. Suppose we have an integrable function of n variables, $h(x_1, \cdots, x_n) = h(\boldsymbol{x})$. If $\boldsymbol{x}(t)$ is a solution to (2.3.1), the *time average* of this function is

$$\bar{h}(\boldsymbol{x}_0) = \lim_{T \to \infty} T^{-1} \int_0^T h(\boldsymbol{x}(t)) \, dt \ . \tag{2.3.5}$$

In terms of the phase flow, (2.3.5) can also be written in the form

$$\bar{h}(\boldsymbol{x}_0) = \lim_{T \to \infty} T^{-1} \int_0^T h(\boldsymbol{F}^t \boldsymbol{x}_0) \, dt \ . \tag{2.3.6}$$

Suppose that the trajectories of a dynamical system do not go to infinity and the motion is limited to some bounded region D of volume V_D in phase space. The *phase average* of function $h(\boldsymbol{x})$ can be defined as

$$\langle h \rangle = V_D^{-1} \int_{(D)} h(\boldsymbol{x}) \, dV \ , \tag{2.3.7}$$

where $dV = dx_1 \cdots dx_n$.

The dynamical system (2.3.1) is *ergodic* if, for any integrable function $h(x)$ and almost all initial conditions x_0, the equality

$$\bar{h}(x_0) = \langle h \rangle \tag{2.3.8}$$

holds. This definition implies that, for an ergodic system, the time average does not depend on the choice of the initial condition x_0.

Let us take some region Ω inside D and consider the function $h_\Omega(x)$, such that $h_\Omega(x) = 1$ when x belongs to Ω and $h_\Omega(x) = 0$ for x outside Ω. Evidently, this function is integrable:

$$V_D^{-1} \int_{(D)} h_\Omega(x)\, dV = \frac{V_\Omega}{V_D} . \tag{2.3.9}$$

Moreover, the phase average of $h_\Omega(x)$ is simply the "relative volume" of region Ω.

The function $h_\Omega(x)$ can be chosen as $h(x)$ in the definition (2.3.8) of ergodic systems. This definition then requires that the relative time spent by a trajectory inside any region Ω should be equal to the relative volume of this region, and cannot depend on the initial conditions. In other words, a phase trajectory of an ergodic system uniformly and densely covers all of the region Ω.

Note that the phase flow of Hamiltonian systems cannot be ergodic in the strict sense of the above definition. Since such systems conserve energy, the phase trajectory must lie on some energetic hypersurface $H = E$ and the flow can be ergodic only within this hypersurface. It there are K additional integrals of motion, the motion of such a system will be restricted to a hypersurface of $n - K - 1$ dimensions. It is this hypersurface G that represents then the attainable "region" of space. With this notion, definition (2.3.9) can be reformulated as

$$\lim_{T \to \infty} T^{-1} \int_0^T h(F^t x_0)\, dt = S_G^{-1} \int_{(G)} h(x)\, dS , \tag{2.3.10}$$

where the last integration is performed over the hypersurface G and S_G is the area of this hypersurface.

Therefore, when the motion of a Hamiltonian system is ergodic, its phase-space trajectory uniformly and densely covers, with time, the entire hypersurface specified by the integrals of motion.

The relative volume V_Ω / V_D (or the relative area S_Ω / S_D) of some region Ω is effectively the *measure* $\mu(\Omega)$ of this region[3]. Using the concept of measure, (2.3.8) can be written as

$$\lim_{T \to \infty} T^{-1} \int_0^T h(F^t x_0)\, dt = \int_{(D)} h(x)\, d\mu , \tag{2.3.11}$$

---

[3] This is only a special case of the measures that are rigorously introduced and discussed in [2.19]. However, for our purposes it is sufficient to interpret the measure simply as a relative volume.

where the integration is performed over the entire attainable region of the phase space and

$$\mu(D) = \int_{(D)} d\mu = 1 \ . \tag{2.3.12}$$

An example of ergodic behavior is provided by the quasiperiodic motion of a Hamiltonian system which takes place, for instance, in the model of two harmonic oscillators with an irrational frequency ratio (Sect. 2.1). In this case, the phase trajectory densely covers the surface of a two-dimensional torus.

Ergodicity is a necessary, but not sufficient, condition for chaotic motion. Indeed, the above example with two harmonic oscillators corresponds to a perfectly regular mapping F^t. This mapping produces steady drift of any small initial region Ω_0 along the torus surface, changing neither the total area nor the form of this region (Fig. 2.15). It is obviously inappropriate to call such motion chaotic.

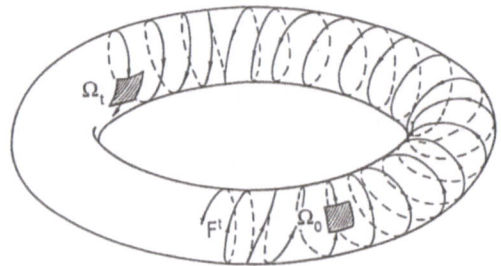

Fig. 2.15. Evolution of a small region Ω_0 under the phase flow of quasiperiodic motion

There are Hamiltonian dynamical systems with much more complicated regimes of motion. In these systems, any initial small region Ω_0 becomes strongly deformed as it drifts over the energetic hypersurface. In the process of such drift, the region ejects long irregular tentacles but never breaks (Fig. 2.16). If we wait sufficiently long, the initial region would spread over the entire hypersurface, so that its pieces can be found in any part of this hypersurface. Such dynamical systems are said to possess the property of mixing.

Mixing can be illustrated by the classical example first discussed by *Gibbs*. Suppose that a vessel contains 30% ink and 70% water, and that initially the ink and water are not mixed. Now let us shake the vessel. After some time, we would find that the liquid looks homogeneous: each part of the mixture consists of 30% ink and 70% water. Dynamical mixing in the phase space resembles this process, the only exception being that the initial region is not broken into separate drops but always remains connected.

To formalize the concept of mixing, we follow here *Balescu* [2.20], and consider two arbitrarily small regions A and B with measures $\mu(A)$ and $\mu(B)$. Let us assume that region B stays immobile while A evolves with time according to $A_t = F^t A$, obeying the continuous mapping F^t produced by a dynamical system (2.3.1). Then the intersection $A_t \cap B$ would consist of all the pieces of A that had

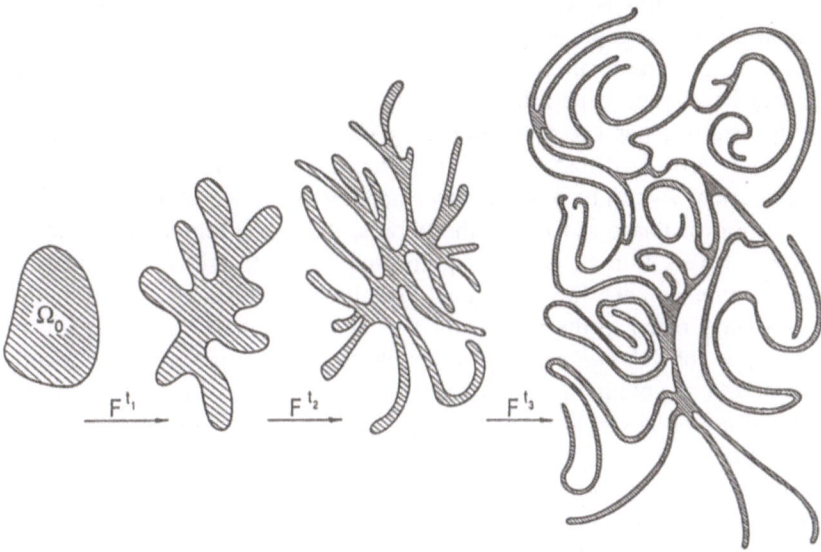

Fig. 2.16. The evolution of an initial region Ω_0 under the phase flow of a mixing motion

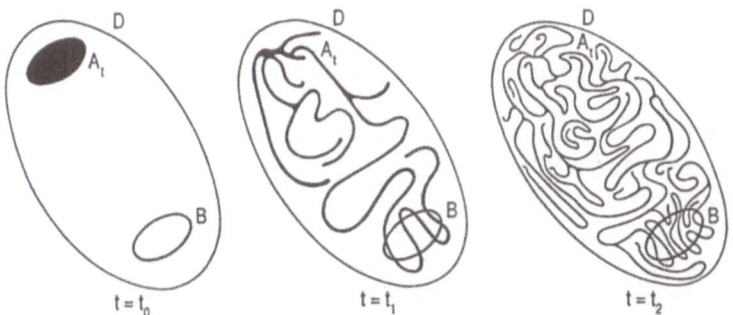

Fig. 2.17. Evolution of an initial region in the case of mixing

come, by time t, into the immobile region B (Fig. 2.17). The dynamical system is said to be *mixing*, if there is a limit at $t \to \infty$ which is equal to

$$\lim_{t \to \infty} \frac{\mu(A_t \cap B)}{\mu(B)} = \mu(A) . \tag{2.3.13}$$

One may comment on this definition as follows. Since the considered motion is conservative, it leaves invariant the measure of region A, i.e. $\mu(A_t) = \mu(A)$. Furthermore, $\mu(A) = \mu(A_t)/\mu(D)$ is the relative fraction of the volume occupied by A in the entire attainable region D. On the other hand, $\mu(A_t \cap B)/\mu(B)$ represents the relative fraction of the volume occupied in region B by the pieces of A_t which had reached it by time t. Equation (2.3.13) states that in the limit

$t \rightarrow \infty$ these two ratios coincide, independent of the original sizes, forms, and relative positions of the regions A and B.

There are several important implications of the above definition. The immobile region B can be chosen to be arbitrarily small and can be positioned anywhere. Despite this, after the passage of a sufficiently long time t, we will find inside B pieces of region $A_t = F^t A$ that evolved from the initial region A. This means that at $t \rightarrow \infty$ pieces of A_t can be found in any small neighborhood of any point in the attainable region of phase space. In other words, the initial region A transforms with time into a very thin all-penetrating cobweb whose threads can be detected inside *any* volume element, however small. Yet the total volume occupied by this cobweb is always equal to the volume of the initial region A. Moreover, in the limit $t \rightarrow \infty$ this cobweb *uniformly* spans the phase space: its pieces would occupy the same fraction of volume in any arbitrarily chosen region of the phase space.

The initial region A can also be chosen arbitrarily small and placed anywhere in the attainable part of the phase space. Nevertheless, at $t \rightarrow \infty$ it would always evolve into the all-penetrating cobweb, as discussed above. This implies that the phase-space trajectories of such a dynamical system with mixing are always absolutely unstable with respect to small perturbations and that they diverge with the passage of time.

The divergence of trajectories means that the behavior of this dynamical system is *unpredictable*. If we know the initial position of a phase-space point only with some finite accuracy, i.e. we know only that this point belongs to some small region Ω_ε of a characteristic size ε, we cannot tell where it would be found after a sufficiently long time.

Actually, mixing brings about an effective *irreversibility*. We see that after enough time has elapsed, any attainable part of phase space will contain the pieces that come from many distant initial regions (and, in the limit $t \rightarrow \infty$, from almost any initial region). Consequently, if we know only that, at a final moment in time, the phase-space point is found in a small region of size ε, we cannot uniquely determine where the point was initially. Since in any physical experiment we can measure the properties only with some finite accuracy, the deterministic description of dynamical systems with mixing is meaningless. In this situation we are well advised to look for statistical theories. Thus, the concept of mixing corresponds well to our intuitive understanding of what should be called chaotic motion.

The consequence of mixing is *fading of temporal correlations*. Consider any two integrable functions f and g in a closed region D of phase space. According to (2.3.7), their phase averages are

$$\langle f \rangle = V_D^{-1} \int_{(D)} f(\boldsymbol{x}) \, d\boldsymbol{x} \,, \quad \langle g \rangle = V_D^{-1} \int_{(D)} g(\boldsymbol{x}) \, d\boldsymbol{x} \,. \tag{2.3.14}$$

Suppose that the initial state of a system is given by $\boldsymbol{x}(0) = \boldsymbol{x}_0$. Then, after some time t, the system will move into another state $\boldsymbol{x} = \boldsymbol{x}(t, \boldsymbol{x}_0)$. Let us define the correlator of f and g by

$$\langle f(\boldsymbol{x}(t, \boldsymbol{x}_0)) g(\boldsymbol{x}_0) \rangle = V_D^{-1} \int_{(D)} f(\boldsymbol{x}(t, \boldsymbol{x}_0)) g(\boldsymbol{x}_0) \, d\boldsymbol{x}_0 \,. \tag{2.3.15}$$

This expression can also be written as

$$\langle f(x)g(x_0(t,x))\rangle = V_D^{-1} \int_{(D)} f(x)g(x_0(t,x))\,dx \ . \tag{2.3.16}$$

We now want to show that, if the system is mixing, we shall obtain, in the limit $t \to \infty$,

$$\langle f(x)g(x_0)\rangle \to \langle f\rangle\langle g\rangle \tag{2.3.17}$$

for all functions f and g. This is the property of fading of temporal correlations.

To prove this property, we first consider the following special choice of the functions f and g,

$$g_A(x) = \begin{cases} 1, & x \in A \\ 0, & x \notin A \end{cases}, \quad f_B(x) = \begin{cases} 1, & x \in B \\ 0, & x \notin B \end{cases}, \tag{2.3.18}$$

where A and B are some subregions of D. Suppose that B remains immobile, while A evolves with time as $A_t = F^t A$. Then, at large times t, the mixing condition (2.3.13) yields the approximate relation

$$\mu(A_t \cap B) \approx \mu(A)\mu(B) \ . \tag{2.3.19}$$

However, $\mu(A_t \cap B)$ is the relative volume of the intersection of A_t with region B. It can be written as

$$\mu(A_t \cap B) = V_D^{-1} \int_{(D)} f_B(x)g_{A_t}(x)\,dx$$
$$= V_D^{-1} \int_{(D)} f_B(x)g_A(x_0(x,t))\,dx \ . \tag{2.3.20}$$

On the other hand, we have

$$\mu(A)\mu(B) = \left(V_D^{-1} \int_{(D)} g_A(x)\,dx\right)\left(V_D^{-1} \int_{(D)} f_B(x)\,dx\right) \ . \tag{2.3.21}$$

If we now take into account definition (2.3.16) of the correlator and definition (2.3.14) of phase averages, it can clearly be seen that at $t \to \infty$ equations (2.3.19–21) imply

$$\langle f_B(x)g_A(x_0)\rangle \approx \langle f_B\rangle\langle g_A\rangle \ . \tag{2.3.22}$$

Let us further divide entire region D of the phase space into a set of nonintersecting subregions A_j. Then, for sufficiently fine graining, we can approximate any smooth function $h(x)$ in D as

$$h(x) = \sum_j \gamma_j h_{A_j}(x) \ , \quad \text{where} \tag{2.3.23}$$

$$h_{A_j}(\boldsymbol{x}) = \begin{cases} 1, & \boldsymbol{x} \in A_j \\ 0, & \boldsymbol{x} \notin A_j \end{cases} . \tag{2.3.24}$$

If we approximate $f(\boldsymbol{x})$ and $g(\boldsymbol{x})$ by (2.3.23), we obtain, using (2.3.22),

$$\begin{aligned} \langle f(\boldsymbol{x})g(\boldsymbol{x}_0)\rangle &= \sum_{j,k} \gamma_j\gamma_k' \langle f_{A_j}(\boldsymbol{x})g_{A_k}(\boldsymbol{x}_0)\rangle \\ &\approx \sum_{j,k} \gamma_j\gamma_k' \langle f_{A_j}(\boldsymbol{x})\rangle\langle g_{A_k}(\boldsymbol{x}_0)\rangle \\ &= \langle f\rangle\langle g\rangle , \end{aligned} \tag{2.3.25}$$

which proves the statement.

Fading of correlations means that, in time, the dynamical system "forgets" its initial conditions.

It can be shown (see [2.13]) that ergodicity follows from mixing. The reverse statement is generally not true: ergodicity does not imply mixing. This is already seen in the above example of two harmonic oscillators with an irrational frequency ratio. For such a system, the phase flow $\boldsymbol{F}^t$ simply moves the initial region Ω_0 over the surface of a torus without changing its form (Fig. 2.15). Then, at different moments in time, the measure of the intersection of $\Omega_t = \boldsymbol{F}^t\Omega_0$ with some small fixed region B alternates between zero and some constant value, so that $\mu(\Omega_t \cap B)$ has no definite limit at $t \to \infty$. This example indicates that there are systems which have ergodic behavior but are not mixing.

Hence, mixing represents a much stronger property than ergodicity. If a dynamical system is mixing, it is natural to consider it as chaotic. Proceeding from the property of mixing, one can show that the behavior of such a system is irreversible and unpredictable; thus it cannot be described in a deterministic way.

For a long time it was believed that mixing (and chaotic behavior) is possible only in systems with very many degrees of freedom. However, *Sinai* [2.21,22] proved in 1962 that this property is possessed even by a system consisting of two rigid flat discs on a table with rigid reflecting borders. Moreover, it is even possible to fix one of the discs in a certain position (Fig. 2.18a). If we replace the immobile disc by a disc of twice the radius and consider a moving material

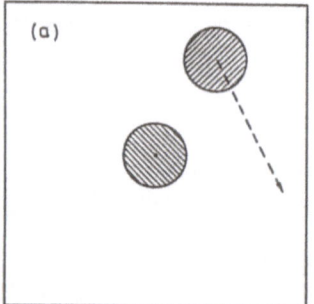
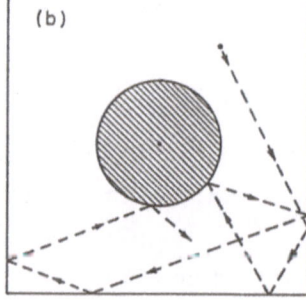

Fig. 2.18a,b. The simplest form of the Sinai billiard

particle instead of the mobile disc, this would lead us to the problem of motion inside a *billiard* (Fig. 2.18b). The mathematical "billiard" describes a dynamical system where the point particle reflects elastically from the borders of some closed region and performs free straight motion between the reflections. In the above case one can rigorously prove that the system is mixing for almost any choice of initial conditions. This result can be further generalized to the case when both discs are mobile.

Although the rigorous results of *Sinai* apply only to two interacting particles, it can be expected that the same property holds for ensembles of many elastically colliding particles. In fact, the property of mixing is always implicitly assumed in the construction of the kinetic gas theory.

The presence of mixing strongly implies chaoticity in the behavior of a system. However, not every chaotic motion is obliged to be mixing. The definition (2.3.13) imposes very severe requirements of ideal uniformity that are not always fulfilled in realistic systems.

3. Strange Attractors

In this chapter we begin the analysis of dissipative dynamical systems. In contrast to the Hamiltonian systems discussed earlier, these systems compress the phase volume. In dissipative dynamical systems (almost) all trajectories are asymtotically attracted with time to some limit sets of zero volume which are called attractors. Simple attractors are points, lines and (hyper)surfaces in the phase space. However, much more complex strange attractor are also possible.

While attractive fixed points correspond to stable steady states, all other attractors give rise to self-sustained oscillations. When an attractor is a closed line or a (hyper)surface, such oscillations are periodic or quasiperiodic. Strange attractors produce chaotic self-oscillations.

3.1 Dissipative Systems and Their Attractors

The behavior of dissipative dynamical systems is described by a set of first-order differential equations

$$\dot{x} = v(x) \quad , \quad x(0) = x_0 \quad , \tag{3.1.1}$$

where $x = \{x_1, \cdots, x_n\}$ is some n-dimensional vector with components $x_1, \cdots, x_n$ and $v = \{v_1, \cdots, v_n\}$ is a vector function which has components $v_1, \cdots, v_n$. We will assume that the function v does not depend explicitly on time, i.e. that system (3.1.1) is autonomous. Note that any differential equation, solvable for the highest-order derivative, can be reduced to the form (3.1.1) by an appropriate change of variables. Moreover, if some system is not autonomous, we can introduce a new independent variable τ and, by adding the equation $d\tau/dt = 1$, convert it to the autonomous form.

The phase space of system (3.1.1), which we continue to denote by M, is formed by the components $x_1, \cdots, x_n$ of vector x and therefore has dimensionality n. Any solution $x_i(t) = F_i(t, x_0)$, $i = 1, 2, \cdots, n$, of (3.1.1) defines a certain line, or *phase-space trajectory*, in the space M. The set of all trajectories $x(t)$ for a given dynamical system constitutes its *n-dimensional flow*. If v is a smooth function, a solution $x(t) = F(t, x_0)$ exists at all times t and every point of phase space corresponds to a single state of the system.

The characteristic property of the Hamiltonian system studied in the previous chapter consists in the conservation of phase-space volume. In contrast to this, dissipative dynamical systems compress the phase-space volume.

Let us estimate, following *Lichtenberg* and *Lieberman* [3.1], the rate of change for a small element of the phase-space volume

$$\Delta\Omega = \prod_i \Delta x_i \quad . \tag{3.1.2}$$

We have

$$\frac{d(\Delta\Omega)}{dt} = \frac{d(\Delta x_1)}{dt}\prod_{i\neq 1}\Delta x_i + \frac{d(\Delta x_2)}{dt}\prod_{i\neq 2}\Delta x_i + \cdots$$
$$= \left(\prod_i \Delta x_i\right)\frac{d(\Delta x_1)}{dt}\Delta x_1^{-1} + \left(\prod_i \Delta x_i\right)\frac{d(\Delta x_2)}{dt}\Delta x_2^{-1} + \cdots . \tag{3.1.3}$$

Therefore, taking into account (3.1.2), we find

$$\frac{d(\Delta\Omega)}{dt} = \Delta\Omega\sum_i \Delta x_i^{-1}\frac{d(\Delta x_i)}{dt} \quad . \tag{3.1.4}$$

Noting that

$$\Delta x_i(t) = \frac{\partial x_i(t)}{\partial x_i(t_0)}\Delta x_i(t_0) \quad , \tag{3.1.5}$$

where t_0 is some previous moment of time, we obtain

$$\frac{d(\Delta x_i)}{dt} = \Delta x_i(t_0)\frac{d}{dt}\frac{\partial x_i(t)}{\partial x_i(t_0)} \quad . \tag{3.1.6}$$

Substitution of (3.1.6) into (3.1.4) yields

$$\frac{d(\Delta\Omega)}{dt} = \Delta\Omega\sum_i \frac{\Delta x_i(t_0)}{\Delta x_i(t)}\frac{d}{dt}\frac{\partial x_i(t)}{\partial x_i(t_0)}$$
$$= \Delta\Omega\sum_i \frac{\Delta x_i(t_0)}{\Delta x_i(t)}\frac{\partial v_i(t)}{\partial x_i(t_0)} \quad . \tag{3.1.7}$$

In the limit $t_0 \to t$ and thus $\Delta x_i(t_0) \to \Delta x_i(t)$, this equation reduces to

$$\frac{d(\Delta\Omega)}{dt} = \Delta\Omega\sum_i \frac{\partial v_i}{\partial x_i} \quad . \tag{3.1.8}$$

Thus we find that, if the condition

$$\text{div } \boldsymbol{v} = \sum_i \frac{\partial v_i}{\partial x_i} < 0 \tag{3.1.9}$$

is satisfied at (almost) every point in phase space, the dynamical system compresses its initial phase-space volume in the course of time, and we should call it dissipative.

In the long time limit ($t \to \infty$), all solutions to the dynamical equations of a dissipative system are absorbed by some subset with a vanishing volume in phase space. This subset is called the attractor. A formal definition was given by *Lanford* [3.2] (see also [3.3–9]), who stated that an attractor represents a bounded subset A of the phase space M that satisfies the following three conditions:

i) A is invariant under the flow;
ii) there exists some neighborhood U which shrinks to A under the action of the flow;
iii) A cannot be decomposed into two nonintersecting invariant sets.

A few comments concerning this definition are appropriate. The solution $x(t) = F(t, x_0)$ can be viewed as a rule which relates any initial point x_0 to some point $x(t)$ at time t. In other words, it specifies some mapping $F^t : M \to M$ which acts in the phase space. The invariance of the attractor A under the flow, assumed by the first condition, means that $F^t A = A$. The second condition implies that A is a subset of the set U and is a limit of $F^t U$ when $t \to \infty$. The third condition is necessary if we want to discriminate between different attractors in the phase space when several are present[1].

The *basin* of attractor A is the set of all initial points x_0, such that trajectories originating from these points converge at $t \to \infty$ to the attractor A.

Points x^0 at which the right-hand side of dynamical equations vanish, are called the *fixed points* of system (3.1.1); these fixed points can be either stable or unstable. Close to a fixed point, we can often use, instead of (3.1.1), a system of linearized equations for the deviations $\xi(t) = x(t) - x^0$, i.e.

$$\dot{\xi}_i = \sum_{k=1}^{n} a_{ik} \xi_k \quad , \tag{3.1.10}$$

where $a_{ik} = \partial v_i / \partial x_k$ are the elements of the linearization matrix A. If all eigenvalues λ_i of this matrix satisfy the condition $\text{Re}\{\lambda_i\} < 0$, then the fixed point x^0 is asymptotically stable.

When a system consists of two differential equations ($n = 2$) there are four possible types of fixed points. If both eigenvalues λ_1 and λ_2 are real and have the same sign, the fixed point is a *node* (Fig. 3.1). If λ_1 and λ_2 are real and have opposite signs, this unstable fixed point is a *saddle* (Fig. 3.2). Trajectories I and II, passing through the saddle, are *separatrices*. When both eigenvalues are complex (but not purely imaginary), the fixed point is a *focus* (Fig. 3.3). Finally, when both eigenvalues are purely imaginary we have a *center* (Fig. 3.4). Note that the last type of fixed point is permitted only in conservative systems.

---

[1] Usually a dynamical system has a finite number of attractors. However, examples of systems possessing infinitely many attractors are also known [3.10].

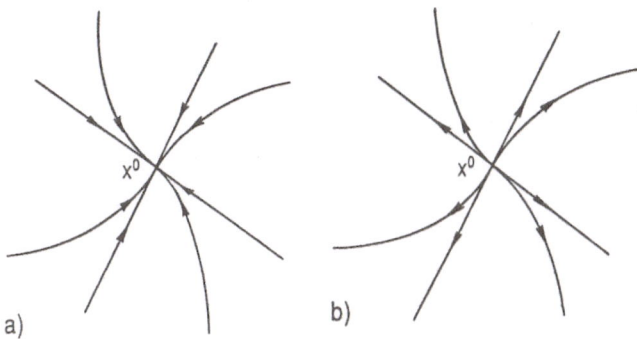

Fig. 3.1. Stable (**a**) and unstable (**b**) nodes

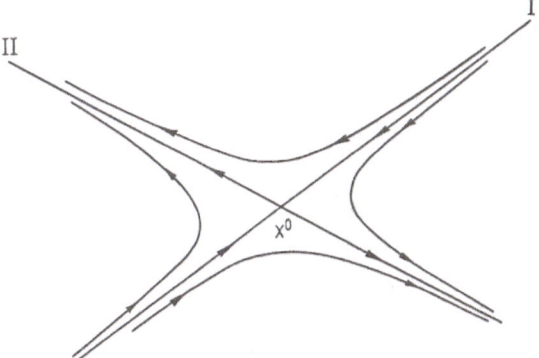

Fig. 3.2. A saddle point

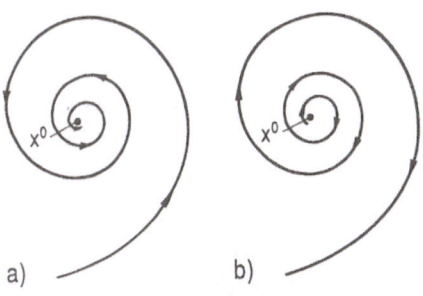

Fig. 3.3. Stable (**a**) and unstable (**b**) focuses

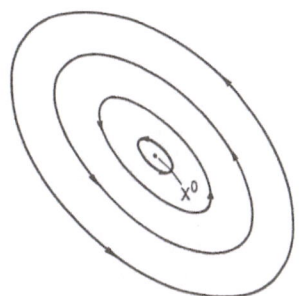

Fig. 3.4. Center

In phase spaces of higher dimensions, more complicated types of fixed points are possible. For instance, one can find combinations of the first and second, or of the second and third types. In these cases, the fixed points are the *saddle-node* (Fig. 3.5) and the *saddle-focus* (Fig. 3.6), respectively.

Obviously, attractive fixed points, such as a stable node and a stable focus, are the attractors of dynamical systems.

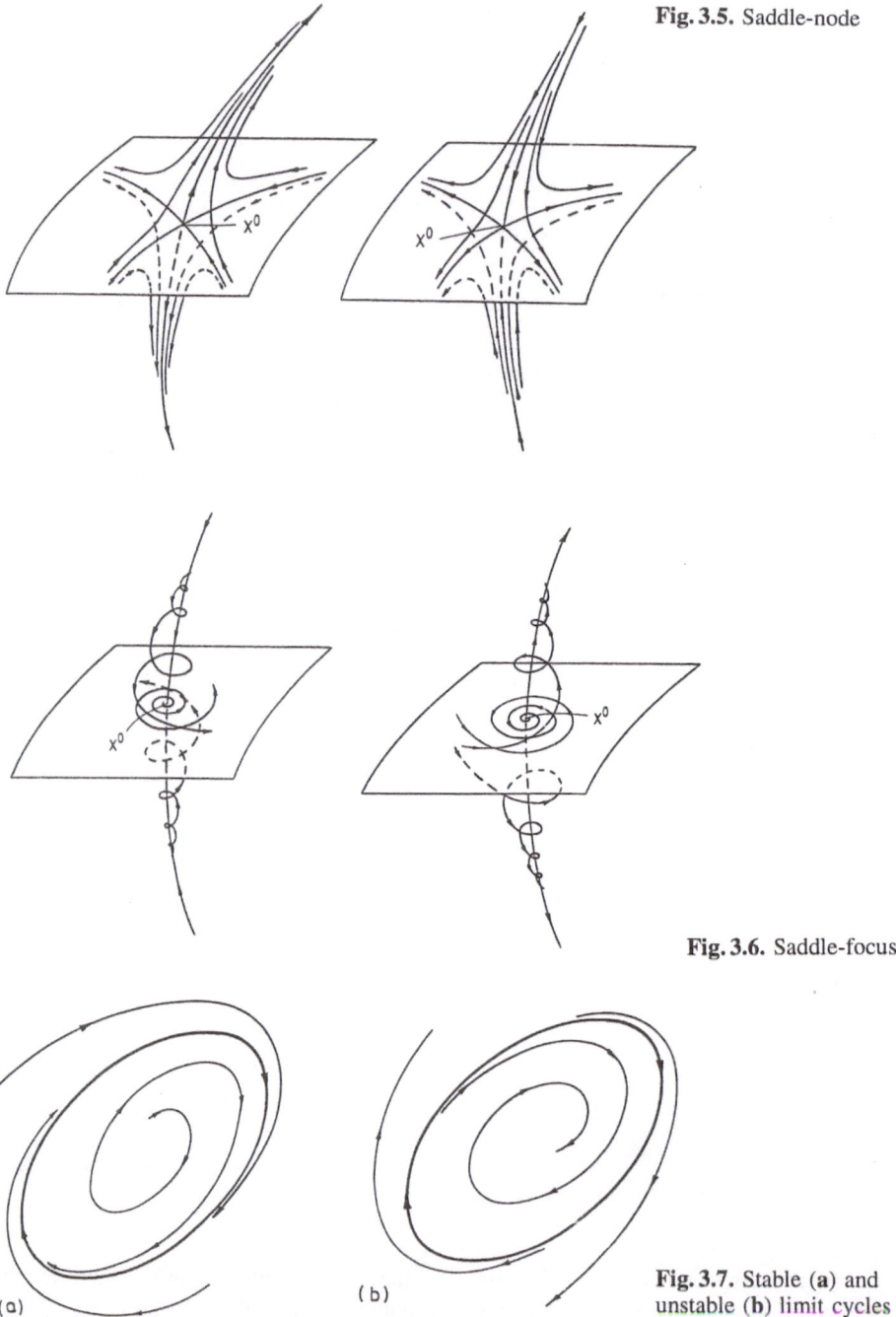

Fig. 3.5. Saddle-node

Fig. 3.6. Saddle-focus

(a)

(b)

Fig. 3.7. Stable (**a**) and unstable (**b**) limit cycles

The attractors of dissipative systems can consist of closed phase-space curves corresponding to periodic motion. Such isolated closed trajectories are called *limit cycles*. Stable limit cycles are the attractors. They have the property that, in their sufficiently close neighborhood, there are no other closed trajectories and all phase-space curves from this neighborhood wind with time onto a single closed trajectory of the limit cycle (Fig. 3.7a).

In the opposite case, when all trajectories unwind from the limit cycle trajectory, it is absolutely unstable (Fig. 3.7b). Saddle limit cycles are also possible. In the vicinity of such a cycle some phase-space trajectories wind onto it as $t \rightarrow \infty$, forming its stable manifold W^s, while others, forming the unstable manifold W^u, unwind from the limit cycle (Fig. 3.8). Neither absolutely unstable nor saddle limit cycles can be attractors.

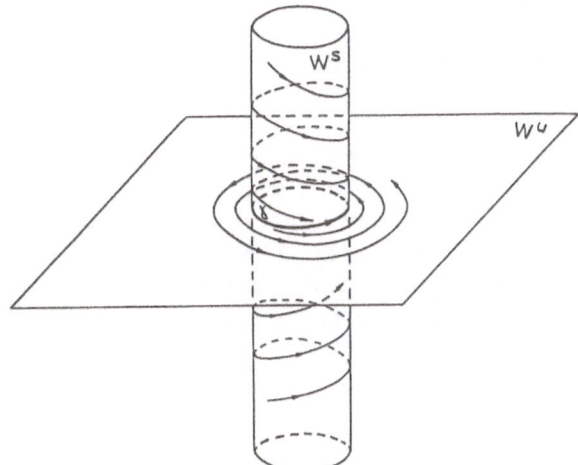

Fig. 3.8. Stable (W^s) and unstable (W^u) manifolds of a saddle limit cycle

In two-dimensional dissipative dynamical systems there can be no attractors other than stable fixed points and stable limit cycles, which immediately follows from the fact that the phase trajectories do not intersect. But already in systems with a three-dimensional phase space ($n = 3$), more complicated attractors can exist. For instance, one can imagine two-dimensional *invariant tori* which correspond to some quasiperiodic motion with two rationally independent frequencies. If $n > 3$ such invariant tori might have higher dimensions, corresponding to more complex quasiperiodic motions with many different frequencies. Usually, however, such multidimensional attractive tori are not structurally stable: they are easily destroyed even by small perturbations to the right-hand sides of the dynamical equations (3.1.1). These effects, known as *frequency-locking* of self-oscillations, result in the emergence of a simple attractive periodic motion (see Chap. 6). Hence, attractive multidimensional tori represent something of an exception.

All the above-mentioned attractors (i.e. stable fixed points, stable limit cycles and invariant tori) are called *simple*, since the dynamics of a system possessing such attractors is not chaotic (it can at most be ergodic).

Any simple attractor represents a submanifold in the phase space of a dynamical system[2]. However, dissipative dynamical systems with a larger number of variables ($n \geq 3$) could also possess bounded attractive sets which are not submanifolds. These attractors, if they are not reducible to a finite superposition of submanifolds, are called *strange*[3].

The above definition of a strange attractor is rather formal. In practical applications, a strange attractor is usually understood in the following way. Any strange attractor is a limit set which absorbs with time the phase-space trajectories of a dissipative dynamical system. Having entered the neighborhood of an attractor set, the trajectories remain there forever. The distinguishing property is that the motion on a strange attractor is intrinsically unstable: any two trajectories diverge exponentially with time while still belonging to the same attractor. This implies that the behavior of a dissipative dynamical system with a strange attractor is characterized by a combination of global compression of the phase-space volume with local instability of trajectories.

To illustrate a possible structure of the set representing a strange attractor, we consider an example suggested by *Shaw* [3.12]. Let us take an infinitely thin rectangular sheet, consisting of an infinite number of tightly pressed nonintersecting surfaces, and fold it at one end only, as shown in Fig. 3.9a. After that, we stretch the folded edge to double its length and paste it to the edge on the opposite side of the sheet (Fig. 3.9b). The resulting object would represent a set of zero volume which can contain diverging trajectories.

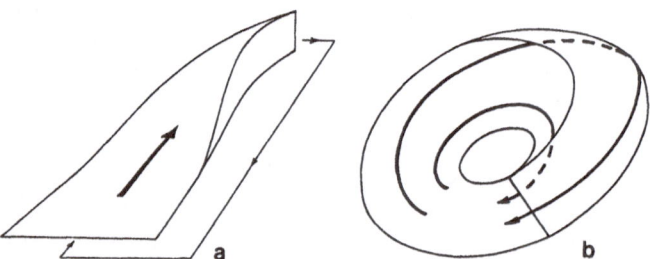

Fig. 3.9a,b. Construction of a geometric set which can contain an infinite number of diverging trajectories

---

[2] A submanifold in space M is a subset W in space $M' \leq M$ which locally looks like a "piece" of space M' and has at every point a unique tangent hyperplane, i.e. W is smoothly embedded in M. A limit cycle and a two-dimensional invariant torus are examples of the one-dimensional and two-dimensional submanifolds.

[3] The term "strange attractor" was initially introduced by *Ruelle* and *Takens* [3.11] to denote an attractor which differs from both a fixed point and a limit cycle.

Since phase-space trajectories diverge on a strange attractor, the dynamics of a system with such an attractor should be similar to the dynamics of a conservative system with mixing, i.e. it is chaotic. Usually, strange attractors are stable in the sense that, although small perturbations of a dynamical system can change the internal structure of an attractor, they do not completely destroy it, so that the motion remains chaotic.

In the physics literature one can sometimes find a wider definition of a strange attractor, which allows it to include stable periodic trajectories which extremely large periods and very small attraction basins. *Afraimovich* and *Shilnikov* [3.13] suggested that one refer in these cases to quasi-attractors. Since small fluctuations can easily throw a trajectory from one long-period limit cycle to another, dynamical systems with quasi-attractors can display apparently chaotic properties in computer simulations and in real experiments.

3.2 The Lorenz Model

The *Lorenz model* [3.14] proposed in 1963 was the first realistic model of a dynamical process with intrinsically chaotic behavior. It was derived using a number of drastic simplifications to the hydrodynamical problem of thermal convection in a horizontal layer of liquid heated from below. It is given by three equations

$$\dot{x} = \sigma y - \sigma x \quad , \quad \dot{y} = rx - y - xz \quad , \quad \dot{z} = xy - bz \quad . \tag{3.2.1}$$

Here σ is the Prandtl number, r is reduced Rayleigh number and b a parameter specifying the linear dimension of the physical system. Below we shall use this extensively investigated system to illustrate the typical properties of a strange attractor.

The Lorenz system is dissipative because it satisfies the condition

$$\operatorname{div} v = \sum_{i=1}^{3} \frac{\partial v_i}{\partial x_i} = -(b + \sigma + 1) < 0 \quad . \tag{3.2.2}$$

Hence, it uniformly compresses the phase-space volume. The system has no trajectories which go to infinity [3.14]. Each solution of (3.2.1) enters a sphere of some radius $R = R(\sigma, r, b)$ and remains there forever. It should be also noted that the Lorenz system is invariant with respect to the transformation $x \to -x$, $y \to -y$, $z \to z$.

Depending on the value of the parameter r, the Lorenz system can exhibit various forms of dynamical behavior. When $r < 1$ it has a stable fixed point $O = \{0, 0, 0\}$, the origin of the coordinates, which is an attractive node. At $r = 1$ this fixed point loses its stability and transforms into a saddle-node. At the same value of the parameter r, two additional fixed points

$$\begin{aligned}
O_1 &= \{[b(r-1)]^{1/2} \quad , \quad [b(r-1)]^{1/2} \quad , \quad (r-1)\} \quad , \\
O_2 &= \{-[b(r-1)]^{1/2} \quad , \quad -[b(r-1)]^{1/2} \quad , \quad (r-1)\}
\end{aligned} \tag{3.2.3}$$

appear. At $r > 1$ the Lorenz system has only these three fixed points, O, O_1 and O_2.

The stability of points O_1 and O_2 is determined by the roots of the characteristic equation

$$\lambda^3 + (\sigma + b + 1)\lambda^2 + (r + \sigma)b\lambda + 2\sigma b(r - 1) = 0 \quad . \tag{3.2.4}$$

Analysis of this equation shows that O_1 and O_2 are stable if $\sigma > b + 1$ and $1 < r < r^*$, where

$$r^* = \sigma(\sigma + b + 3)(\sigma - b - 1)^{-1} \quad . \tag{3.2.5}$$

When $r > r^*$, points O_1 and O_2 become unstable. In this case the characteristic equation (3.2.4) has one real negative root and two complex conjugate roots with positive real parts, which implies that O_1 and O_2 are saddle-focuses.

Further investigation of the properties of the phase-space trajectories of (3.2.1) can be performed only by means of computer simulations, since a local analysis in the neighborhoods of the unstable fixed points O_1 and O_2 does not provide information about the global properties of motion in the system.

Numerical investigations of the Lorenz system have been carried out by many authors (reviewed in [3.15–17]). Below we summarize the results of investigation of this system at $\sigma = 10$, $b = 8/3$ in the interval $10 \leq r \leq 28$.

1) When $10 \leq r < r_1$, where $r_1 \approx 13.926$, the system has three fixed points O, O_1 and O_2. Point O is unstable and represents a saddle-node with a two-dimensional stable manifold W^s and two separatrices Γ_1 and Γ_2. Two other fixed points O_1 and O_2 are stable. The structure of the phase space in this case is shown in Fig. 3.10.
2) At $r = r_1$ each of the curves Γ_1 and Γ_2 transforms into a closed loop; points O_1 and O_2 remain stable (Fig. 3.11).

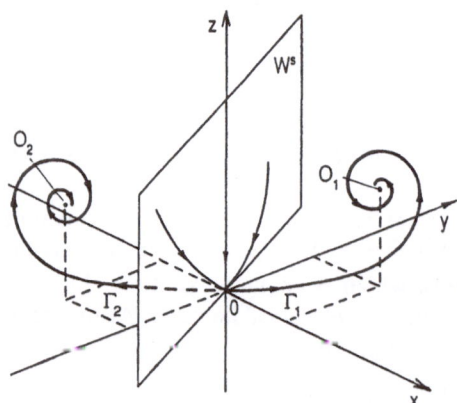

Fig. 3.10. Phase space of the Lorenz model at $\sigma = 10$, $b = 8/3$, $10 < r < 13.926$

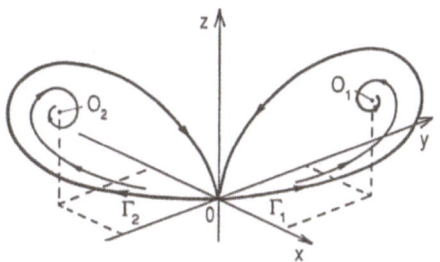

Fig. 3.11. Phase space of the Lorenz model at $\sigma = 10$, $b = 8/3$, $r = 13.926$

3) In the interval $r_1 < r < r_2$ where $r_2 \approx 24.06$, points O_1 and O_2 are still stable, but each of the previous loops Γ_1 and Γ_2 gives birth to a saddle periodic trajectory L_1 and L_2. The separatrices Γ_1 and Γ_2 now go to the points O_2 and O_1, respectively. Moreover, new trajectories appear which go from the saddle limit cycle L_1 and L_2 and from L_2 to L_1. These curves form a one-dimensional set B_1 that is not, however, attracting. The stable manifolds of the saddle periodic trajectories L_1 and L_2 are the boundaries of the attraction basins of the fixed points O_1 and O_2. The phase curve, which originated outside these regions, can oscillate from the neighborhood of L_1 to the neighborhood of L_2 and vice versa, until it comes into the basin of one of the attractors O_1 or O_2. The number of oscillations performed greatly increases when r approaches r_2. This behavior is called *metastable chaos* [3.18]. The general structure of the phase space in this case is shown in Fig. 3.12.

4) At $r = r_2$ the fixed points O_1 and O_2 remain stable, but the separatrices Γ_1 and Γ_2 no longer go to them. Instead they wind on the saddle trajectories L_1 and L_2, respectively (Fig. 3.13). Set B_1 is replaced by set B_2, which becomes stable and attractive for $r > r_2$.

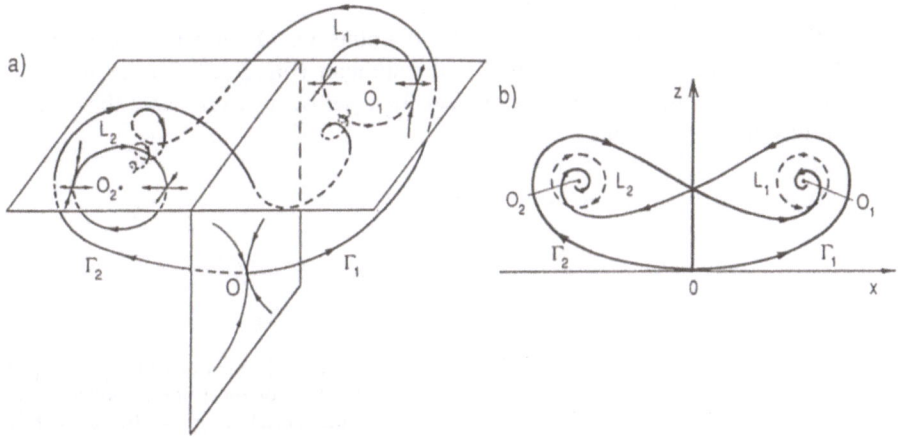

Fig. 3.12. Basic elements (**a**) of the phase space structure of the Lorenz model at $\sigma = 10$, $b = 8/3$, $13.926 < r < 24.06$ and their projection (**b**) on the coordinate plane (x, z)

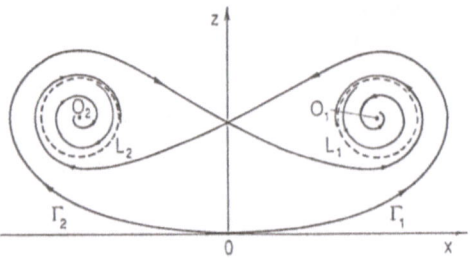

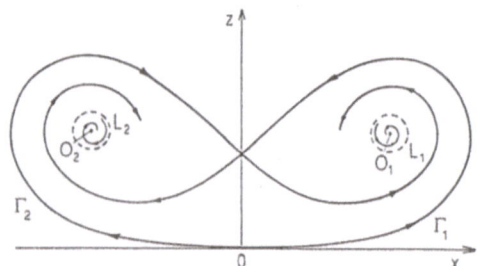

Fig. 3.13. Projection of the phase space structure of the Lorenz model on the plane (x, z) at $\sigma = 10$, $b = 8/3$, $r = 24.06$

Fig. 3.14. Projection of the phase space structure of the Lorenz model on the plane (x, z) at $\sigma = 10$, $b = 8/3$, $24.06 < r < 27.74$

5) In the interval $r_2 < r < r^*$ where $r^* \approx 27.74$, points O_1 and O_2 are still stable. However, the phase space contains also a limit set B_2 which is called the Lorenz attractor. This set B_2 consists of integral curves going from L_1 to L_2 and vice versa; it also includes the saddle point O and its separatrices Γ_1 and Γ_2. Hence, within this interval, the Lorenz system has three attractors, i.e. the fixed points O_1 and O_2 and the Lorenz attractor B_2. The basin of B_2 is bounded by the stable manifolds of the saddle limit cycles L_1 and L_2 (Fig. 3.14). Depending on the initial conditions, the phase trajectories of this system either go to points O_1 and O_2 or perform oscillations, wandering at random from revolutions around O_1 to revolutions around O_2 and vice versa. Consequently, for different initial conditions, the system can approach a steady state or perform chaotic motion.

6) When r tends to r^*, the saddle limit cycles L_1 and L_2 shrink to the fixed points O_1 and O_2. At $r = r^*$ they merge with such points and disappear. At this value of r the fixed points lose their stability.

7) In the interval $r^* < r < 28$, all fixed points O, O_1 and O_2 are unstable. The only attractive set is B_2, the Lorenz attractor (Fig. 3.15). Therefore, under any choice of initial conditions (except for a set of measure zero) this system exhibits chaotic motion.

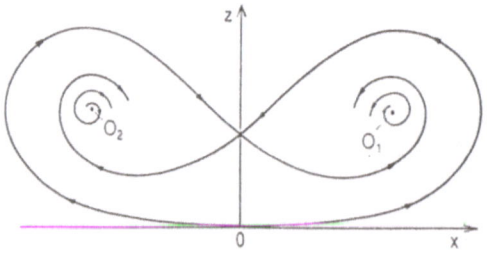

Fig. 3.15. Projection of the phase space structure of the Lorenz model on the plane (x, z) at $\sigma = 10$, $b = 8/3$, $27.74 < r < 28$

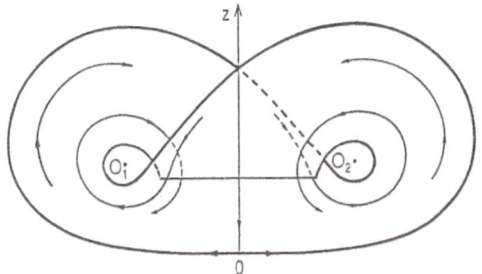

Fig. 3.16. Model of the Lorenz attractor

In order to understand this motion more clearly, let us use a geometric model, as suggested in [3.14]. This model represents a branching surface S, which lies near the vertical plane $x = y$, contains fixed points O_1 and O_2, and is symmetric with respect to this vertical plane that includes the line $0z$ (Fig. 3.16). The phase trajectory which starts on this surface on the left of the vertical plane rotates clockwise around the point O_1, unwinding as a spiral until it crosses the vertical plane. After that it begins to rotate around O_2, but in the opposite direction (counterclockwise). The phase trajectory proceeds to curl out from point O_2 until it once more crosses the vertical plane. Then the rotation will be performed around O_1, etc. In this process, the subsequent numbers of rotations around points O_1 and O_2 change in an irregular way and the motion appears chaotic.

Note that this motion cannot be realized on a single simple surface because the trajectories should never intersect. Therefore the surface S must consist of an infinite number of distinct leaves, linked and tightly pressed together, so that the different leaves do not merge but are separated by vanishingly small distances. The phase-space trajectory belongs to such an infinitely branching surface.

Figure 3.17 shows the results of numerical integration of equations (3.2.1) for $\sigma = 10$, $b = 8/3$, $r = 28$ and the initial conditions $x(0) = y(0) = z(0) \approx 0$.

Thus, the Lorenz attractor can be pictured neither as a two-dimensional surface, nor as a simple two-dimensional manifold. It turns out that, if we take a planar cross section through this attractor, the plane will be intersected by a trajectory at multiple points which form a Kantor set (Chap. 4).

To conclude this section, we briefly discuss another example of a limit set that combines local instability with global compression of the phase-space volume, i.e. the *Smale-Williams solenoid* [3.20–21]. This famous example is purely speculative and does not correspond to any real system.

The Smale-Williams solenoid can be constructed in a space of three or more dimensions. Consider a toroidal region D representing the interior of a two-dimensional torus (Fig. 3.18a). Let us stretch out the torus while simultaneously contracting its diameter, then twist and fold it, and place it again into the region of space occupied by the initial torus (Fig. 3.18b). After such a transformation the initial region D will be mapped into another smaller region D' lying within D. We can now repeat the same transformation applying it this time to the torus that contains D'. Then we will obtain yet another region D'' which lies inside D'. The

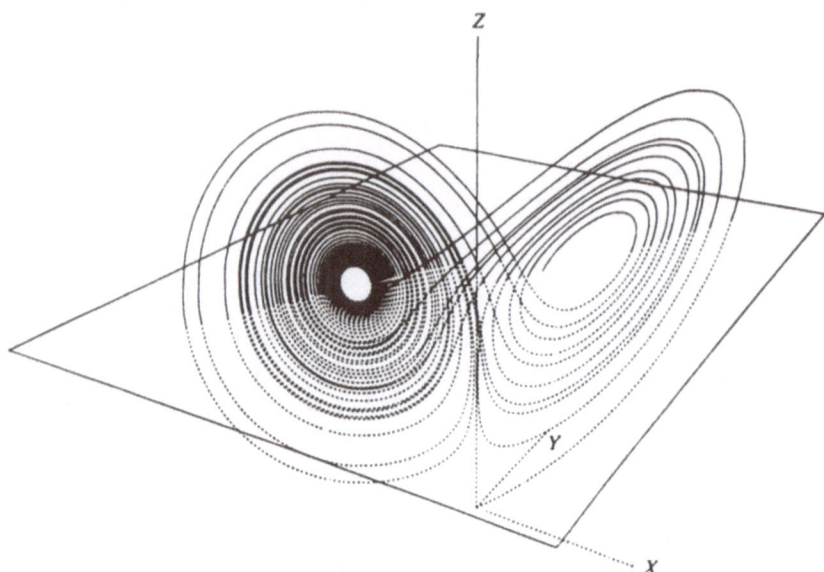

Fig. 3.17. Chaotic phase trajectory on the Lorenz attractor. From [3.19]

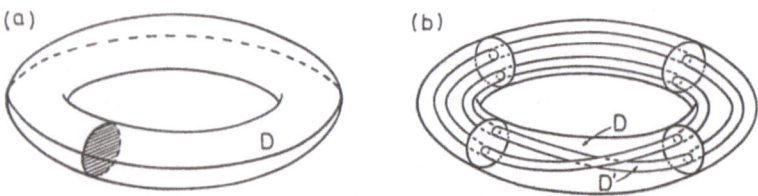

Fig. 3.18a,b. Construction of the Smale-Williams solenoid

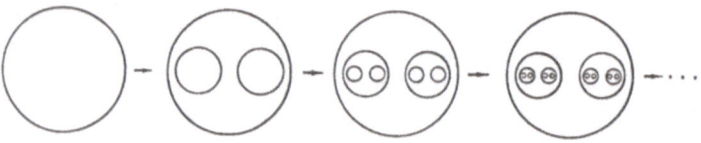

Fig. 3.19. Cross-section of a hierarchical structure produced constructing the Smale-Williams solenoid

infinite series of such transformations produces a hierarchy of regions that will look in cross-section as shown in Fig. 3.19. If we consider any two initially close points, we would find that in the process of such transformations they will become separated by an exponentially increasing distance. On the other hand, the volume of successive regions will tend to zero. The limit set produced by the sequence of these transformations represents an open, infinitely thin filament that pierces any vertical cross-section at an infinitely large number of points.

Thus, we have demonstrated in this section that even very simple dissipative dynamical systems, such as the Lorenz model, can support extremely complicated and irregular forms of motion. The mathematical description of such chaotic behavior of dissipative systems is based on the concept of strange attractors, which represent the attractive sets of a complex structure. It is natural now to turn to a discussion of the practical criteria that allow us to distinguish regular and chaotic motion.

3.3 Lyapunov Exponents

We already know that there are several possible types of attractors of dissipative dynamical systems. These are stable fixed points, stable limit cycles, and invariant tori (all of these belonging to the class of simple attractors), and also strange attractors. Simple attractors correspond to some regular motion of a system (which, however, might appear rather complex, as in the case of quasiperiodic motion of an invariant torus). On the other hand, the dynamics of systems with strange attractors is intrinsically chaotic. Therefore, it is important to develop some criteria to distinguish between regular (but apparently complex) and true chaotic motion.

Since chaos is intimately related to the instability of phase-space trajectories, which leads to fast divergence of any two initially close trajectories in phase space, it seems natural to base such a criterion on measuring this divergence.

Consider again a dynamical system defined by a set of ordinary differential equations

$$\dot{x} = v(x) \quad , \quad x(0) = x_0 \quad . \tag{3.3.1}$$

Let us take two neighboring points $x_1(0)$ and $x_2(0)$ in phase space and observe the behavior of the trajectories originating at $t = 0$ from these two points. In particular, we can measure the separation $d(t) = |\xi(t)| = |x_2(t) - x_1(t)|$ between these trajectories at any time moment t. Then, if the behavior is chaotic, $d(t)$ will increase exponentially with time, so that in the long time limit we would have

$$d(t) \approx d(0)\,e^{kt} \quad . \tag{3.3.2}$$

Using (3.3.2), we can estimate the mean rate of exponential divergence of the trajectories as

$$k \approx \frac{1}{t} \ln \left[\frac{d(t)}{d(0)} \right] \quad . \tag{3.3.3}$$

This estimate is not, however, generally acceptable. Indeed, for finite motion (which we are considering here), the distance $d(t)$ cannot grow indefinitely. Therefore, at large times, k will approach zero regardless of whether the motion is regular or chaotic. However, the smaller we choose the initial separation $d(0) = |\xi(0)|$ the longer we can observe the growth of $d(t)$ before it reaches its maximal value. Consequently, we should take the limits $d(0) \to 0$ and $t \to \infty$ in (3.3.3) and write

$$
h = \lim_{\substack{d(0) \to 0 \\ t \to \infty}} \frac{1}{t} \ln \left[\frac{d(t)}{d(0)} \right] \quad . \tag{3.3.4}
$$

In the literature h is sometimes called[4] the *Kolmogorov-Sinai entropy* (e.g. [3.25]). Using this property, one can distinguish between chaotic and regular behavior. For instance, if a system has periodic or quasiperiodic dynamics, the distance $d(t)$ does not increase exponentially with time and h is equal to zero. If the motion consists of attraction to some fixed point, $d(t)$ vanishes in the long time limit and therefore $h < 0$. In the case of chaotic motion h is always positive.

Note that h is a dimensional entity (it has the dimension of inverse time) and therefore it gives not only the qualitative, but also a quantitative specification of a dynamical regime. When $h > 0$, the inverse quantity $t_{mix} = h^{-1}$ determines the characteristic mixing time for the system considered. After a time $t \gg t_{mix}$, any initial region Ω_0 will spread over the entire limit set in phase space, i.e. over the entire strange attractor[5]. Only a probabilistic description of the system dynamics is justified at $t \gg t_{mix}$. On the other hand, for $t \ll t_{mix}$ the behavior of a system can be predicted with sufficiently good accuracy (which obviously cannot exceed the accuracy ε with which the initial position of the point is specified).

Therefore, the "entropy" h defined by (3.3.4) is an important property of a dynamical system. Today it is actively used as a practical criterion for chaotic motion. Detailed numerical recipes for the computation of h can be found in [3.1, 26–27].

Generally, the value of h might also depend on the choice of the initial point $x_1(0)$ of the trajectory. A dynamical system can simultaneously possess several different attractors with their own basins and, depending on the particular basin into which $x_1(0)$ falls, we will obtain different values of h. For example, the Lorenz system (3.2.1) has two simple and one strange attractor in the interval of r from 24.06 to 24.74 for $\sigma = 10$ and $b = 8/3$. One can also imagine a situation in which the phase space of some dissipative system has several strange attractors.

At first glance, the definition (3.3.4) seems to leave some ambiguity. Indeed, is it not possible that the value of the limit depends on the direction of the initial shift $\xi(0)$? Below we show that, although such dependence exists, it is not essential. If the direction of the vector of the initial shift is chosen at random, the obtained limit value of h will almost always be the same. These properties are closely related to the Lyapunov exponents which play a fundamental role in the theory of dynamical systems.

---

[4] Actually, it does not coincide with the entropy proposed by *Kolmogorov* [3.22–23] and *Sinai* [3.24]. It will be shown that h represents the maximal Lyapunov exponent. Under some conditions, it can give an estimate of the true entropy [3.26].

[5] This quantity can also be defined for the Hamiltonian systems studied in Chap. 2. Then t_{mix} specifies the characteristic time during which Ω_0 spreads over the energetic hypersurface.

Let us take a trajectory $x(t)$ of a dynamical system (3.3.1), originating from point $x(0)$, and another neighboring trajectory $\tilde{x}(t) = x(t) + \xi(t)$. Consider the function

$$\Lambda(\xi(0)) = \lim_{t \to \infty} \frac{1}{t} \ln \left[\frac{|\xi(t)|}{|\xi(0)|} \right] \tag{3.3.5}$$

defined for the vectors $\xi(0)$ of the initial shift, such that $|\xi(0)| = \varepsilon$ and $\varepsilon \to 0$. We claim that, under all possible rotations of the vector $\xi(0)$, the function Λ will change by jumps and will acquire a finite set of values $\{\lambda_j\}$, $j = 1, 2, \cdots, n$. These values are called the *(global) Lyapunov exponents*.

The meaning of this definition can be explained as follows. If two trajectories remain close in the course of time, the evolution of $\xi(t)$ obeys a linearized equation

$$\dot{\xi} = A\xi \quad, \tag{3.3.6}$$

where A is a matrix with elements

$$a_{ik} = \left. \frac{\partial v_i}{\partial x_k} \right|_{x=x(t)} \quad . \tag{3.3.7}$$

In general the matrix A will depend on time, i.e. $A = A(t)$. However, to simplify the argument we first analyze the case where it is constant. We also assume that all eigenvalues of A are real. The general solution to (3.3.6) can then be written in the form

$$\xi(t) = \sum_j C_j e_j \exp(\lambda_j t) \quad, \tag{3.3.8}$$

where C_j are coefficients determined by the initial conditions, and e_j are the eigenvectors of A corresponding to different eigenvalues λ_j.

Each eigenvector e_j defines a certain invariant direction in the space of all vectors ξ. If the vector $\xi(0)$ of the initial shift was directed along one of the vectors e_j, the vector $\xi(t)$ will remain parallel to it and, as time elapses, the distance between the two trajectories considered will increase or decrease exponentially, $|\xi(t)| \sim \exp(\lambda_j t)$. With such a choice of initial shift vector $\xi(0)$ we have

$$\Lambda(\xi(0)) = \lambda_j \quad . \tag{3.3.9}$$

Generally, the initial shift vector $\xi(0)$ would have components along several or all vectors e_j (Fig. 3.20). In this case is determined at large times, as seen from (3.3.8), by the term with the largest exponent in the decomposition of $\xi(t)$. This implies that, in the long time limit, the distance is always proportional to one of the exponents $\exp(\lambda_j t)$. Consequently, the function $\Lambda(\xi(0))$ can only take values λ_j and, when the vector $\xi(0)$ is rotated, this function will change its value by abrupt jumps from one λ_j to another.

Besides real eigenvalues, the matrix A may also have complex eigenvalues $\alpha = \lambda + i\omega$. Since the matrix A is real, the complex conjugate $\alpha^* = \lambda - i\omega$ of

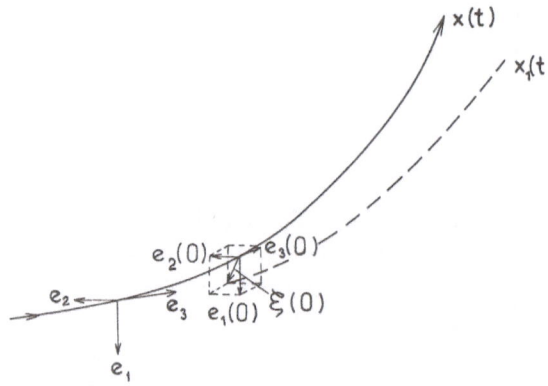

Fig. 3.20. Two close trajectories of a dynamical system

any such eigenvalue should again represent a valid eigenvalue of A. This pair of complex-conjugate eigenvalues specifies a certain invariant plane which is spanned by two vectors e and e'. On such a plane, the solution to (3.3.6) has the form

$$\boldsymbol{\xi}(t) = (Ce \cos \omega t + C'e' \sin \omega t) \exp(\lambda t) \quad . \tag{3.3.10}$$

Therefore, if we choose an initial shift $\boldsymbol{\xi}(0)$ belonging to this plane, the distance between these two trajectories will change with time, on average, as $|\boldsymbol{\xi}(t)| \sim \exp(\lambda t)$. Then (3.3.5) implies that $\Lambda(\boldsymbol{\xi}(0)) = \lambda$. We see that any pair of complex-conjugate eigenvalues α and α^* gives rise to a Lyapunov exponent $\lambda = \mathrm{Re}\{\alpha\}$. This exponent should be considered as doubly degenerate because it corresponds to the motion on an invariant plane, not along some invariant direction. Hence, for a constant matrix A, all Lyapunov exponents coincide with the real parts of its eigenvalues.

Actually, the elements of the matrix A depend on time. The theory of linear differential equations with time-dependent coefficients [3.28] states that a general solution to (3.3.6) can be written as a superposition of n fundamental partial solutions $\{\boldsymbol{\xi}_j(t)\}$, i.e.

$$\boldsymbol{\xi}(t) = \sum_j C_j \boldsymbol{\xi}_j(t) \quad , \tag{3.3.11}$$

where the coefficients C_j are determined from the decomposition of the initial shift $\boldsymbol{\xi}(0)$ in terms of the vectors $\boldsymbol{\xi}_j(0)$. Generally speaking, the time dependence of the fundamental partial solutions $\{\boldsymbol{\xi}_j(t)\}$ can be arbitrary. However, if we take into account that the matrix $A(t)$ does not increase indefinitely with time, for large times we can write [3.28]

$$|\boldsymbol{\xi}_j(t)| = \Phi_j(t) \exp(\lambda_j t) \quad , \tag{3.3.12}$$

where the function $\Phi_j(t)$ grows more slowly than any time exponential. Therefore, when we choose the initial shift vector $\boldsymbol{\xi}(0)$ directed along one of the vectors $\boldsymbol{\xi}_j(0)$, (3.3.5) yields $\Lambda(\boldsymbol{\xi}(0)) = \lambda_j$, which formally coincides with the result (3.3.9). Note, however, that λ_j is no longer the real part of an eigenvalue of A.

For an arbitrary initial shift $\boldsymbol{\xi}(0)$, the limit (3.3.5) almost certainly yields the maximal Lyapunov exponent. Indeed, if we take the initial shift vector at random, it would most probably have nonvanishing projections on all the vectors $\boldsymbol{\xi}_j$, including the one corresponding to the maximal value of λ_j. The term with this value λ_j would dominate in (3.3.11) at large times, and therefore the limit (3.3.5) would yield $\lambda_{\max}$. We see that the quantity h, introduced above by (3.3.4) and called the Kolmogorov-Sinai entropy, is essentially the maximal Lyapunov exponent.

Below we list, without proof, the principal properties of the Lyapunov exponents in dissipative and conservative dynamical systems.

i) The total number of exponents is equal to the dimensionality n of the phase space of a particular dynamical system. It is convenient to enumerate them in order of decreasing magnitude, so that $\lambda_1 \geq \lambda_2 \geq \cdots \geq \lambda_n$. If a trajectory does not end in a fixed point, one of the Lyapunov exponents (corresponding to the shift along the trajectory) is always equal to zero.

ii) In Hamiltonian systems [3.11] these exponents satisfy the condition $\lambda_j = -\lambda_{2k-j+1}$, where k is the number of degrees of freedom ($n = 2k$). Therefore, in such systems at least two exponents should be equal to zero. In completely integrable systems all Lyapunov exponents are equal to zero.

iii) The sum of all Lyaspunov exponents gives the value of div v, averaged along a trajectory, i.e.

$$\lim_{t \to \infty} \frac{1}{t} \int_0^t \operatorname{div}[\boldsymbol{v}(\boldsymbol{x}(t))]\, dt = \sum_j \lambda_j \quad . \tag{3.3.13}$$

It was shown in Sect. 3.1 that the evolution of the phase-space volume is determined by div v. When div $v < 0$, the system is dissipative, while div $v = 0$ implies that the phase-space volume is conserved. For Hamiltonian systems we thus have $\sum_j \lambda_j = 0$, whereas for dissipative systems $\sum_j \lambda_j < 0$.

A dynamical system is called *hyperbolic* if it has both positive and negative Lyapunov exponents. Hyperbolic dissipative systems possess strange attractors.

Usually, the analytical calculation of Lyapunov exponents (and of $h = \lambda_{\max}$) is not possible. Nvertheless, there are very efficient and reliable numerical algorithms which allow one to compute all Lyapunov exponents [3.1, 29–31]. For instance, *Shimada* and *Nagascima* [3.29] found that the Lorenz system (3.2.1) at $\sigma = 16$, $b = 4$ and $r = 40$ has exponents $\lambda_1 = 1.37$, $\lambda_2 = 0$, $\lambda_3 = -22.37$.

3.4 The Autocorrelation Function

Another characteristic property, used along with the Lyapunov exponents in studies of complex dynamical systems, is the autocorrelation function. This function contains the essential information about the behavior of a system. For example, periodic and quasiperiodic regimes are characterized, respectively, by periodic and quasiperiodic autocorrelation functions. On the other hand, if the autocorrelation function fades with time and the system has no attractive fixed points, we can expect that its dynamics is chaotic. Hence, fading of the autocorrelation function can be used as a criterion for dynamical chaos.

In experimental and computational studies it is more convenient to estimate, instead of the autocorrelation function, the *spectral density* (or simply the spectrum) which is closely related to the former. The form of the spectral density provides one of the simplest criteria for classification of dynamical regimes. Using it, we can easily distinguish between chaotic, periodic and quasiperiodic behavior.

When a system displays periodic behavior (which corresponds to a limit cycle) with a time period T_0, it produces a discrete spectrum consisting of narrow lines at the frequency $\omega_0 = 2\pi/T_0$ and its integer multiples $2\omega_0, 3\omega_0, \cdots$, etc. If the dynamics of a system is quasiperiodic with some incommensurable frequencies $\omega_1, \cdots, \omega_k$ (this corresponds to a k-dimensional nonresonant torus), its spectrum consists of separate lines at the frequencies $\omega_1, \cdots, \omega_k$ and at their integer combinations. We see that for quasiperiodic motion the spectrum remains discrete (although the lines in the spectrum can be very close if the number k of different basic frequencies is large enough). In contrast to this, for genuinely chaotic regimes the spectrum should be *continuous*.

Below we discuss in more detail the properties of the autocorrelation function and spectral density. This discussion is preceded by their formal definitons.

Suppose $x(t) = \{x_1(t), \cdots, x_n(t)\}$ is a solution of the dynamical system (3.3.1). We will consider only one component of this solution, denoted simply as $x(t)$. The *autocorrelation function* $g(\tau)$ represents the average over some large time T of the product of the values of x, taken with a constant time shift τ, i.e.

$$g(\tau) = \lim_{T \to \infty} \frac{1}{T} \int_0^T x(t)\,x(t+\tau)\,dt \quad . \tag{3.4.1}$$

The *spectral density* $S(\omega)$ is defined by

$$S(\omega) = \frac{1}{2\pi} \int_{-\infty}^{\infty} \exp(-\mathrm{i}\,\omega\tau) g(\tau)\,d\tau \quad . \tag{3.4.2}$$

Suppose that $x(t)$ is a periodic function with period T_1. Then we can expand it in the Fourier series

$$x(t) = \sum_{n=-\infty}^{\infty} C_n \exp(-\mathrm{i}\,n\omega_1 t) \quad , \tag{3.4.3}$$

where $\omega_1 = 2\pi/T_1$ and C_n are the Fourier coefficients. In this case the autocorrelation function is

$$g(\tau) = \frac{1}{T} \int_0^T x(t)x(t+\tau)\,dt$$

$$= \sum_{n=-\infty}^{\infty} \sum_{n=-\infty}^{\infty} C_n C_{n'} \exp(-in'\omega_1\tau)$$

$$\times \frac{1}{T} \int_0^T \exp[-i(n+n')\omega_1 t]\,dt$$

$$= \sum_{n=-\infty}^{\infty} C_n C_{-n} \exp(-in\omega_1\tau) = \sum_{n=-\infty}^{\infty} |C_n|^2 \exp(-in\omega_1\tau) \quad , \qquad (3.4.4)$$

where we have taken into account that $C-n = C_n^*$ because the function $x(t)$ is real. We see that $g(\tau)$ is periodic with period T_1.

Using (3.4.4) and (3.4.2) we can find the spectral density

$$S(\omega) = \sum_{n=-\infty}^{\infty} |C_n|^2 \frac{1}{2\pi} \int_{-\infty}^{\infty} \exp[-i(n\omega_1 - \omega)\tau]\,d\tau$$

$$= \sum_{n=-\infty}^{\infty} |C_n|^2 \delta(\omega - n\omega_1) \quad . \qquad (3.4.5)$$

Hence the spectrum of a periodic function $x(t)$ corresponding to the periodic motion of a dynamical system (3.3.1) is discrete and contains, besides the principal frequency ω_1, all its integer multiples $n\omega_1$.

As an illustration, we consider the example of spectra obtained by *Turner* et al. [3.32] in experiments with the Belousov-Zhabotinskii chemical reaction in a stirred tank flow reactor. This reaction was described in Chap. 3 of *Foundations of Synergetics I*, where it was noted that, in a thin nonstirred layer, it can develop very complicated spatio-temporal patterns. When a stirred tank is used, the solution remains homogeneous but develops complex temporal oscillations. If new portions of fresh reagents are permanently supplied to the reactor, such oscillations can persist indefinitely. Depending on the concentration ratio in the solution, one can observe either periodic or chaotic oscillations. Figure 3.21 shows periodic oscillations in such a system together with their spectrum.

Suppose now that $x(t)$ is a quasiperiodic function,

$$x(t) = x(\phi_1(t), \cdots, \phi_k(t)) \quad , \qquad (3.4.6)$$

where x is 2π-periodic with respect to each phase ϕ_i,

$$x(\phi_1 + 2\pi, \cdots, \phi_k + 2\pi) = x(\phi_1, \cdots, \phi_k) \quad , \qquad (3.4.7)$$

and $\phi_i(t) = \omega_i t$, $i = 1, 2, \cdots, k$. Quasiperiodicity implies that different frequencies ω_i are not rationally related, i.e. there is no common frequency ω_0 such all ω_i are some integer multiples of ω_0.

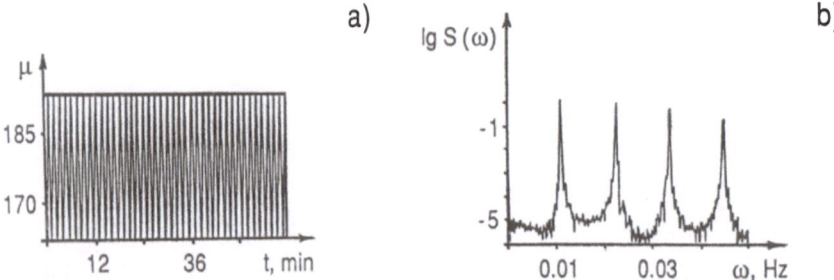

Fig. 3.21. Periodic oscillations of the concentration (**a**) in the Belousov-Zhabotinskii solution and the corresponding spectral density (**b**). From [3.32]

Let us expand $x(t)$ in a Fourier series with respect to phases ϕ_i,

$$x(t) = \sum_{n_1,\cdots,n_k} C_{n_1} \cdots C_{n_k} \exp[-i(n_1\phi_1 + \cdots + n_k\phi_k)] \quad . \tag{3.4.8}$$

Substitution of $\phi_i = \omega_i t$ into (3.4.8) yields

$$x(t) = \sum_{n_1,\cdots,n_k} C_{n_1} \cdots C_{n_k} \exp[-i(n_1\omega_1 + \cdots + n_k\omega_k)t] \quad . \tag{3.4.9}$$

Using the last decomposition, we can find the autocorrelation function and the spectral density

$$g(\tau) = \sum_{n_1,\cdots,n_k} \left| C_{n_1} \cdots C_{n_k} \right|^2 \exp[-i(n_1\omega_1 + \cdots + n_k\omega_k)\tau] \quad , \tag{3.4.10}$$

$$S(\omega) = \sum_{n_1,\cdots,n_k} \left| C_{n_1} \cdots C_{n_k} \right|^2 \delta(\omega - n_1\omega_1 - \cdots - n_k\omega_k) \quad . \tag{3.4.11}$$

Therefore, the spectrum of a quasiperiodic function contains all possible linear combinations $n_1\omega_1 + \cdots + n_k\omega_k$. If the number of principal frequencies ω_i is large, lines in this spectrum might be very closely spaced, but would nevertheless remain discrete.

Finally, if the dynamics of a system is chaotic, the function $x(t)$ is aperiodic and its spectrum is *continuous*. In experiments [3.32] with the Belousov-Zhabotinskii reaction, the transition from periodic to chaotic oscillations, which occurs with the change in solution composition, is manifested by a characteristic transformation of the spectrum, as shown in Fig. 3.22.

Besides the two above-mentioned criteria for dynamical chaos based on the Lyapunov exponents and the autocorrelation function, there is also a simple criterion which uses the Poincaré map. A detailed description of the map for Hamiltonian systems was given in Chap. 2. Below we briefly outline its application to dissipative dynamical systems.

Suppose that the phase-space flow generated by a dynamical system (3.3.1) is intersected by some surface S. Let us mark on this surface successive points

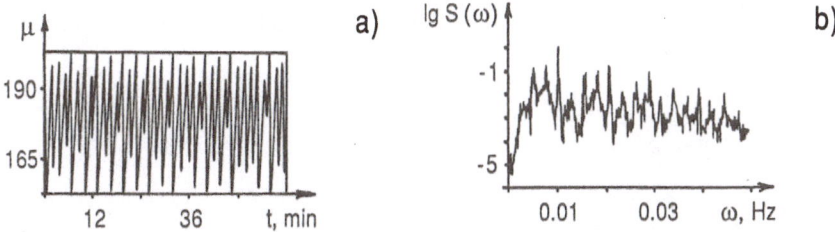

Fig. 3.22. Chaotic oscillations (**a**) in the Belousov-Zhabotinskii reaction and their spectrum (**b**). From [3.32]

where the trajectory intersects the surface in a specified direction (Fig. 3.23). For simplicity, we now assume that the phase space of the system is three-dimensional; generalization to the n-dimensional case is the same as in Chap. 2. Then, if we obtain in the cross-section S a finite set of points that are subsequently mapped into one another under the system evolution (Fig. 3.23a), this would indicate the presence of a limit cycle. When successive intersection points of a trajectory fill an entire closed curve on the surface S, this corresponds to an invariant torus (Fig. 3.23b). Finally, if we find that a trajectory hits the surface S at a set of points that are irregularly scattered over some region (Fig. 3.24), this indicates the presence of a strange attractor and the regime of chaotic motion.

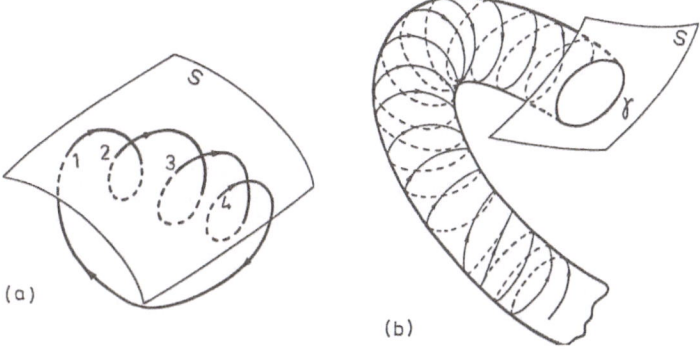

Fig. 3.23. Poincaré maps of periodic (**a**) and quasiperiodic (**b**) motion

To conclude this section we note that, due to the finite duration of any experimental observation (or computer simulation), it is never possible to rigorously distinguish periodic motion with a very large period and genuinely chaotic motion. Indeed, when the period is extremely large, distances between the lines in the spectrum of motion are very small. On the other hand, the finite time T of observation gives rise to a broadening of each line, so that it occupies an interval of frequencies of about $2\pi/T$. When the time of observation is comparable to the period of motion, the broadened lines overlap and the resulting spectrum is

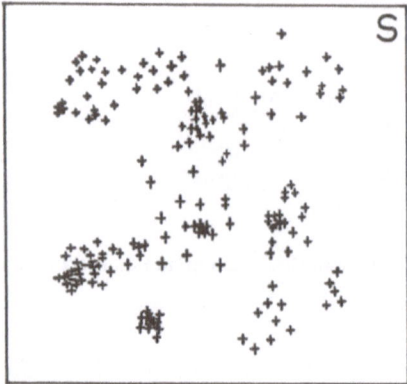

Fig. 3.24. Example of a Poincaré map for chaotic motion

practically indistinguishable from a continuous one. The same remark applies to the Poincaré map. When the period of motion is very large, the trajectory would hit the cross-sectional surface at a very large number of different points before the cycle is completed.

However, the distinction between genuinely chaotic motion and periodic motion with extremely long periods is not essential from a practical point of view. If the motion is indistinguishable from chaotic motion over a long time of observation, it is sufficient to treat it as chaotic for all practical purposes.

4. Fractals

Strange attractors represent sets in phase space which are different from points, lines or pieces of (hyper)surfaces. To give an adequate description of these complex patterns, calls for the use of special geometric objects known as fractal sets.

The chaotic dynamics of dissipative systems is not the only field of science where fractals appear. Such geometric objects as jagged coastlines, trajectories of Brownian motion or clusters of particles obtained by diffusion-limited aggregation are also examples of fractal sets.

An important property of an ideal fractal is its self-similar nature. If we magnify some fragment of such a pattern, we would see precisely the same geometric structure reproduced on a new scale. Obviously, this implies that such objects should differ from the smooth manifolds which represent lines or pieces of surfaces.

Since the fractals occupy an intermediate position between standard geometric objects with integer dimensions (1 for a line, 2 for a surface, etc.), they can be conveniently characterized by their fractal dimensions (if such an object has a dimension 1.9, it is "almost" a piece of a surface). Fractal dimensions can be introduced in various distinct ways, each emphasizing a different geometric aspect of a pattern.

4.1 Self-Similar Patterns

A favorite example of many school textbooks is the trajectory of the Brownian motion of a particle wandering diffusively in a liquid. Although this trajectory is extremely irregular, it still displays some remarkable invariance: If we take a small region containing part of the trajectory, and enlarge it to the full scale, the result would look very similar to the original whole trajectory. The procedure can be repeated many times with the same effect, until one finally reaches the molecular scale of distances.

This example demonstrates the important property of *self-similarity* or *scale invariance*, which characterizes patterns found in a wide variety of different problems. If this property is possessed (at least approximately) by some pattern, it is called a *fractal* [4.1–3].

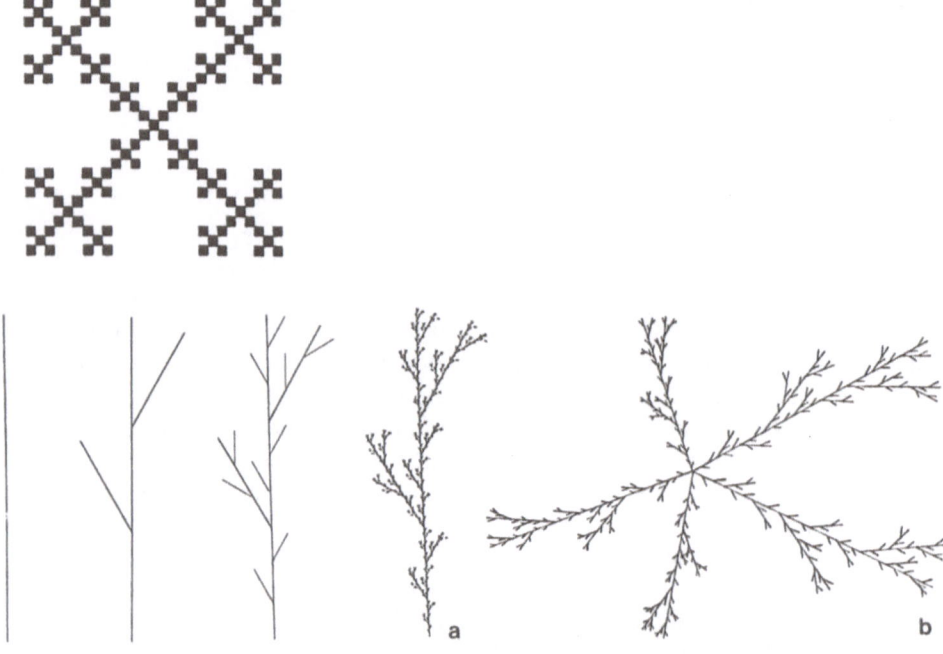

Fig. 4.2. Construction of a fractal tree (**a**) and the final tree (**b**) with several initial trunks

Consider, for instance, an idealized geometric "snowflake" (Fig. 4.1) as suggested by *Vicsek* [4.4]. Each element of this fractal pattern consists of five identical elements of a smaller size, each with the same structure. Clearly, this pattern is strictly scale-invariant: if we enlarge any of its elements three times, we would obtain the entire initial pattern, and so on.

Another geometric self-similar pattern [4.5] represents an idealized "tree" (Fig. 4.2). It is constructed as follows. We start from a bare trunk and add two branches, each three times shorter than the trunk. Next we add two new branches, also three times shorter, to each previous branch and to the trunk of the tree. If the initial "tree" had several trunks pointing in different directions, repetition of this procedure would yield the pattern shown in Fig. 4.2b.

Analogous forms are often detected in nature. Figure 4.3 shows the so-called "viscous fingers" [4.6] observed in a hydrodynamical experiment with the Hele-Shaw cell. This cell is bounded by two plates, and the space between them is occupied by liquid. If we inject some less viscous liquid into the middle of the cell, a pattern of "viscous fingers" develops.

Similar patterns are formed in the process of diffusion-limited aggregation. Suppose that we have a thin layer of liquid where some particles wander randomly. Inside the liquid there is initially a nucleus with a smooth surface. When one of the wandering particles touches the nucleus, it sticks to it. Further particles can stick either to the bare surface of the nucleus or to particles already attached

Fig. 4.3. "Viscous fingers" in the Hele-Show cell. From [4.3]

Fig. 4.4. Fractal cluster obtained in diffusion-limited aggregation. From [4.2]

to it. After this process has continued for some time, no bare space remains on the initial nucleus and the aggregating particles produce a large, highly irregular cluster (Fig. 4.4). Remarkably, this cluster looks very much like the fractal tree shown in Fig. 4.2.

The aggregation process can be described [4.7] by a system of partial differential equations. The formation of fractal patterns then results from instabilities inherent in the dynamics of such a system.

Strange attractors of the dissipative dynamical systems analyzed in the previous chapter, are often found to possess a self-similar structure. As an illustration, consider the *Henon attractor* [4.8]. This attractor is obtained in a dynamical model that is described by some discrete map. However, in Chap. 5 it will be shown that discrete maps are very closely related to dynamical systems with continuous time. In principle, the Henon map can be interpreted as a Poincaré map for some continuous dynamical system. When a dynamical system has a strange attractor,

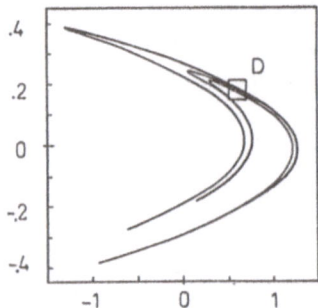

Fig. 4.5. The Henon attractor. From [4.6]

its phase-space trajectories leave a complicated set of points in the cross section. For the Henon map, this set is shown in Fig. 4.5.

Investigating Fig. 4.5, we are led to suspect that the Henon attractor consists of "thick" curves with some internal structure. To further check this, we can enlarge some region D of the attractor. It would then be seen (Fig. 4.6) that this part is indeed resolved into a set of several close curves. If we repeat this procedure, we come to the conclusion that each of the "curves" is made up of an infinite number of components. The structure of this set is repeated at each new scale level. Hence, the Henon attractor is self-similar.

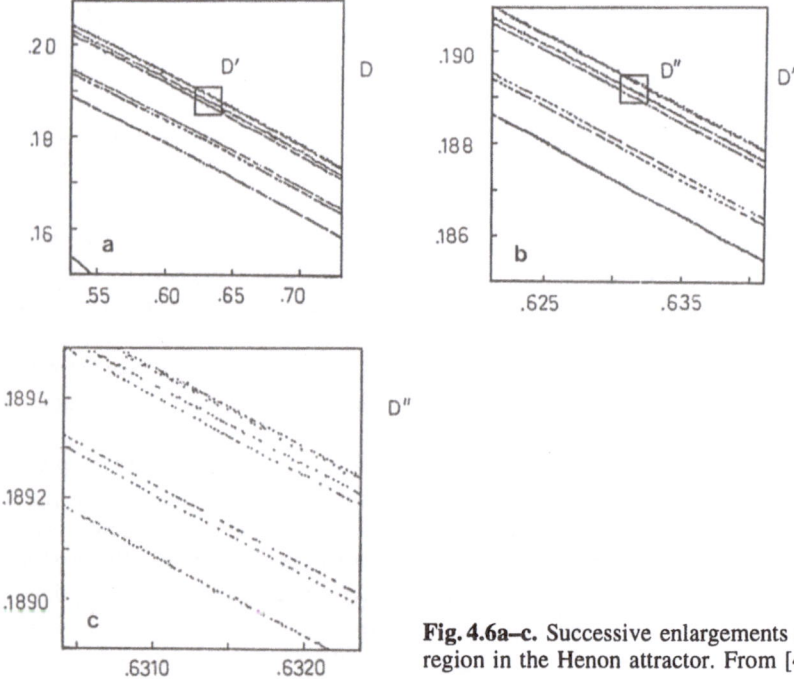

Fig. 4.6a–c. Successive enlargements of a small region in the Henon attractor. From [4.6]

We clearly need some mathematical concepts to provide a qualitative description and measure of complexity, applicable to self-similar patterns of any origin. These concepts were initially developed for the *Kantor sets* which are sets with scale invariance.

The simplest example of a Kantor set can be constructed in the following way. Take a line segment of unit length [0,1], divide it into three equal parts and delete the central interval $(1/3, 2/3)$. Repeat the procedure iteratively with each of the remaining parts (Fig. 4.7). The parts which remain after an infinite number of iterations form the Kantor set.

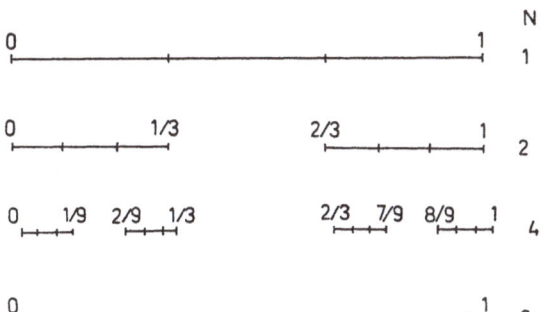

Fig. 4.7. Construction of the simplest Kantor set

Let us estimate the total length of the intervals deleted in the process of construction. At the kth step we delete 2^{k-1} intervals, each of length 3^{-k}. Therefore, the total length of deleted intervals is

$$\frac{1}{3} + \frac{2}{9} + \frac{4}{27} + \cdots = \frac{1}{2} \sum_{k=1}^{\infty} \left(\frac{2}{3}\right)^k = \frac{(1/3)}{(1 - 2/3)} = 1 \quad . \tag{4.1.1}$$

Therefore the measure (i.e. the length) of the remaining parts of the initial line segment which form the Kantor set, is strictly zero.

The two-dimensional generalization of the above example yields the *Serpinski gasket* (Fig. 4.8). We now take a square with sides of unit length and divide it into nine equal squares. Then, at the first iteration ($k = 1$) we delete the interior of the central square. Later, the procedure is repeated iteratively for all remaining squares. One can again verify that as $k \to \infty$ the total area of deleted squares is equal to 1, i.e. to the area of the initial square. What is left after an infinite sequence of deletions represents a Kantor set known as the Serpinski gasket. Similar constructions can be realized in a space of any dimensionality.

The strange attractors of dynamical systems have the local structure of a Kantor set and, in this respect, they differ from "simple" attractors which represent either single points or entire pieces of (hyper)surfaces.

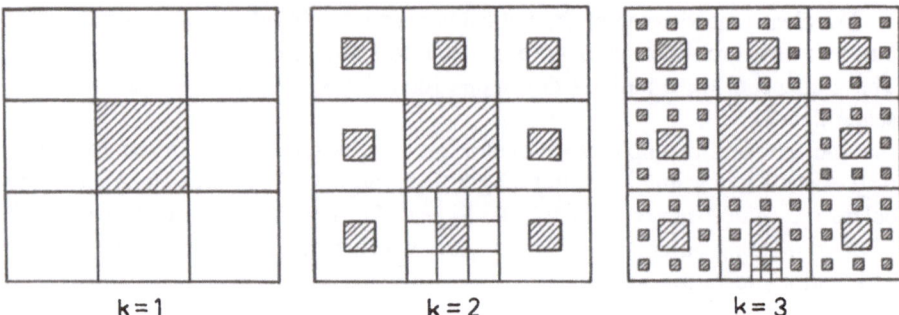

k = 1   k = 2   k = 3

Fig. 4.8. Construction of the Serpinski gasket (the shaded regions are deleted)

4.2 Fractal Dimensions

From a geometric point of view, fractal patterns occupy an intermediate position. For instance, the Serpinski gasket is apparently "more" than a mere line, and yet it is not a true surface because its area is equal to zero. The same can be said about other examples of such patterns. A *fractal dimension* is designated to measure the degree of proximity between a fractal pattern and the nearest smooth manifold.

Since such a "degree of proximity" may be defined in a number of different ways, there is an entire family of different fractal dimensions. Below we give definitions of several dimensions, together with a discussion of the relationship between them.

Consider some set A in n-dimensional space. Let us cover this set by a system of n-dimensional cubic cells with sides ε, in such a way that these cells contain all the points of A. Suppose $N(\varepsilon)$ is the minimum number of cells necessary to cover A. Then we can define the limit

$$D_F = \lim_{\varepsilon \to 0} \frac{\ln N(\varepsilon)}{\ln(1/\varepsilon)} \quad . \tag{4.2.1}$$

The quantity D_F is the *capacity* of A [4.9–10]. However, following *Mandelbrot* [4.1, 2], it is also often called simply the fractal dimension of set A.

For lines, surfaces or pieces of a higher-dimensional space, the capacity D_F takes integer values and coincides with the ordinary dimensionality. Indeed, the number of small cubes of side ε required to cover a continuous volume in a d-dimensional space is equal to this volume divided by ε^d, i.e. $N(\varepsilon) \sim K\varepsilon^{-d}$. Substitution of this expression into (4.2.1) yields $D_F = d$.

The capacity of self-similar patterns is non-integer. As an illustration, we determine below the capacities of the two Kantor sets defined in Sect. 4.1. In the first example, after an iteration step k, we have $N_k = 2^k$ undeleted segments, each of length $\varepsilon_k = 3^{-k}$. Therefore, if we choose the side of a "one-dimensional cube" (i.e. the length of a line segment) equal to ε_k, it would be sufficient to use N_k

"cubes" to cover the entire Kantor set (since all its points lie inside the undeleted segments). Consequently, we find

$$D_F = \lim_{k \to \infty} \frac{\ln N_k}{\ln(1/\varepsilon_k)} = \lim_{k \to \infty} \frac{\ln 2^k}{\ln 3^k} = \frac{\ln 2}{\ln 3} \approx 0.631 \quad . \tag{4.2.2}$$

For the Serpinski gasket, after a step k we have $N_k = 8^k$ undeleted squares, each of size $\varepsilon_k = 3^{-k}$. Therefore we obtain

$$D_F = \lim_{k \to \infty} \frac{\ln 8^k}{\ln 3^k} = \frac{\ln 8}{\ln 3} \approx 1.893 \quad . \tag{4.2.3}$$

Homogeneous fractal sets, such as the Serpinski gasket, are something of a geometric idealization. Fractals found in realistic situations are often highly inhomogeneous, i.e. their elements are positioned much more "thickly" in some areas than in others. The simplest fractal dimension D_F is not sensitive to the presence of such local variations. To take them into account, more sophisticated dimensions are constructed.

Suppose we have covered the entire fractal pattern by $N(\varepsilon)$ cubic cells of side ε. Then, for each of the cells we can specify the probability p_i that a point from this pattern, chosen at random, would belong to the ith cell. The resulting probability distribution can be used to construct the *generalized dimension* [4.11–12]

$$D_q = \lim_{\varepsilon \to 0} \frac{\ln \sum_{i=1}^{N(\varepsilon)} p_i^q}{(q-1)\ln \varepsilon} \quad , \tag{4.2.4}$$

where q is some real number, $0 \le q < \infty$.

If a fractal set is completely homogeneous, the probability p_i is the same for all cells, i.e. $p_i = 1/N(\varepsilon)$. Therefore,

$$\sum_{i=1}^{N(\varepsilon)} p_i^q = N(\varepsilon) \left[\frac{1}{N(\varepsilon)} \right]^q = N(\varepsilon)^{1-q} \tag{4.2.5}$$

and the generalized dimension D_q, given by (4.2.4), coincides with the capacity D_F [cf. (4.2.1)]. However, when homogeneity is lacking, D_q differs from D_F. *Hentschel* and *Procaccia* [4.12] showed that generally $D_q \ge D_{q'}$ for any $q < q'$; the equality is realized only if a fractal set is homogeneous.

Let us examine the meaning of D_q for several integer values of q. If $q = 0$ we have $p_i^q = 1$. Then the sum in (4.2.4) is simply $N(\varepsilon)$ and therefore (4.2.4) reduces to (4.2.1). Hence, $D_0 = D_F$ and we obtain the capacity.

The case when $q = 1$ is special, because both the denominator and the numerator in (4.2.4) vanish (note that the sum of all probabilities p_i is equal to unity). Nonetheless, we can still consider the limit $q \to 1$. Introducing $\kappa = q - 1$, we transform (4.2.4) as

$$D_1 = \lim_{\substack{\varepsilon \to 0 \\ \kappa \to 0}} \frac{\ln\left[\sum_{i=1}^{N(\varepsilon)} p_i \exp(\kappa \ln p_i)\right]}{\kappa \ln \varepsilon}$$

$$= \lim_{\substack{\varepsilon \to 0 \\ \kappa \to 0}} \frac{\ln\left(1 + \kappa \sum_{i=1}^{N(\varepsilon)} p_i \ln p_i\right)}{\kappa \ln \varepsilon}$$

$$= \lim_{\varepsilon \to 0} \frac{\sum_{i=1}^{N(\varepsilon)} p_i \ln p_i}{\ln \varepsilon} \quad . \tag{4.2.6}$$

Since $S(\varepsilon) = -\sum_i p_i \ln p_i$ is the information content of the proability distribution, D_1 shows how this content increases when we make the graining increasingly fine (i.e. when $\varepsilon \to 0$). Consequently, D_1 is called the *information dimension* of fractal sets.

If $q = 2$, the definition (4.2.4) yields

$$D_2 = \lim_{\varepsilon \to 0} \frac{\ln\left(\sum_{i=1}^{N(\varepsilon)} p_i^2\right)}{\ln \varepsilon} \quad . \tag{4.2.7}$$

The meaning of this expression can be clarified by noting that p_i^2 gives the probability of finding at least *two* points inside the ith cell. Therefore, D_2 is sensitive to binary correlations in the probability distribution. It is called the *correlation dimension*. This dimension is very useful in studies of turbulent flows; we return to its discussion in Chap. 7.

Higher generalized dimensions D_q with $q = 3, 4, 5, \cdots$ are related to correlations of higher orders and provide additional information about the geometric structure of a fractal.

To characterize complex inhomogeneous fractal sets, other criteria such as the α-spectrum [4.13] and γ-spectrum [4.14, 15] have also been proposed. Moreover, examples of *fat* fractals representing Kantor sets but still having an integer capacity D_F are known. Such objects are characterized by introducing a *metadimension* [4.16]. A detailed discussion of the problems related to fractal dimensions can be found in [4.1–3, 17–20].

4.3 Dimensions of Strange Attractors and Fractal Basin Boundaries

In the previous chapter, we have seen that strange attractors are extremely complicated geometric structures. Actually, *Ruelle* and *Takens* [4.21] introduced the concept of a "strange" attractor to emphasize its difference from smooth manifolds in phase space. Moreover, the example of the Henon attractor demonstrated that these objects might be self-similar. Therefore one can try to determine the fractal dimensions of strange attractors.

To numerically estimate the capacity D_F, using its definition (4.2.1) directly, one should divide the phase space of a dynamical system into cells of side ε. The phase-space trajectory should be integrated for a sufficiently long time, in order to exclude the transient process. After that time, one can begin to mark the cells when they are visited by a trajectory. If the trajectory is very long, the number of marked cells tends to $N(\varepsilon)$. This procedure may be repeated for different values of ε. The definition (4.2.1) of D_F implies that $\ln N(\varepsilon)$ should depend linearly on $\ln(1/\varepsilon)$ for small ε. Therefore, if we plot the dependence of $\ln N(\varepsilon)$ on $\ln(1/\varepsilon)$, which is obtained in a numerical experiment, the slope of this curve far from the origin of the coordinate system would give the fractal dimension (i.e. the capacity) D_F.

For example, the numerically determined [4.22] fractal dimension D_F of a strange attractor in the Lorenz model (3.2.1) at $\sigma = 10$, $b = 8/3$ and $r = 28$ is $D_F = 2.05 \pm 0.01$. Therefore, at these values of the parameters, the Lorenz attractor is very close to being a two-dimensional surface.

Russel et al. [4.23] numerically determined the fractal dimension of a strange attractor in the model described by the equations

$$
\begin{aligned}
\dot{x} &= y(z - 1 + x^2) + \gamma x \quad , \\
\dot{y} &= x(3z + 1 - x^2) + \gamma y \quad , \\
\dot{z} &= -2z(\nu + xy)
\end{aligned}
\tag{4.3.1}
$$

at $\nu = 1.1$ and $\gamma = 0.87$. They found that in this case $D_F = 2.318 \pm 0.002$, i.e. the attractor occupies an intermediate position between a surface and a three dimensional volume.

A simple estimate of the fractal dimension of a strange attractor can be obtained using the hypothesis of *Kaplan* and *Yorke* [4.24]. This hypothesis assumes that the fractal dimension D_F can be well approximated by the *Lyapunov dimension*

$$
D_L = j + \frac{\sum_{i=1}^{j} \lambda_i}{|\lambda_{j+1}|} \quad .
\tag{4.3.2}
$$

Here λ_i are the Lyapunov exponents ordered according to their values, i.e. $\lambda_1 \geq \lambda_2 \geq \cdots \geq \lambda_n$, and j is determined from the conditions

$$
\begin{aligned}
\lambda_1 + \lambda_2 + \cdots + \lambda_j &\geq 0 \quad , \\
\lambda_1 + \lambda_2 + \cdots + \lambda_j + \lambda_{j+1} &< 0 \quad .
\end{aligned}
\tag{4.3.3}
$$

For dynamical systems with $n = 3$ the Lyapunov dimension is given by

$$
D_L = 2 + \frac{\lambda_1}{|\lambda_3|} \quad .
\tag{4.3.4}
$$

This follows from the fact that, if such a system has a strange attractor, it has one positive (λ_1) and one negative (λ_3) Lyapunov exponent. Furthermore, one Lyapunov exponent should vanish ($\lambda_2 = 0$). The compression of phase-space volume implies that $\lambda_1 + \lambda_3 < 0$ (Sect. 3.1). Therefore, $\lambda_1 < |\lambda_3|$, and consequently the Lyapunov dimension D_L of a strange attractor in such a system should lie between 2 and 3.

The approximate correspondence between D_L and the fractal dimension D_F of a strange attractor can be explained using the example of a system with $n = 3$. We take a small volume element V_0 of the phase space at an initial time $t = 0$ and consider its evolution for $t > 0$. Since the dynamical system is dissipative and, we assume, the motion takes place on a strange attractor, this volume element becomes progressively compressed and deformed with time (Fig. 4.9a). Let us (mentally) straighten the deformed volume element, covering the band at each time moment t by three-dimensional cubes of side $\varepsilon(t)$, as shown in Fig. 4.9b,c. We know (cf. Sect. 3.3) that the Lyapunov exponents determine the degree of compression of a volume element along some directions and its degree of stretching along others. Using the Lyapunov exponents, we can estimate the time dependence of a volume element as

$$V(t) \sim V_0 \exp[(\lambda_1 - |\lambda_3|)t] \quad . \tag{4.3.5}$$

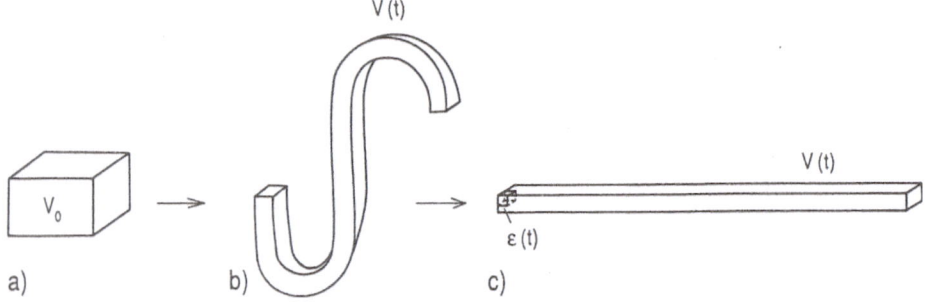

Fig. 4.9a–c. Illustration of the Kaplan-York hypothesis

The side $\varepsilon(t)$ of the cubes covering the deformed volume element at time t can be chosen to be equal to the minimum thickness of this band, which exponentially decreases with time. This yields

$$\varepsilon(t) \sim \exp(-|\lambda_3|t) \quad \text{and} \tag{4.3.6}$$

$$\varepsilon(t)^3 \sim \exp(-3|\lambda_3|t) \quad . \tag{4.3.7}$$

Since $\varepsilon(t)^3$ is the volume of cubes that should be taken to cover the volume element $V(t)$ at time moment t, the total required number of such cubes would be

$$N(t) \sim \frac{V(t)}{\varepsilon(t)^3} \sim \exp(\lambda_1 t + 2|\lambda_3|t) \quad . \tag{4.3.8}$$

Hence,

$$D_F = \lim_{t\to\infty} \frac{\ln N(t)}{\ln(1/\varepsilon(t))} = 2 + \frac{\lambda_1}{|\lambda_3|} \quad . \tag{4.3.9}$$

This result coincides with expression (4.3.4) for the Lyapunov dimension.

Although the above arguments indicate a strong correlation between the Lyapunov dimension and the fractal dimension D_F of strange attractors, they are not rigorous. The two dimensions coincide only in some special cases [4.23, 25–26]. Generally, the Kaplan-Yorke hypothesis does not hold [4.27]. It can be proved only that D_L provides an upper estimate for D_F [4.27–29].

Thus we have seen that fractal dimension is an important property of strange attractors. In the remainder of this section we consider only simple attractors, such as fixed points and limit cycles. The geometrical properties of such attractors themselves are, of course, trivial. However, as it turns out, the boundaries of attraction basins for these *simple* attractors may still have fractal properties. The presence of fractal basin boundaries can have important implications for the dynamics of a system. When they occur, small perturbations of the initial conditions produce substantial changes in the subsequent behavior of the system.

Fractal curves have been known since the 19th century – they were considered by K. Weierstrass as early as 1871. The question of when basin boundaries of an attractor would be a nondifferentiable curve was first addressed in 1918 by *Julia* [4.30] and one year later by *Fatou* [4.31]. They studied the basin boundaries for rational mappings of a complex plane into itself. However, their work remained almost unknown until quite recently. Today, a basin boundary for a rational map is called a *Julia set*.

A simple example of a rational map possessing such a property is given by

$$z_{n+1} = z_n^2 + c \,, \tag{4.3.10}$$

where z is a complex variable and c is a complex constant. The attractor of this map has a fractal basin boundary, i.e., a fractal Julia set (Fig. 4.10).

A two-dimensional map with fractal boundaries between nonchaotic attractors has been investigated by *Grebogi* et al. [4.33]:

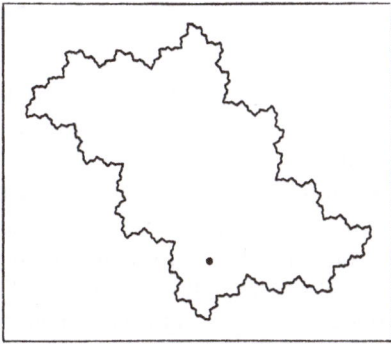

Fig. 4.10. Julia set for $c = -0.12375+0.56508i$ (from [4.32])

$$x_{n+1} = 2x_n \; ,$$

$$(4.3.11)$$

$$y_{n+1} = \lambda y_n + \cos x_n \; ,$$

where the parameter λ lies inside an interval $2 > \lambda > 1$ and the variable x varies between 0 and 2π. They have found that for this map almost all initial conditions generate sequences of y_n such that either $y_n \to +\infty$ or $y_n \to -\infty$ in the limit $n \to \infty$. Hence, $+\infty$ and $-\infty$ represent two different attractors of this system. A function $f(x)$ can be further defined such that $y_n \to +\infty$ if the initial values of x and y satisfy the condition $y > f(x)$, and $y_n \to -\infty$ otherwise. This function can be analytically determined [4.33]. It is given by

$$f(x) = -\sum_{k=1}^{\infty} \frac{1}{\lambda^{k+1}} \cos (2^k x) \; . \qquad (4.3.12)$$

Since $\lambda > 1$ the sum converges. However, the derivative of this function is

$$\frac{df}{dx} = \frac{1}{2} \sum_{k=1}^{\infty} \left(\frac{2}{\lambda}\right)^{k+1} \sin (2^k x) \; . \qquad (4.3.13)$$

When $\lambda < 2$ this derivative diverges and, hence, the function $f(x)$ is not differentiable. The fractal dimension of this curve is $D_F = 2 - \ln \lambda / \ln 2$.

As shown in Sect. 2.1, iterative maps are closely related to dynamical systems with continuous time. Indeed, if we consider only the points where a phase trajectory crosses a certain (hyper)plane in the phase space of the dynamical system, a corresponding Poincaré map can be constructed.

Thus, such simple attractors as fixed points or limit cycles might still possess fractal boundaries separating their attraction basins. *Moon* and *Li* [4.34] have studied a model that describes the motion of a particle in a two-well potential in the presence of an external force f,

$$\dot{x} = y \; ,$$

$$(4.3.14)$$

$$\dot{y} = -\gamma y + \tfrac{1}{2}x(1 - x^2) + f \cos \omega t \; .$$

This system has two fixed points $x = \pm 1$, $y = 0$. In the interval $0.1 > f > 0.05$ for $\omega = 0.833$ and $\gamma = 0.15$, they are unstable and the attractors represent two limit cycles surrounding each of these fixed points.

To determine the attraction basins of the attractors, many integrations of equations (4.3.14) with varying initial conditions $(x_0, \dot{x}_0 = y_0)$ have been performed. For low values of the force (e.g., for $f = 0.05$) the basin boundary looks smooth (Fig. 4.11). However, when f is increased to $f = 0.1$, a complex structure of the boundary develops (Fig. 4.12). To confirm the fractal properties of the basin boundary in this case, a sequence of successive enlargements of increasingly smaller regions in the phase space near such a boundary has been carried out. Each of

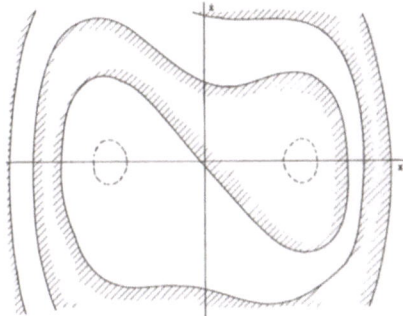

Fig. 4.11. Smooth basin boundary (*solid lines*) for periodic motion about fixed points of the system (4.3.14) at $f = 0.05$, $\omega = 0.833$, and $\gamma = 0.15$ (from [4.34])

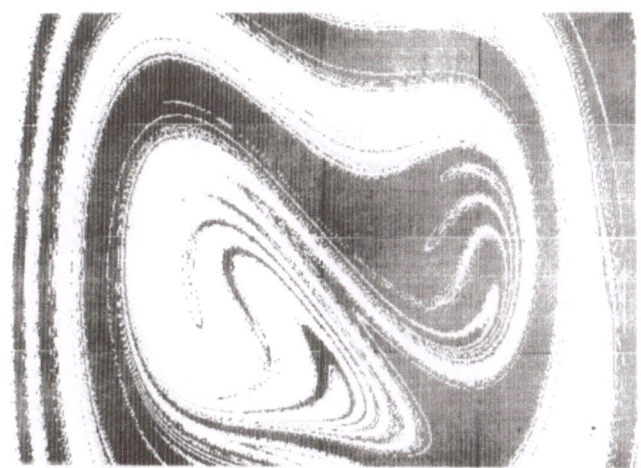

Fig. 4.12. An attraction basin with a fractal boundary for the periodic motion in the system (4.3.14) at $f = 0.1$, $\omega = 0.833$, and $\gamma = 0.15$ calculated from 160×10^3 initial conditions (from [4.34])

the enlargements shows finer and finer structure (Fig. 4.13). Thus we see that a simple mechanical system with a two-well potential can already exhibit a fractal boundary between the basins of two nonchaotic attractors.

Fig. 4.13. Finer-scale enlargement of the fractal boundary shown in Fig. 4.12 (from [4.34])

5. Iterative Maps

The mathematical models known as iterative maps are closely related to dynamical systems with continuous time. They can arise naturally in problems where the state of a system is allowed to change only at some prescribed instants in time. In fact, iterative maps are a special case of an automaton with instantaneous states described by continuous variables.

Despite their greater simplicity, iterative maps possess many properties in common with dynamical systems; particularly, they have almost the same classes of attractors and the same kinds of bifurcations. Maps can also exhibit chaotic dynamics.

The relative simplicity of iterative maps facilitates their analysis and allows us to proceed further with the investigation of chaotic dynamics. This investigation provides an elegant analytical description of the transition to chaos through a sequence of period-doubling bifurcations.

5.1 Fixed Points and Cycles

Consider the Poincaré map for some system of differential equations

$$\dot{x} = v(x) . \tag{5.1.1}$$

It relates each point in the cross-sectional surface to another point in this surface, i.e. the point where the trajectory of (5.1.1) will next interesect it. This implies that there is some function ϕ which relates the positions of any two successive intersection points. Let us denote the coordinates of the nth intersection point on the surface by x_n. Then the coordinates of the next intersection point can be obtained as

$$x_{n+1} = \phi(x_n) . \tag{5.1.2}$$

We say that (5.1.2) defines an *iterative map*. Applying this map, an infinite sequence of points

$$x_0, x_1, \ldots, x_n, x_{n+1}, \ldots \tag{5.1.3}$$

can be obtained. This sequence is uniquely determined by the initial point x_0.

The use of an iterative map instead of the related differential equations is very efficient in studies of dynamical systems, because it allows a reduction of their dimensionality. It turns out that many properties of a full dynamical system are already determined by the respective map.

Obviously, maps such as (5.1.2) can also be defined in the absence of any connection with continuous dynamical systems, as may be seen from the following simple ecological example. Suppose we have a population of butterfiles which reproduce only in a fixed season of a year. Let us denote the number of individuals in this population in the nth year by x_n. If we neglect environmental fluctuations, the number of individuals in the next year x_{n+1} is uniquely determined by the size of the population in the previous years, i.e. by x_n. Consequently, the time evolution of the population number x_n will follow some iterative map (5.1.2). In this particular example, such a map is one dimensional. However, in a similar manner it is possible to construct examples where x_n would represent a set of several variables.

A *fixed point* x^* of a map ϕ satisfies the equation

$$x^* = \phi(x^*) . \tag{5.1.4}$$

Fixed points play essentially the same role in the theory of iterative maps as the fixed points of differential equations. Let us find the stability condition of a fixed point x^* for the map (5.1.2).

Let $x = x^* + \delta x$ where δx is small compared with x^*. Then (5.1.2) takes the form

$$x^* + \delta x_{n+1} = \phi(x^* + \delta x_n) . \tag{5.1.5}$$

Linearizing the right-hand side of (5.1.5) and taking into account the definition (5.1.4) of a fixed point, we find

$$\delta x_{n+1} = A \delta x_n . \tag{5.1.6}$$

Here A is the matrix with components $a_{ik} = \partial \phi_i / \partial x_k$ taken at $x = x^*$.

The eigenvalues α_j of matrix A are called *multiplicators*. A fixed point x^* is *stable* only if all its multiplicators α_j satisfy the condition $|\alpha_j| < 1$. Indeed, in this case any intial deviation δx from the fixed point will fade with time, as follows from the linearized map (5.1.6). Unstable fixed points with multiplicators of magnitude $|\alpha_j| < 1$ and $|\alpha_j| > 1$ are called the *saddle* points.

As an example, we consider the two-dimensional *Henon map* [5.1] mentioned in the previous chapter. This map is given by

$$x_{n+1} = \phi_1(x_n, y_n) = y_n + 1 - a x_n^2 ,$$
$$y_{n+1} = \phi_2(x_n, y_n) = b x_n . \tag{5.1.7}$$

Fixed points of the Henon map are found from the equations

$$x = y + 1 - a x^2 , \qquad y = b x . \tag{5.1.8}$$

When $a > a_0$, where $a_0 = -(1 - b)^2/4$, and $a \neq 0$, the coordinates of the fixed points are

$$x_{\pm} = \frac{1}{2a}\left\{-(1 - b) \pm \left[(1 - b)^2 + 4a\right]^{1/2}\right\},$$ (5.1.9)

$$y_{\pm} = bx_{\pm}.$$

The point (x_+, y_+) is stable only for $a < a_1$, where $a_1 = (\frac{3}{4})(1 - b)^2$. The point (x_-, y_-) is always unstable.

It can be shown that the Henon map is dissipative, i.e. that sequences of points produced by iterations of this map converge to some attracting set of zero measure (to the attractor). This set is very complex; it has a fractal structure (Figs. 4.5, 6).

Below we consider only one-dimensional iterative maps

$$x_{n+1} = \phi(x_n).$$ (5.1.10)

The behavior of a one-dimensional map can be easily visualized. Let us take the plane (x_{n+1}, x_n), plot there the function $x_{n+1} = \phi(x_n)$ and draw the line $x_{n+1} = x_n$. Subsequent iterations of the map (5.1.10) can then be constructed using the *Lamerey diagram* (Fig. 5.1).

With the help of this construction, one can easily determine the fixed points of (5.1.10) and analyze their stability. Indeed, the fixed points are simply the points of intersection of the curve $x_{n+1} = \phi(x_n)$ with the straight line $x_{n+1} = x_n$. Figure 5.1 shows that the sequence x_1, x_2, x_3, ... converges to a fixed point x_1^* and therefore x_1^* is stable. From the same construction one sees that x_2^* is unstable: after a small perturbation it gives rise to a diverging sequence x_1', x_2', x_3', ...

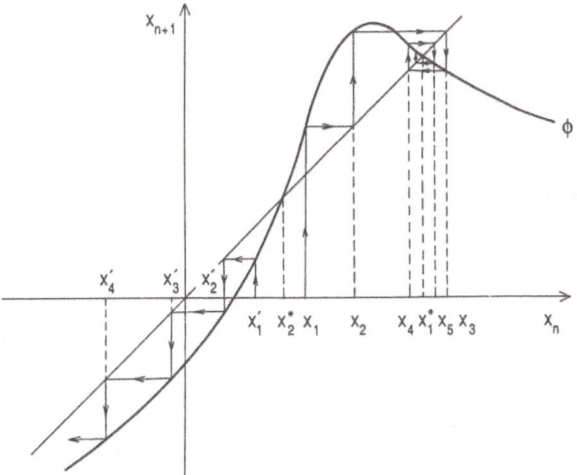

Fig. 5.1. The Lamerey diagram for a map ϕ. The fixed point x_1^* is stable, and the fixed point x_2^* is unstable

The analytic condition of stability of a fixed point x^* reduces for (5.1.10) to the inequality $|\phi'(x^*)| < 1$, which has a simple geometric interpretation; namely, the tangent to the curve $x_{n+1} = \phi(x_n)$ at the point x^* should be less steep than the bisectrix $x_{n+1} = x_n$.

Besides fixed points, a one-dimensional map might have cycles. An m-*cycle* of a map (5.1.10) is a finite sequence of points $x_1^*, x_2^*, \ldots, x_m^*$, such that

$$x_2^* = \phi(x_1^*), \quad x_3^* = \phi(x_2^*), \ldots, \quad x_1^* = \phi(x_m^*) \tag{5.1.11}$$

and no elements in this sequence coincide. Sometimes $x_1^*, x_2^*, \ldots, x_m^*$ are also called m-*multiple fixed points*. Figure 5.2 shows a stable 2-cycle of a one-dimensional map, formed by points x_1^* and x_2^*; the fixed point x^* is unstable.

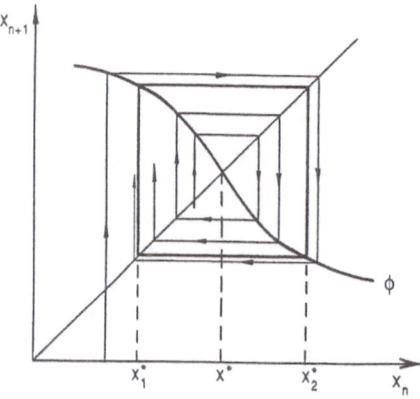

Fig. 5.2. Stable 2-cycle of a one-dimensional map

Since for any point x_1^* of an m-cycle we have

$$x_i^* = \phi^m(x_i^*) = \phi(\phi(\phi \ldots \phi(x_i^*) \ldots)), \tag{5.1.12}$$
$$m \text{ times}$$

each of its m points is simultaneously a fixed point of the map ϕ^m. The analysis of the stability of an m-cycle is thus reduced to the investigation of the stability of the respective fixed points of the map ϕ^m. The m-cycle (5.1.11) is stable when the condition

$$|\phi'(x_1^*)\phi'(x_2^*) \ldots \phi'(x_m^*)| < 1 \tag{5.1.13}$$

is satisfied. This condition is easily derived using the rule for the derivative of a compound function, i.e. $(f(g(x)))' = f'(g(x))g'(x)$.

A map (5.1.10) yields a *one-to-one correspondence* if, for any x and y, there is no equality $\phi(x) = \phi(y)$. One-to-one correspondence implies that the function $\phi(x)$ either grows monotonically or decreases monotonically with x.

Let us consider a map with the monotonically increasing function $\phi(x)$, shown in Fig. 5.3. This map has three fixed points x_1^*, x_2^*, x_3^*. It is easily seen from the

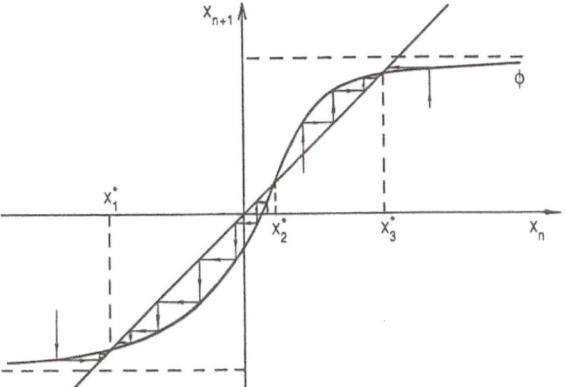

Fig. 5.3. The Lamerey diagram for a map with the monotonically increasing function $\phi(x)$

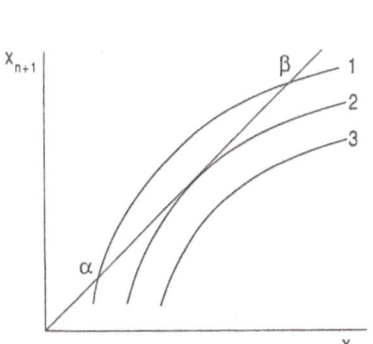

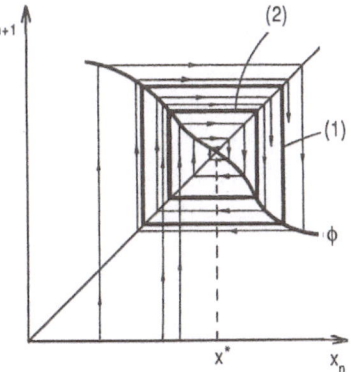

Fig. 5.4. Creation (or disappearance) of a pair of stable (β) and unstable (α) fixed points

Fig. 5.5. Stable (*1*) and unstable (*2*) cycles, and a stable fixed point x^*, of a iterative map with the monotonically decreasing function $\phi(x)$

construction that x_2^* is unstable, while x_1^* and x_3^* are stable. Each initial point x_0 gives rise to a sequence of points that converge either to x_1^* if $x_0 \in (-\infty, x_2^*)$ or to x_3^* if $x_0 \in (x_2^*, \infty)$. Therefore, the entire line of x is divided into the attraction basins of two fixed points x_1^* and x_3^*.

A similar property holds for a general one-to-one map with the monotonically increasing function $\phi(x)$. The entire line of x is divided into the attraction basins of different stable fixed points. Each fixed point x_i^* lies between two unstable points x_{i-1}^* and x_{i+1}^*; the interval (x_{i-1}^*, x_{i+1}^*) represents an attraction basin of the point x_i^*.

A special situation arises when the curve $x_{n+1} = \phi(x_n)$ touches the bisectrix $x_{n+1} = x_n$. This corresponds to a *bifurcation* of the map that separates two quali-

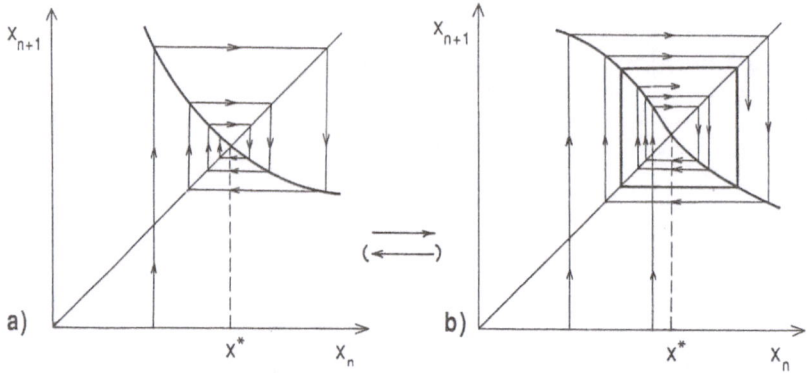

Fig. 5.6a, b. Creation (disappearance) of a 2-cycle from a fixed point

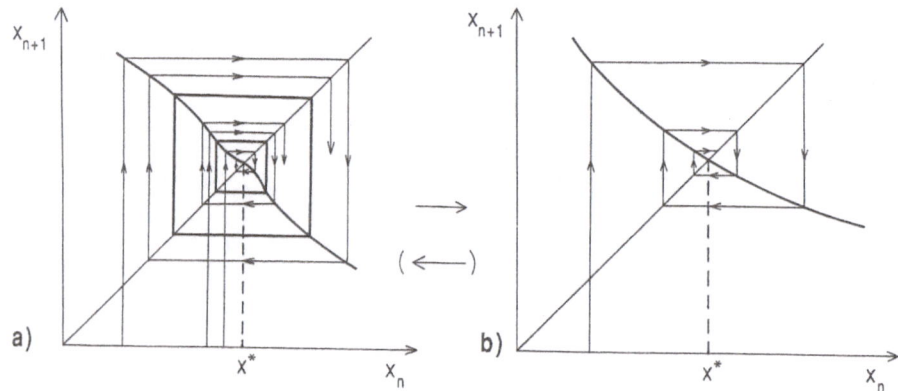

Fig. 5.7a, b. Disappearance (creation) of a pair of 2-cycles

tatively different regimes. For example, the bifurcation in Fig. 5.4 corresponds to the merging and disappearance of stable and unstable fixed points, if we go from curve 1 to curve 3. The reverse transition from 3 to 1 results in the emergence of two fixed points. If we interpret (5.1.10) as a Poincaré map for some dynamical system, this bifurcation would correspond to the birth or disappearance of a pair of limit cycles, one stable and another unstable.

Now we can consider a one-to-one map with a monotonically decreasing function $\phi(x)$. In this case there is a single fixed point x^* which, depending on the form of the function $\phi(x)$, can be stable or unstable. To investigate this map, we note that the function $\phi^2(x) = \phi(\phi(x))$ would give a one-to-one map with a monotonically *increasing* function. Indeed, we have

$$\left(\phi^2(x)\right)' = \phi'(\phi(x))\phi'(x) > 0 \ . \tag{5.1.14}$$

Each fixed point x_i^* of a twice-iterated map ϕ^2, which differs from the only fixed point x^* of the original map $\phi(x)$, corresponds to some 2-cycle of the map

$\phi(x)$. Consequently, the entire line of x is divided into the attraction basin of point x^* (if it is stable) and the attraction basins of stable 2-cycles (Fig. 5.5).

There are two possible types of bifurcations for a one-to-one map with the decreasing function $\phi(x)$: creation or disappearance of a 2-cycle from the fixed point (Fig. 5.6) and creation or disappearance of a pair of 2-cycles (Fig. 5.7).

5.2 Chaotic Maps

Now we consider maps without a one-to-one correspondence (Fig. 5.8). They may have very complex dynamics, which generates aperiodic sequences $x_0, x_1, x_2, \ldots$. In this case a map has no stable fixed points or cycles of any period. This possibility is related to the fact that the inverse mapping ϕ^{-1} is now not unique (for instance, it might have two branches, as in Fig. 5.8). Therefore, if we try to reconstruct the sequence of points in the backward direction, we would have to decide each time which of the branches should be chosen.

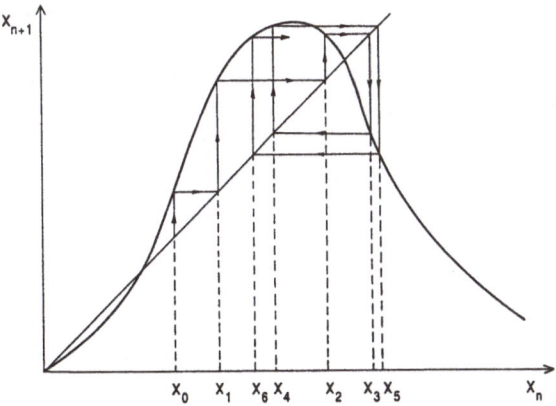

Fig. 5.8. An example of a map without a one-to-one correspondence

The following arguments [5.2] show that nonmonotonic maps possess infinite sets of different fixed points and cycles. Suppose the inverse mapping ϕ^{-1} has p branches, i.e. it consists of p one-to-one maps ϕ_i^{-1}, $i = 1, 2, \ldots, p$. Consider a product of m one-to-one maps

$$\Phi = \phi_{i_m}^{-1}\phi_{i_{m-1}}^{-1} \cdots \phi_{i_2}^{-1}\phi_{i_1}^{-1} \tag{5.2.1}$$

where i_m are any integers from 1 to p. This map is monotonic and possesses at least one fixed point $x^*_{i_1 i_2 \ldots i_m}$. However, such a fixed point will also be a fixed point of ϕ^{-m} and, consequently, of the map ϕ^m. A fixed point of ϕ^m corresponds either to a fixed point of the original map ϕ or to a cycle of this map. By choosing different integers $i_1 i_2 \ldots i_m$ and different numbers m, we can produce an infinite set of different maps (5.2.1). Since the fixed points of these maps are generally different, this construction would yield an infinite number of fixed points and cycles.

Chaotic maps have *only* unstable fixed points and cycles. As an example, consider a simple *logistic map* [5.2, 3]

$$x_{n+1} = \mu x_n (1 - x_n) \tag{5.2.2}$$

with $\mu = 4$. This map transforms the segment $[0, 1]$ into itself (Fig. 5.9) and, as shown below, for $\mu = 4$ it possesses an infinite number of different cycles, all of which are unstable. An m-cycle $x_1^*, x_2^*, \ldots, x_m^*$ of the map (5.2.2) is unstable if

$$|\phi'(x_1^*)||\phi'(x_2^*) \ldots \phi'(x_m^*)| > 1 . \tag{5.2.3}$$

It follows from (5.2) that $\phi'(x) = -8(x - \frac{1}{2})$ and, therefore, $|\phi'(x)| \leq 1$ for $|x - \frac{1}{2}| \leq 1/8$ and $|\phi'(x)| \geq 1$ for $|x - \frac{1}{2}| \geq 1/8$. If all the factors in (5.2.3) satisfy the condition $|\phi'(x_i^*)| > 1$, this inequality is obviously true. Suppose that some of these factors are smaller than one, e.g. $|\phi'(x_1^*)| \leq 1$ and hence $|x_1^* - \frac{1}{2}| \leq 1/8$. Then, by rather tedious estimates using (5.2.2), it is possible to show (see [5.2]) that this factor should be followed by a number of other factors, each greater than one, so that the product of $|\phi'(x_i^*)|$ and this group of factors is nevertheless greater than one. Actually, any small factor in (5.2.3) is followed by such a group of factors, so that the inequality (5.2.3) is satisfied. Since (5.2.3) holds for an arbitrary cycle, this proves the instability of *all* cycles for the map (5.2.2). The resulting dynamics is evidently very complex and irregular.

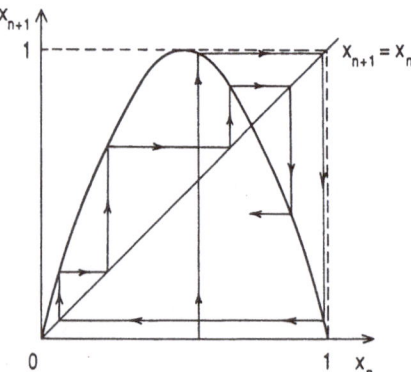

Fig. 5.9. Graph of the logistic map (5.2.2)

Two other well-known examples of chaotic maps are piece-linear. The *tent map* (Fig. 5.10a) is given by

$$x_{n+1} = \begin{cases} 2x_n , & 0 \leq x \leq \frac{1}{2} , \\ 2(1 - x_n), & \frac{1}{2} < x \leq 1 . \end{cases} \tag{5.2.4}$$

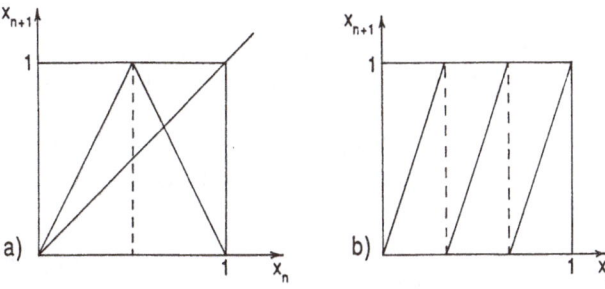

Fig. 5.10. Graphs of the tent map (**a**) and the saw map (**b**)

It can be considered as a piece-linear approximation of the logistic map. The *saw map* (Fig. 5.10b) is

$$x_{n+1} = \{\gamma x_n\} , \tag{5.2.5}$$

where x_n belongs to the segment $[0, 1]$, we have $\gamma > 1$, and $\{\dots\}$ denotes taking the fractional part of the argument.

Let us calculate the Lyapunov exponent of map (5.2.5):

$$\lambda = \lim_{n \to \infty} \frac{1}{n} \ln \left| \frac{\delta x_n}{\delta x_0} \right| , \tag{5.2.6}$$

where $\delta x_n = x_n - x_n'$ is the present distance between points that were initially separated by a vanishingly small distance $\delta x_0 = x_0 - x_0'$. Since $\delta x_0 \to 0$, we can write

$$\delta x_{n+1} = \phi'(x_n) \delta x_n = \gamma \delta x_n \tag{5.2.7}$$

and therefore

$$\delta x_n = \gamma^n \delta x_0 . \tag{5.2.8}$$

Substitution of (5.2.8) into (5.2.7) yields $\lambda = \ln\gamma > 0$. We see that the map (5.2.5) stretches the distance between any two initially close points. Since the dynamics is restricted to the finite segment $[0, 1]$, this property leads to some sort of mixing.

Because of their apparent simplicity and their relationship to the sets of differential equations, iterative maps represent attractive objects for mathematical studies. In 1964 *Sharkovskii* [5.4] (see also [5.5]) proved a theorem which allows one to order all possible m-cycles into the sequence

$$3, 5, 7, \dots, 2 \cdot 3, 2 \cdot 5, 2 \cdot 7, \dots, 2^m \cdot 3, 2^m \cdot 5, 2^m \cdot 7, \dots,$$
$$2^n, \dots, 2^3, 2^2, 2, 1 . \tag{5.2.9}$$

If a map with a continuous function $\phi(x)$ has a cycle of some period m_1, it must also possess cycles of periods m which stay further to the right in this sequence. For instance, if a map has a 2-cycle it should also possess a 1-cycle (which is simply a fixed point). Note that the sequence is opened with 3. Therefore, if a map has a 3-cycle, it must have the cycles of *all* periods. This property was discovered independently by *Li* and *Yorke* [5.6] who thereupon coined the phrase "period 3 implies chaos".

It is interesting to note that, for some chaotic maps, one can even find an analytic solution for the iteration series. If we again take the logistic map (5.2.2) with $\mu = 4$, we can see that its exact solution is

$$x_n = \sin^2\left(2^n x_0\right) , \tag{5.2.10}$$

where x_0 is any number from the segment $[0, 1]$. Indeed, from (5.2.2) we obtain

$$\begin{aligned} x_{n+1} &= 4\sin^2\left(2^n x_0\right)\left[1 - \sin^2\left(2^n x_0\right)\right] \\ &= 4\sin^2\left(2^n x_0\right)\cos^2\left(2^n x_0\right) = \sin^2\left(2^{n+1} x_0\right) . \end{aligned} \tag{5.2.11}$$

Hence, the simple analytic expression (5.2.10) is able to generate an infinite chaotic sequence of numbers!

5.3 Feigenbaum Universality

In the previous section, we showed that some iterative maps have chaotic dynamics. However, with a different choice of parameters, the same maps might have regular behavior which is generally characterized by attractive cycles. When we gradually change the parameters of a map, it can undergo a transition from regular to chaotic dynamics. We discuss below the properties of such a transition, using the example of a logistic map (5.2.2). We will see that the final results are also applicable to a large class of other maps.

Let us investigate the behavior of the logistic map (5.2.2) when the parameter μ is gradually increased from 0 to 4.

a) If $0 < \mu \leq 1$ the map has only the fixed point $x = 0$ which is stable (Fig. 5.11a).
b) When $\mu \geq 1$, this fixed point $x = 0$ loses its stability, since now $\phi'(0) > 1$, and another fixed point $x_1 = 1 - 1/\mu$ appears inside the segment $[0, 1]$ (Fig. 5.11b). Its multiplicator is $\alpha(x_1) = 2 - \mu$ and, therefore, this fixed point remains stable until $\mu = 3$.
c) When $1 + \sqrt{6} \geq \mu > 3$, the point x_1 is unstable. By means of a bifurcation it gives birth to a stable 2-cycle (Fig. 5.11c) formed by

$$x_2^{(1),(2)} = \left(\frac{1}{2\mu}\right)\left[\mu + 1 \pm \left(\mu^2 - 2\mu - 3\right)^{1/2}\right] , \tag{5.3.1}$$

which are the fixed points of the twice-iterated map ϕ^2 (Fig. 5.11c).
d) When the value $\mu = 1 + \sqrt{6} \approx 3.45$ is exceeded, the next bifurcation occurs. The 2-cycle $\{x_2^{(1)}, x_2^{(2)}\}$ loses its stability, but a new stable 4-cycle appears. At $\mu \approx 3.54$ this cycle also becomes unstable, giving birth to a stable 8-cycle,

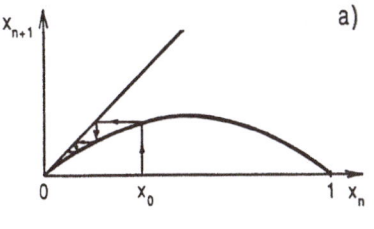
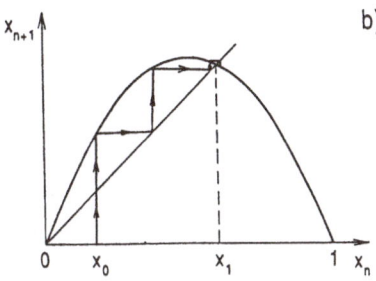
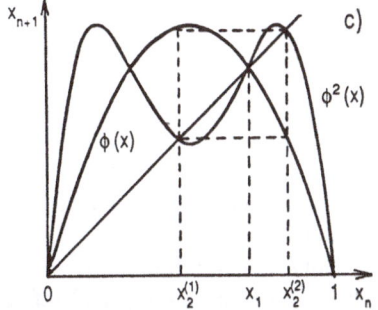

Fig. 5.11. Graphs of the logistic map for (**a**) $0 < \mu \leq 1$, (**b**) $1 < \mu \leq 3$, (**c**) $3 < \mu \leq 1+\sqrt{6}$

etc. The subsequent period-doubling bifurcations of an attractive cycle take place until the value $\mu_\infty = 3.5699 \ldots$ is reached. At $\mu = \mu_\infty$ the period of the attractive cycle diverges and all finite cycles of periods 2^m, $m = 1, 2, 3, \ldots$, are repulsive.

e) When $\mu_\infty < \mu \leq 4$, the logistic map (5.2.2) may have cycles of any period and aperiodic sequences which give rise to chaotic behavior.

Hence, for $\mu < \mu_\infty$, the map (5.2.2) has a single stable cycle of period 2^m, with m depending on μ, which attracts (almost) all points from the segment $[0, 1]$. When $\mu > \mu_\infty$, the dynamics of the map (5.2.2) becomes much more complicated: it includes aperiodic trajectories that are not attracted to the cycles.

The qualitative changes in the dynamics of the map (5.2.2) upon variation of the parameter μ can be visualized using a so-called *bifurcation diagram* (Fig. 5.12). Along the vertical axis we plot (multiple) fixed points of the map which form the stable cycle of period 2^m, born at $\mu = \mu_m$ from the cycle of period 2^{m-1}. The values of μ are plotted along the horizontal axis. The sequence of values μ_m, at which the period-doubling bifurcations occur, obey a simple law (found by *Feigenbaum* [5.7–9], and *Grossmann* and *Thomae* [5.10])

$$\lim_{m \to \infty} \frac{\mu_m - \mu_{m-1}}{\mu_{m+1} - \mu_m} = \delta = 4.6692 \ldots. \tag{5.3.2}$$

Let us introduce the distance d_m between the line $x = 1/2$ and the nearest element of the cycle with the period 2^m at $\mu = \mu_m$ (Fig. 5.12). It turns out [5.7–9] that the ratio d_m/d_{m+1} has the limit

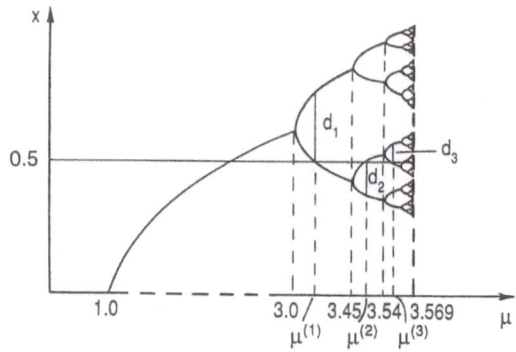

Fig. 5.12. The bifurcation diagram of the logistic map (5.3.1)

$$\lim_{m \to \infty} \frac{d_m}{d_{m+1}} = \alpha = 2.5029 \ldots . \qquad (5.3.3)$$

The parameters δ and α are called the *universal Feigenbaum numbers*. Note that δ specifies the rate of convergence of the sequence of μ_m, $m = 1, 2, 3, \ldots$, while α determines the change of scales under subsequent bifurcations. If we expand the bifurcation diagram by a factor α near $x = 1/2$ at $\mu = \mu_m$ and reflect it in the horizontal axis, it would look precisely like the corresponding part of the bifurcation diagram at $\mu = \mu_{m-1}$.

Although (5.3.2) and (5.3.3) were derived for the logistic map (5.2.2), the same asymptotic laws are valid for other iterative maps with a maximum that can be approximated by a quadratic parabola. Moreover, even the values of δ and α remain the same for all such maps. This phenomenon is known as *Feigenbaum universality* [5.7–9]. We will explain this below using the renormalization theory [5.9, 11].

Let us plot (Fig. 5.13a,b) the logistic function $\phi(\mu, x) = \mu x (1 - x)$ and its second iterate $\phi^2(\mu, x) = \phi(\phi(\mu, x))$ a the point $\mu = \mu^{(1)}$ (Fig. 5.12). Note that $x = \frac{1}{2}$ is a fixed point of ϕ^2 and a part of a 2-cycle of the map ϕ. Next we repeat the construction (Fig. 5.13c,d) for the second and fourth iterates $\phi^2(\mu, x)$ and $\phi^4(\mu, x)$ at the point $\mu = \mu^{(2)}$. Then $x = \frac{1}{2}$ will again be a fixed point of ϕ^4 and a part of a 2-cycle of ϕ^2.

Compare the function ϕ in Fig. 5.13a with the part of ϕ^2 inside the outlined square in Fig. 5.13c. We can see that these two functions almost coincide after an appropriate scale transformation and two reflections in the horizontal and vertical axes. In fact, the same property can be observed for any two iterated maps ϕ^k and ϕ^{2k} with $k = 2^m$. Moreover, when $m \to \infty$ the difference between the two functions, after a proper transformation, becomes vanishingly small (the scale transformation factor also tends to a certain limit value). In other words, a universal limit function $g(y)$ must exist which remains invariant under this operation, i.e.

$$g(g(y)) = -\frac{1}{\alpha} g(-\alpha y) , \qquad (5.3.4)$$

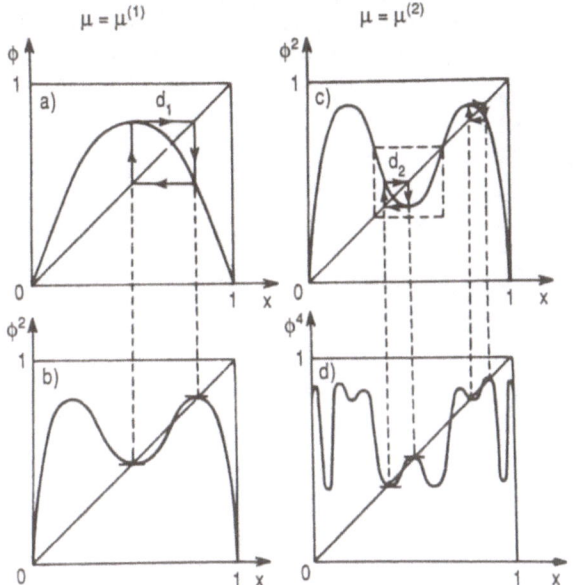

Fig. 5.13. The functions $\phi(\mu, x)$ and $\phi^2(\mu, x)$ for $\mu = \mu^{(1)}$ (**a, b**); and the functions $\phi^2(\mu, x)$ and $\phi^4(\mu, x)$ for $\mu = \mu^{(2)}$ (**c, d**)

where α is an unknown scale factor and $y = x - \frac{1}{2}$. The parameter α specifies how strongly we should stretch the plot of the function $g(y)$ near the point $y = 0$ before the inversion. In the limit $m \rightarrow \infty$ the iterated maps $\phi^k(\mu_{m-1}, y)$ with $k = 2^m$ approach the functions

$$g(g(\ \dots\ g(y)\ \dots)) = (-\alpha)^{-m} g\big((-\alpha)^m y\big) \ . \tag{5.3.5}$$
$$\text{m times}$$

Equation (5.3.4) is effectively a functional equation for the universal function $g(y)$. Since $y = 0$ (i.e. $x = 1/2$) should always be the fixed point of the 2^m times iterated map $\phi(\mu_{m-1}, y)$, we require that $g(0) = 1$. Then (5.3.4) yields

$$\alpha = -\frac{1}{g(1)} \ . \tag{5.3.6}$$

Feigenbaum [5.7–8] (see also [5.12]) found numerically a polynomial approximation for the solution of the functional equation (5.3.4):

$$g = 1 - 1.52763y^2 + 0.104815y^4 - 0.0267057y^6 + \dots \ . \tag{5.3.7}$$

According to (5.3.6) this gives $\alpha = 2.50280787 \dots$, which coincides with the limit value (5.3.3) obtained by direct simulations.

The universal number δ (5.3.2) can be found from renormalization theory if we consider the evolution of small perturbations to $g(y)$. A more detailed discussion of renormalization theory for period-doubling bifurcations is given in [5.7–15].

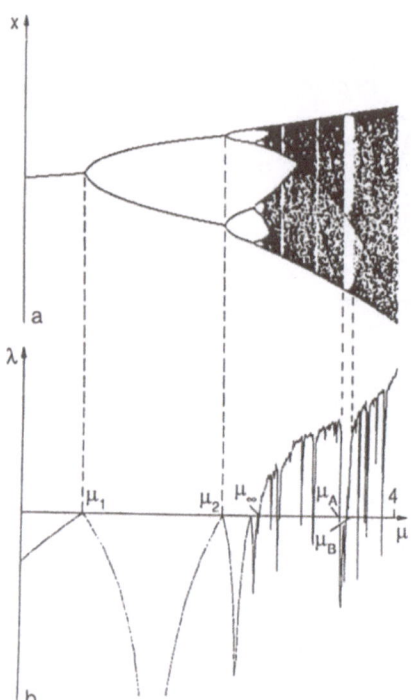

Fig. 5.14. The bifurcation diagram of a logistic map (**a**) and the corresponding behavior of the Lyapunov exponent (**b**). From [5.14]

After an infinite series of period doublings, the period of the cycle becomes infinite. Yet the dynamics of the logistic map is not chaotic at $\mu = \mu_\infty$, because the Lyapunov exponent remains equal to zero at this point, and the exponential divergence of initially close trajectories is absent. Nevertheless, the logistic map already possesses a Kantor invariant set at $\mu = \mu_\infty$.

For higher values of μ, chaotic behavior with a positive Lyapunov exponent is found (Fig. 5.14b). However, in the interval $\mu_\infty < \mu < 4$ bands with chaotic behavior alternate with bands (or *windows*) where the dynamics is regular. For example, at $\mu \approx 3.83$ there is a wide window with a stable cycle of period 3, while at $\mu \approx 3.74$ we have a window with a cycle of period 5. Thorough investigation of the structure in Fig. 5.14a reveals that there are also other narrower windows with stable cycles of higher periods (these regions are clearly seen in the plot of the Lyapunov exponent, Fig. 5.14b).

The emergence of a stable cycle at the left edge μ_A of any window (μ_A, μ_B) is described by a tangent bifurcation, while its disappearance at the right edge μ_B is the result of a crisis (Sect. 6.1). The sequences of these bifurcations also have some universal properties which were discussed in [5.16, 17].

6. Routes to Temporal Chaos

Temporal chaos sets in after the breakdown of long-range time order and the disappearence of coherent temporal behavior. In the previous chapter we outlined one of the possible transitions to chaos in the special case of models with discrete time. Now we want to discuss the principal scenarios leading to temporal chaos in general dynamical systems. Before proceeding to this discussion, we briefly describe some of the concepts of bifurcation theory which are used in the analysis.

6.1 Bifurcations

We have seen in Chap. 3 that the character of asymptotic motion in dynamical systems depends on the type of attractors present. Steady states, periodic and quasiperiodic motions correspond to simple attractors (a fixed point, a limit cycle and an invariant torus, respectively). Chaotic motion is produced by a strange attractor.

Suppose a dynamical system includes some control parameter μ, i.e.

$$\dot{x} = v(x, \mu) \ . \tag{6.1.1}$$

When the control parameter μ is varied, the attractors of this system also change. As a rule, these changes are smooth and continuous. A small variation of the control parameter usually results in a slight shift of a fixed point, or in a small change in the form and period of a limit cycle, etc. Only at some critical values of this parameter will the attractor experience a radical change, accompanied by a sharp modification of the system dynamics. This occurs, for example, when a limit cycle gives rise to an invariant torus, so that the periodic motion is replaced by a quasiperiodic one. In general, any discontinuous qualitative change in the behavior of a system is called a *bifurcation.*

A bifurcation can produce either more complex or a simpler form of motion. The latter case is realized when, for instance, an invariant torus collapses into a limit cycle. A bifurcation can also lead to the disappearance of an attractor – it then represents the *crisis* of an attractor.

Suppose a dynamical system (6.1.1) has a fixed point x^0 which depends, generally, on a control parameter, i.e. $x^0 = x^0(\mu)$. Let us assume that this fixed point is stable for $\mu < \mu_0$ and unstable for $\mu > \mu_0$. Then at $\mu = \mu_0$, the real

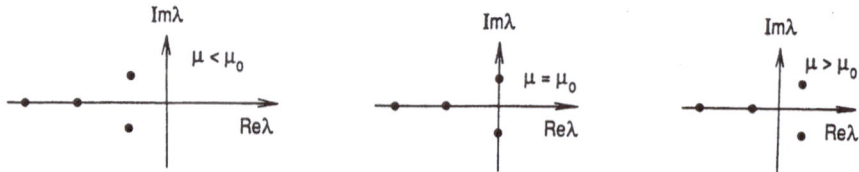

Fig. 6.1. Behavior of eigenvalues of the linearization matrix under the Andronov-Hopf bifurcation

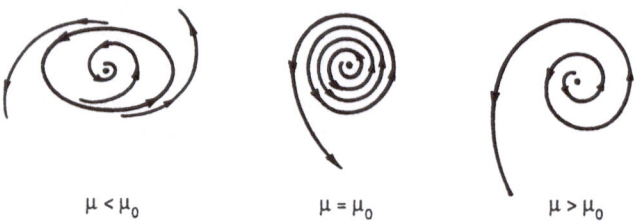

Fig. 6.2. The inverted Andronov-Hopf bifurcation

part of some eigenvalues of the linearization matrix (Sect. 3.1) vanishes; these eigenvalues cross the imaginary axis.

The best-known example of a bifurcation from the fixed point is the *Andronov-Hopf bifurcation* [6.1–2]. In this case, the pair of complex conjugate eigenvalues $\lambda_{1,2} = \xi \pm i\omega$ of the linearization matrix moves into the right half-plane, as shown in Fig. 6.1. This results in the excitation of oscillations with the period $T = 2\pi/\omega$, i.e. in the creation of a stable limit cycle.

Another common mechanism for the loss of stability of a fixed point is *hard excitation* (or the *subcritical* Andronov-Hopf bifurcation), when a shrinking unstable limit cycle merges with the stable fixed point. In this case the attraction basin of the fixed point diminishes when μ approaches μ_0 and disappears at $\mu = \mu_0$ (Fig. 6.2). Since the attraction basin becomes very small near the bifurcation, even weak external fluctuations may throw the phase trajectory out of the attraction basin of x_0. In other words, although the fixed point is formally stable until the bifurcation value $\mu = \mu_0$ is reached, it actually loses its stability with respect to finite-amplitude perturbations already at $\mu < \mu_0$. After this bifurcation the attractor, i.e. the stable fixed point, disappears; hence it is a crisis.

Consider a stable limit cycle born as a result of the Andronov-Hopf bifurcation. What further bifurcation can it undergo with variation of the control parameter? There is no unique answer to this question, because different types of evolution are possible. A detailed review may be found in [6.3–7]. Below, we describe only two possible further bifurcations which lead eventually to the formation of a strange attractor.

The first possibility consists in the creation of a two-dimensional attractive torus and the simultaneous loss of stability of the limit cycle (Fig. 6.3). Let us take a cross-section of the phase flow of such a dynamical system in a plane S.

Fig. 6.3. Creation of a two-dimensional attractive torus from a limit cycle

Then the original limit cycle would correspond to a fixed point in the plane of cross-section, while the emerging stable torus would be portrayed by a closed curve that surrounds this point. Therefore, in the plane of cross-section, the creation of a two-dimensional attractive torus looks the same as the birth of a limit cycle from a stable focus.

The second principal possibility consists in the creation of a stable limit cycle with a double period from the original stable limit cycle. Subsequent developments, leading eventually to the formation of a strange attractor, are discussed in the following sections.

6.2 The Ruelle–Takens Scenario

In this section, we consider the case of an initial limit cycle that has given birth to an invariant attractive torus in the phase space of a dynamical system. As already noted in the discussion of Hamiltonian systems in Chap. 2, a significant role is then played by the ratio of the frequencies corresponding to the motion along the meridian of the torus and along its equator. If this is irrational (i.e. cannot be expressed as a ratio of two integers) the phase trajectory densely covers the entire torus. Otherwise the trajectory closes after some number of windings and the motion is periodic.

In principle, one can imagine that this two-dimensional torus loses stability at some higher value of the control parameter μ and gives birth to a stable three-dimensional toroidal manifold, so that the asymptotic motion is specified by three independent frequencies. Further increase of the control parameter might result in a sequence of bifurcations which create invariant tori of increasingly higher dimensionalities. After a large number of such bifurcations, we would arrive at a complex quasiperiodic motion with k incommensurable frequencies, which appears chaotic.

Assuming that such a sequence of bifurcations is actually realized, *Landau* [6.8–9] and independently *Hopf* [6.10–11] suggested that the chaotic dynamics of dissipative systems simply represents quasiperiodic motion on a torus of very high dimensionality. This motion has a large number of free parameters (i.e. of the initial phases of all oscillations) which can be interpreted as its effective "degrees of freedom". In the course of time the phase trajectory visits every neighborhood, however small, of a given point on the torus and the motion is ergodic. Although the spectrum is discrete, the spectral lines become ever closer as one approaches the limit $k \to \infty$.

However, as shown in Chap. 2, quasiperiodic motion with even a large number of incommensurable frequencies cannot be called chaotic because it lacks the divergence of phase trajectories which is responsible for the emergence of a really chaotic dynamics.

Another serious deficiency of the Landau-Hopf scenario is that, contrary to their assumptions, the sequence of bifurcations leading to the formation of a multi-dimensional invariant torus is not typical – it is found extremely rarely [6.12]. Even small perturbations of the dynamic equations are usually sufficient to destroy this type of motion. As a result it either degenerates to a stable limit cycle on the torus (the effect of *frequency locking*) or it is completely restructured, giving birth to a strange attractor.

The destruction of an attractive toroidal manifold to produce a strange attractor, was first analyzed by *Ruelle* and *Takens* [6.13] (see also [6.14]). They studied the behavior of dynamical systems (6.1.1) using very general assumptions about the properties of their vector fields. It was found that, if three successive bifurcations (beginning from a fixed point) produce a three-frequency quasiperiodic motion, this motion is very easily destabilized by small structural perturbations. A three-dimensional attractive torus is thereby destroyed, giving way to a strange attractor.

Let us discuss this scenario in more detail. Recall that any dynamical system is defined by its vector field v, i.e. by the right-hand sides of the differential equations (6.1.1). The set of all possible vector fields v forms some functional space Φ. Each point in this functional space corresponds to a possible dynamical system and, reciprocally, each dynamical system corresponds to a unique point in Φ. If we slightly perturb the right-hand sides of (6.1.1), this produces a new dynamical system whose vector field v' is very close to the initial one. All dynamical systems whose vector fields can be obtained by a small variation of the vector field of a given dynamical system constitute its neighborhood in the functional space Φ.

It is said that a vector field v of a dynamical system is *structurally stable* if there is a neighborhood U of this field v, such that, for any vector field v' from this neighborhood, the family of phase trajectories is not qualitatively different from the family of a dynamical system defined by the field v. In other words, if some property of a dynamical system is structurally stable, it is preserved under small variations of the system. The concept of structural stability was first proposed for differential equations by *Andronov* and *Pontryagin* [6.15].

When we construct a mathematical model for any realistic process, we must always resort to some simplifications, neglect some insignificant factors, etc. Consequently, the vector field which enters into the right-hand sides of the dynamical equations is always known only with some finite accuracy, i.e. only within some small neighborhood U in the functional space Φ. Therefore, if a certain property of a dynamical system is not structurally stable, it will probably not be observed in an experiment with a real system.

The theorem by *Ruelle* and *Takens* [6.13, 14] states that, if there is a vector field v on a three-dimensional torus which corresponds to some three-frequency quasiperiodic motion, then in any neighborhood U of the corresponding point in the functional space Φ there will be found the vector fields v' on the three-dimensional torus which have strange attractors. The same statement holds for quasiperiodic motion on tori of higher dimensionalities. In other words, it is generally sufficient to slightly perturb the right-hand sides of the dynamical system (6.1.1) in order to transform the three-frequency quasiperiodic motion into a chaotic one. However, this is not so for *any* vector field v' from the neighborhood U. If the vector field v were strictly structurally unstable, any small perturbations would indeed lead to the destruction of the three-dimensional torus and to the emergence of a motion that is qualitatively different from quasiperiodic motion. However, the Ruelle-Takens theorem does not claim that any small perturbation of v would produce this effect. Actually, inside a small neighborhood U one finds both the fields v' that correspond to some strange attractors, and those which retain the quasiperiodic character of motion (Fig. 6.4). The fraction of the region occupied by vector fields with strange attractors is finite, but they do not fill the entire neighborhood [6.12].

Confirmation of the prediction that a small perturbation of a vector field on a three-dimensional torus does not necessarily lead to the emergence of a strange attractor was provided in [6.16–18]. *Grebogi* et al. [6.16] showed that the addition of a smooth nonlinear perturbation does not always destroy three-frequency motion. *Walden* et al. [6.17] described an experimental study of the Rayleigh-Benard convection in liquids, where quasiperiodic regimes with four or even five incommensurable frequencies were observed.

Some aspects of the Ruelle-Takens scenario can be investigated by means of renormalization methods, switching from the vector field to a discrete map [6.19, 20]. Since this analysis is rather complex, we do not include it in this book. A relatively simple exposition of the renormalization theory for the Ruelle-Takens scenario can be found in [6.21–23].

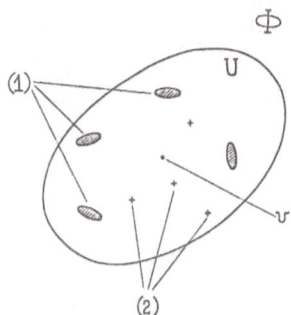

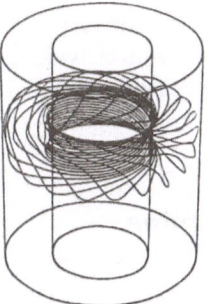

Fig. 6.4. A neighborhood U of a vector field v in the functional space Φ; (*1*) – vector fields with strange attractors, (*2*) – vector fields with regular dynamics. From [6.12]

Fig. 6.5. The trajectory of a particle in a Taylor vortex. From [6.27]

Some experimental data provide firm indications that the Ruelle-Takens scenario of transition to chaos is indeed observed in these systems. For instance, *Gollub* and *Swinney* [6.24] studied the Taylor vortices that appear in a layer of liquid between two concentric cylinders, with the internal cylinder rotating at a constant angular velocity (Fig. 6.5). The dynamics of vortices was investigated using the technique of light scattering. Light that is scattered by a small volume of liquid bears information about the spectrum of frequencies of the radial components of the liquid velocity. Experiments were carried out at different Reynolds numbers. A transition to the chaotic regime in this system was observed after the last bifurcation, resulting in quasiperiodic motion with three independent frequencies.

Similar experiments (for a review see [6.25]) were carried out for the Rayleigh-Benard convection in a horizontal layer of liquid heated from below. When the temperature gradient was increased, a transition to chaotic motion was preceded by the appearance of one and then two independent frequencies in the velocity spectrum (Fig. 6.6a,b). The chaotic regime, characterized by a continuous spectrum, started immediately after the quasiperiodic two-frequency motion (Fig. 6.6c).

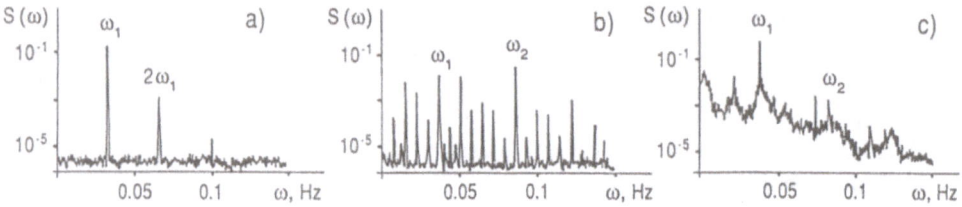

Fig. 6.6. The velocity spectrum in an experiment with the Rayleigh-Benard convection (from [6.25]); (**a**) periodic, (**b**) quasiperiodic and (**c**) chaotic regimes

Chaotic dynamics has been found in studies of the Rayleigh-Benard convection in mercury [6.26] after the successive emergence of three rationally independent frequencies in the spectrum. After emergence of the third frequency, the system was producing noise with a wide continuous spectrum. *Fein* et al. (see [6.23]) experimentally verified the renormalization theory of the transition to chaos via quasiperiodicity.

6.3 Period Doubling

Although the Ruelle-Takens scenario is fairly typical, there are at least two other equally common scenarios. In this section we discuss the scenario in which the transition to chaos is based on a cascade of Feigenbaum bifurcations. This situation corresponds to the second possibility mentioned in Sect. 6.1, i.e. to the case when the original limit cycle of the dynamical system (6.1.1) loses its stability by creating a stable limit cycle with a doubled period. This bifurcation can be multiply repeated

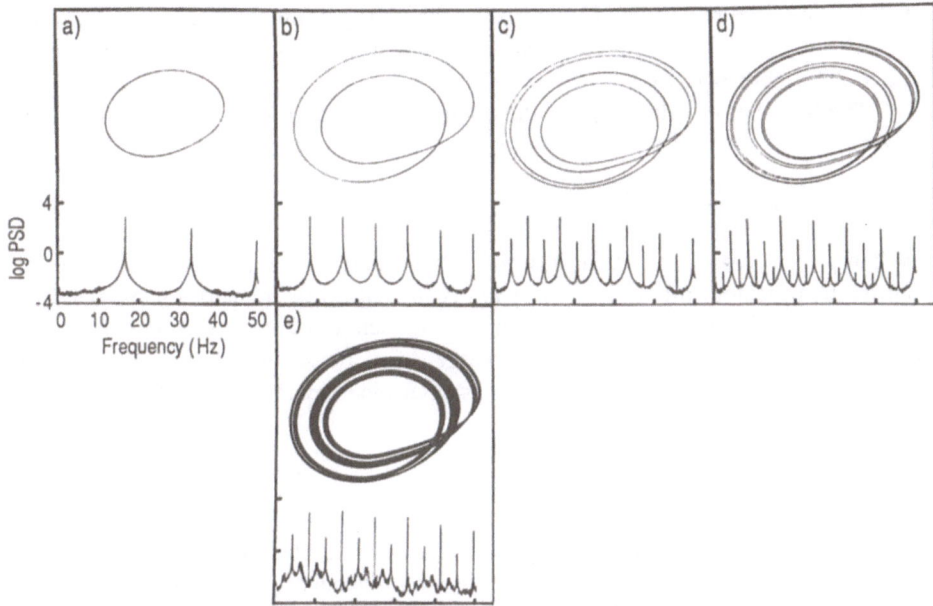

Fig. 6.7a–e. Transition to a strange attractor through the sequence of period-doubling bifurcations in the system (6.3.1). The power spectrum and the form of the attractor are shown
for different values of the control parameter: (**a**) $\mu = 2.6$, (**b**) $\mu = 3.5$, (**c**) $\mu = 4.1$, (**d**)
$\mu = 4.18$ and (**e**) $\mu = 4.23$. From [6.30]

when we further increase the control parameter μ. As a result, the cycle with the
doubled period produces a new stable cycle with the quadrupole period, and so
on. Similar to the phenomena described in Sect. 5.3 for discrete maps, an infinite
period-doubling sequence of bifurcations of the original limit cycle is possible.
This bifurcation sequence can occur within a finite interval of variation of the
control parameter μ, eventually driving the system to the regime with chaotic
dynamics.

This situation can be illustrated by the example of a dynamical system studied
by *Rössler* [6.28]:

$$\dot{x} = -(y + z) \, ,$$
$$\dot{y} = x + \tfrac{1}{5}y \, , \qquad\qquad\qquad (6.3.1)$$
$$\dot{z} = \tfrac{1}{5} + z(x - \mu) \, .$$

Equations (6.3.1) describe the dynamics of an abstract chemical reaction [6.29].
At some value $\mu = \mu_1$ of the control parameter the fixed point gives birth, by an
Andronov-Hopf bifurcation, to the limit cycle of period τ_1 (Fig. 6.7a). This cycle
remains stable until the next bifurcation value $\mu = \mu_2$ is reached. Then the cycle of
period τ_1 transforms into the twice-coiled stable limit cycle of the doubled period
$\tau_2 = 2\tau_1$ (Fig. 6.7b). The spectrum of motion now also includes lines at multiples of

half of the initial frequency. Further period-doubling bifurcations occur at $\mu = \mu_m$, $m = 3, 4, \ldots$, and result in the appearance of stable periodic motions with periods $\tau_m = 2^m \tau_1$ (Fig. 6.7c,d).

The bifurcation values μ_m of the control parameter μ constitute a converging series

$$\lim_{m \to \infty} \mu_m = \mu_\infty \approx 4.20 . \tag{6.3.2}$$

When $\mu = \mu_\infty$ the limit cycle has an "infinite period", i.e. it transforms into an attractive trajectory that never closes up. At $\mu > \mu_\infty$ this trajectory gives rise to a strange attractor (Fig. 6.7e). The dynamics of such a system is then characterized by a continuous spectrum and the divergence of phase trajectories. The rate of convergence of the infinite series of μ_m is determined by the universal *Feigenbaum constant*:

$$\lim_{m \to \infty} \frac{\mu_m - \mu_{m-1}}{\mu_{m+1} - \mu_m} = \delta = 4.6692 \ldots . \tag{6.3.3}$$

Note that this constant is the same as in the case of discrete dynamics (cf. Sect. 5.3).

The transition to a chaotic regime through a sequence of period-doubling bifurcations is also observed in nonautonomous systems. For instance, *Huberman* et al. [6.3] considered the equation of motion of a pendulum with viscous friction under the action of a periodic driving force, i.e.

$$\ddot{\theta} + \beta\dot{\theta} + \omega_0^2 \sin \theta = \nu \cos \omega_1 t . \tag{6.3.4}$$

With the correct choice of the strength of the driving force (such that the pendulum may reach its vertical upper position $\theta = \pi$ with a small remaining velocity), this system exhibits a sequence of period-doubling bifurcations. The role of a control parameter is played here by the ratio ω_1/ω_0 of the frequency of the driving force to the free oscillation frequency of the pendulum. The transition to chaos is realized when this ratio is decreased. A similar behavior was found by *Huberman* and *Crutchfield* [6.32] for the viscous motion of a particle in the "two-hump" potential under the action of a periodic driving force.

Generally, the properties of a period-doubling transition in dissipative dynamical systems are very close to those found for one-dimensional iterative maps (Sect. 5.3). This is explained by the fact that, in the majority of cases, the cross-section Poincaré map of a dynamical system undergoing such a transition is well approximated by some one-dimensional map.

Note that an *infinite* period-doubling sequence can actually be observed only in an ideal system. Small external noises (and the noise produced by a finite computational accuracy) truncate this sequence: motion becomes chaotic after several discernable period-doubling bifurcations. These effects were discussed in [6.12, 23, 33–34].

6.4 Intermittency

The third typical scenario of transition to chaos is connected with the phenomenon of *intermittency*. The intermittent motion of a dynamical system is characterized by the alternation of bursts of apparently chaotic behavior and intervals of almost periodic oscillations (Fig. 6.8). This scenario was described by *Pomeau* and *Manneville* [6.35, 36] who investigated the Lorenz model (Sect. 3.2) at large values of the parameter r. The mathematical aspects of intermittent behavior were also analyzed by *Afraimovich* and *Shilnikov* [6.37] (see also [6.38]).

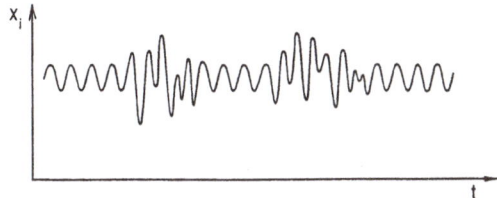

Fig. 6.8. Typical time dependence of a dynamical variable x in the intermittent regime

As already noted, the systems of differential equations are closely related to iterative maps. In particular, each scenario of the transition to chaos in a dynamical system has its counterpart in an iterative map. Bearing this in mind, we explain below the mechanism of intermittent behavior using the example of a one-dimensional map

$$x_{n+1} = \phi(x_n, \mu) \ . \tag{6.4.1}$$

Suppose that the graph of this map touches the bisectrix $x_{n+1} = x_n$ at some critical value μ_c of the control parameter μ. For convenience, we assume that this happens at the point $x = 0$ (Fig. 6.9). Let us expand the function ϕ near this point into a power series of x_n:

$$\phi(x_n, \mu) = 0 + (\mu - \mu_c)\left(\frac{\partial \phi}{\partial \mu}\right)_0 + x_n\left(\frac{\partial \phi}{\partial x_n}\right)_0$$
$$+ \frac{1}{2}x_n^2\left(\frac{\partial^2 \phi}{\partial x_n^2}\right)_0 + \dots \ , \tag{6.4.2}$$

where the derivatives are taken at $x_n = 0$ and $\mu = \mu_c$.

Choosing the appropriate units of measurement for μ and x_n, we can make $(\partial \phi/\partial \mu)_0 = 1$ and $(1/2)(\partial^2 \phi/\partial x_n^2)_0 = 1$.

Furthermore, since the function $\phi(x_n, \mu_c)$ touches the bisectrix at $x_n = 0$, we have $(\partial \phi/\partial x_n)_0 = 1$. Combining these notions and keeping only the first terms of the expansion (6.4.2), we find

$$x_{n+1} = (\mu - \mu_c) + x_n + x_n^2 \ . \tag{6.4.3}$$

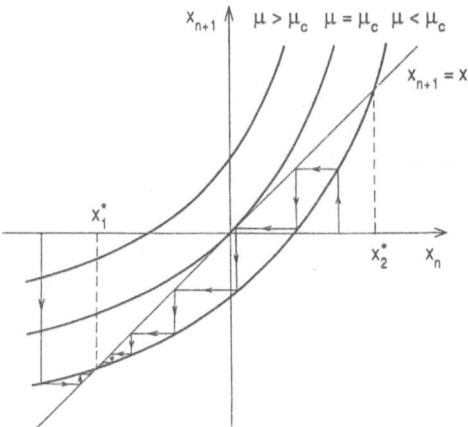

Fig. 6.9. Graph of the map (6.4.3) in the neighborhood of the coordinate origin

When $\mu < \mu_c$, this map has two fixed points

$$x^*_{1,2} = \pm(\mu_c - \mu)^{1/2} , \qquad (6.4.4)$$

the first of which (x^*_1) is stable and the second (x^*_2) unstable (Fig. 6.9). At the critical values μ_c of the control parameter μ, the two points merge and disappear (hence this *tangent bifurcation* represents a crisis).

If the control parameter μ is only slightly larger than μ_c, the distance between the bisectrix $x_{n+1} = x_n$ and the graph of the map (6.4.3) is very small near the point $x = 0$ (Fig. 6.9). Then, as one can easily see, the system can leave the vicinity of the point $x = 0$ only after a very large number of iteration steps. This implies that, within a long discrete time interval, the behavior of the map will be quite regular.

Suppose now that the map (6.4.1) also includes a region that is able to generate a complex dynamics (shown, for example, in Fig. 6.10). If we start from some point x_0 in the interval $[a, b]$ in Fig. 6.10, the map will generate a sequence of points x_0, $x_1, x_2, \ldots$. At some step, the point of this sequence will enter the region near the maximum of the function ϕ and, after performing several irregular oscillations, it will again be thrown out to the region of small values of x and the process will be repeated (Fig. 6.10). The resulting sequence $x_0, x_1, x_2, \ldots, x_k, \ldots$ will not generally include cycles of any finite period.

Let us turn now to the systems of differential equations. We know that, in order to possess a chaotic regime with a strange attractor, the dynamical system should be described by at least three variables. When we introduce the Poincaré map for such a system, phase trajectories are represented by a set of points in the cross-section (hyper)surface. Under a strong compression (dissipation), the Poincaré maps of many systems are almost one-dimensional, so that the behavior of phase trajectories can be approximately described by an appropriate one-dimensional iterative map.

When the phase-space point of such a one-dimensional map spends a long time near some point (e.g. $x = 0$), the respective dynamical system would be

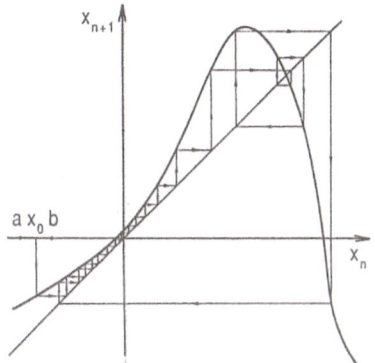

Fig. 6.10. Example of a one-dimensional discrete map with intermittent behavior

characterized within this long time interval by almost periodic behavior. Eventually leaving this region, the system can enter another region of phase space where the dynamics is very complex (this region corresponds to the neighborhood of the maximum for the discrete map shown in Fig. 6.10). After the system has spent some time there, it is pushed out of this region and again visits the region of phase space where almost periodic oscillations occur. Repetitions of such a process give rise to intermittent behavior.

We can estimate the duration of regular oscillations in each interval of almost periodic motion. Note that at very small x_n's the map (6.4.3) can be approximately replaced by a differential equation

$$\frac{dx}{d\tau} = (\mu - \mu_c) + x^2 \tag{6.4.5}$$

(indeed, the value of x then changes very little in any discrete time step $\Delta\tau = 1$). Integration of (6.4.5) yields

$$\tau = (\mu - \mu_c)^{-1/2} \arctan\left[x(\mu - \mu_c)^{-1/2}\right] . \tag{6.4.6}$$

The characteristic number τ_r of discrete time steps necessary for a system to leave the interval $|x| < (\mu - \mu_c)^{1/2}$, is estimated from (6.4.6) as

$$\tau_r = (\mu - \mu_c)^{-1/2} . \tag{6.4.7}$$

The duration T_r of the periodic motion interval can then be obtained if we multiply τ_r by the period T_0 of an individual oscillation, $T_r \sim T_0(\mu - \mu_c)^{-1/2}$.

Hence, the time spent by the system in the region with almost periodic oscillations diverges when the control parameter μ approaches its critical value μ_c. On the other hand, the duration of the intervals with irregular dynamics is not significantly influenced by variations of μ.

Above we described *intermittency of the first kind*, which is produced as a result of a tangent bifurcation. There are, however, two other scenarios producing intermittent behavior. If the function $\phi(x, \mu)$ changes with μ, as shown in

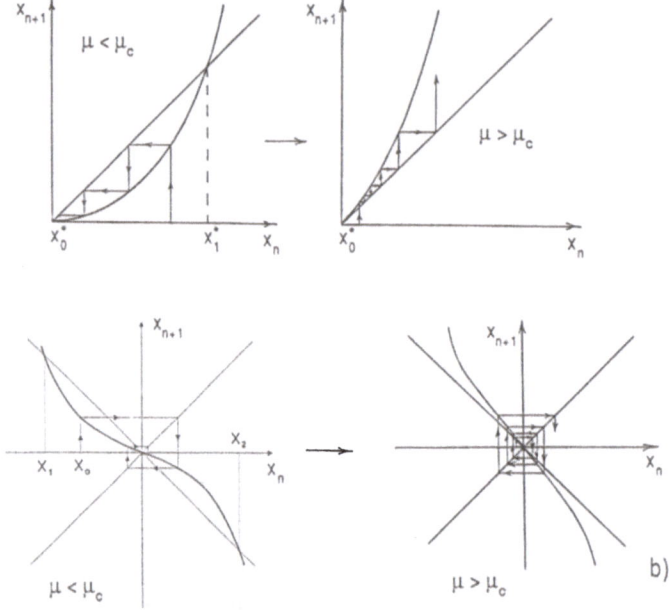

Fig. 6.11. Examples of discrete one-dimensional maps that demonstrate intermittencies of the second kind (**a**) and the third kind (**b**)

Fig. 6.11a, this also results in the intermittent regime. In this case, the fixed point x_0^* loses its stability at $\mu = \mu_c$, merging with the unstable fixed point x_1^*, but still does not disappear. The resulting regime of motion is called *intermittency of the second kind*.

The *third kind of intermittency* is produced when two points x_1 and x_2, where the graph of the function $\phi(x)$ intersects the bisectrix of the II and IV quadrants, merge with the stable fixed point $x_0^* = 0$ (Fig. 6.11b). After this the point $x_0^* = 0$ becomes unstable and an almost periodic motion develops in its vicinity.

Figures 6.11a,b show that in both cases there should be long intervals of almost regular oscillations, during which the phase-space point stays near the coordinate origin. Chaotic bursts are produced when the system enters the outer region with complex dynamics (not shown in Fig. 6.11).

Note that intermittent behavior can be also described by a renormalization theory (see [6.23, 39–41]). Intermittent chaos was experimentally observed in studies of the Rayleigh-Benard convection [6.42, 43], of the Belousov-Zhabotinskii reaction in a continuously stirred flow reactor [6.44, 45], and in other systems [6.46–49].

In this chapter we have outlined three different principal scenarios of the transition to chaos. They do not exhaust the list of all possible variants for such a transition. A strange attractor can be also produced, for instance, from the doubly asymptotic trajectories, as described in Sect. 3.2 for the Lorenz model. Other more rare variants of the transition include the sequence of doubling and destruction of two-dimensional tori [6.50–53], the sequence of period-tripling (or even period-quadrupling) of limit cycles [6.3, 54, 55], torus–chaos intermittency [6.56, 57], and so on.

6.5 Controlling Chaotic Behavior

Chaotic oscillations are a common property of many dynamical systems. In some cases, however, it is desirable that the development of chaos be suppressed. This raises a problem of how to control chaotic dynamical systems, i.e. how to create conditions under which systems that are originally chaotic acquire regular (periodic) dynamics.

The general problem of control is formulated as follows: Suppose we have a dynamical system described by a set (6.1.1) of differential equations. These equations specify a certain flow $F^t(x)$ in the phase space of the system. We want to modify the system in such a way that under the control u its flow $F^t(x, u)$ converges to a prescribed set U in the phase space, i.e., the trajectories approach U and remain in its close vicinity so long as the control is present.

An efficient feedback control method for chaotic dynamical systems has been proposed by *Ott*, *Grebogi*, and *Yorke* [6.58] (see also [6.59]). They note that such systems have many unstable (saddle) limit cycles and these cycles may be found in any neighbourhood of the chaotic (strange) attractor. The idea is to stabilize one of the existing limit cycles.

Let us construct the Poincaré map of our dynamical system, choosing a plane that intersects the given saddle limit cycle (we assume that the cycle is closed after a single revolution of the phase point – more complex cases are considered by *Auerbach* et al. [6.60]). The intersection yields a fixed saddle point x^* of the map which has both stable and unstable separatrixes. In a small neighborhood of x^*, the system's behavior is shown by the dashed trajectories in Fig. 6.12. Initially, the phase point p_1 comes closer to x^*, observing the stable direction. Later, the repulsion prevails and the point leaves the neighborhood of x^*, following the unstable direction.

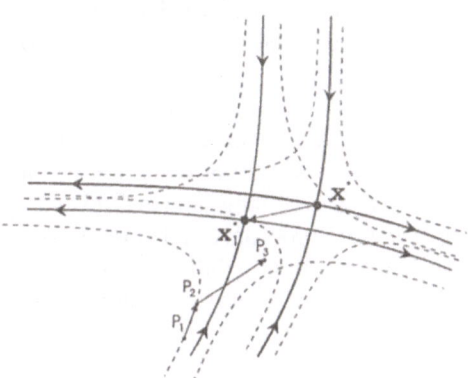

Fig. 6.12. The Poincaré map of a dynamical system

We want to apply perturbations, choosing them in such a way that the phase point would tend towards the stable manifold of x^*. By selecting appropriate perturbations at each step, it can be ensured that the phase point always remains near the stable direction and, hence, that the phase trajectory of the system lies close to the considered limit cycle.

Of course, this procedure can only be used if the phase point p_1 is already located close enough to x^* when the perturbations are started. But, owing to the ergodicity of the system's dynamics, almost all its neigborhoods are accessible. This means that, after waiting for some time, we would see that the point arrives in the vicinity of x^*. When this occurs, the control perturbations can be applied, locking the system in the vicinity of the respective limit cycle.

Now we give a detailed description of this control method. Suppose that the Poincaré map has the form

$$x_{n+1} = f(x_n, \mu) , \tag{6.5.1}$$

where $f = \{f_1, \ldots, f_k\}$ and μ is some parameter. We assume that for $\mu = \mu_0$ this map has a fixed point x^*, i.e., $f(x^*, \mu_0) = x^*$. When the parameter μ is close to μ_0 and x is near to x_0, the linear approximation of the map (6.5.1) could be used, i.e., we can write

$$x_{n+1} - x^* = A(x_n - x^*) + B(\mu - \mu_0) . \tag{6.5.2}$$

Here A is a $k \times k$-dimensional Jacobi matrix and B is a k-dimensional column vector, $A = \partial f / \partial x$ and $B = \partial f / \partial \mu$ at $x = x^*$ and $\mu = \mu_0$.

We exercise the control by adjusting the parameter μ at each successive time step n, so that

$$\mu_n - \mu_0 = -L^T(x_n - x^*) , \tag{6.5.3}$$

where L is some k-dimensional column vector and T denotes the transposition operation. Using (6.5.2), we obtain

$$\delta x_{n+1} = (A - BL^T)\delta x_n , \tag{6.5.4}$$

where $\delta x_n = x_n - x^*$. This linear map has a stable fixed point $\delta x = 0$ if the modulus of all eigenvalues of the matrix $A - BL^T$ is less than one. This condition can be satisfied by a proper choice of the vector L. Then the dynamics of the system becomes stabilized near the saddle limit cycle, corresponding to the point x^* of the Poincaré map.

The method can be used only once the trajectory is near enough to the point x^*, so that the linearization (6.5.2) is justified. As we have remarked, due to the ergodicity of motion, the phase trajectory will eventually reach the required neighborhood. However, one might have to wait a very long time and this can be inconvenient. To eliminate this difficulty, a special procedure called the targeting method has been proposed [6.59, 61, 62]. It offers the possibility of directing the phase trajectory to a certain region of phase space by means of small parametric perturbations.

The method of Ott, Grebogi, and Yorke is very efficient and can be applied to practically any system possessing a chaotic attractor. Its disadvantage, however, is that the deviation of the system from the required state must be computed at each iteration.

An empirical method of *adaptive control* has been proposed by *Huberman* and *Lumer* [6.63]. Suppose that we want to modify the system's dynamics in such a way that it approaches the dynamics of a target system described by

$$\dot{y} = g(y) \ . \tag{6.5.5}$$

To achieve this, one can try to modulate the control parameter, depending on the difference between the actual dynamics of the system and that of the target. Namely, we take

$$\mu - \mu_0 = h(x - y) \ , \tag{6.5.6}$$

where $h(x)$ is a certain function. It turns out that even with the simplest linear choice of $h(x)$, the chaotic dynamics of the original system can be reduced to periodic time behavior.

Besides of the methods employing different variants of feedback, suppression of chaos can often be achieved by applying external fixed control. *Hübler* et al. [6.64, 65] have suggested a method of *resonance stimulations* which is based on the inclusion of an additive external term $F(t)$ to the dynamical equations of the system,

$$\dot{x} = v(x, \mu) + F(t) \ . \tag{6.5.7}$$

If the required (target) dynamics is given by a function $y(t)$ which obeys the equation (6.5.5), the perturbation is chosen as

$$F(t) = g(y(t)) - v(y(t)) \ . \tag{6.5.8}$$

Thus, the perturbation vanishes when dynamics of the controlled system follows that of the target system.

Moreover, it has been shown by *Alexeev* and *Loskutov* [6.66] (see also [6.67, 68]) that, in many cases, periodic temporal variation of a system parameter is already sufficient to suppress the chaotic dynamics and transform it into stable periodic oscillations.

Suppose that the considered system has a chaotic attractor inside a certain interval (μ', μ'') of its parameter μ. Then this parameter can be periodically varied as

$$\mu = \mu_0 + \mu_1 \sin \omega t \ , \tag{6.5.9}$$

where μ_0 and μ_1 should be chosen in such a way that the variations of μ lie within the interval (μ', μ''). Even with this simple choice of the control signal, it is possible to find values of μ_0, μ_1 and ω such that the chaotic behavior is replaced by periodic oscillations. The transition to chaos as the amplitude and the

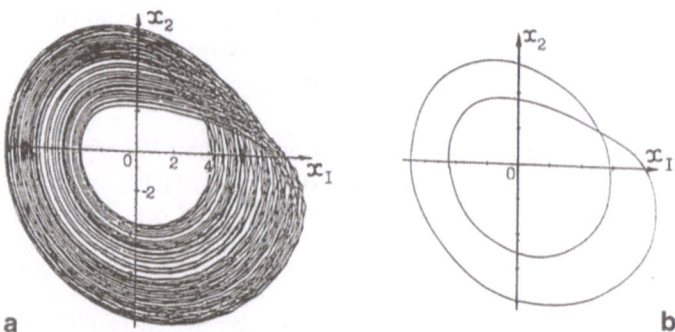

Fig. 6.13. Suppression of chaos in the Rössler model (6.3.1); (**a**) the chaotic phase trajectory at $\mu = 4.46$ and (**b**) the stabilized limit cycle at $\mu_0 = 4.46$, $\mu_1 = 0.21$ and $\omega = 1.62106$. From [6.69]

frequency of the perturbation are gradually changed is then realized via period-doubling bifurcations and intermittency [6.66, 68]. Figure 6.13 shows suppression of chaos in the Rössler system described by the equations (6.3.1).

Thus, the numerical simulations show that application of a weak periodic signal can produce regular periodic oscillations. The analytical proof of this property has been constructed for one-dimensional maps [6.67, 69]. Let us consider a map given by the equation

$$z_{n+1} = z_n \exp\left[a(1 - z_n)\right] . \tag{6.5.10}$$

When the parameter a is increased, this map exhibits in the interval $0 < a < a_\infty$ a sequence of period-doubling bifurcations that gives rise to the chaotic behavior which is observed for some parameter values $a > a_\infty$. We denote the set of these values of a as A_c.

To realize the control, we periodically vary the parameter a in the map (6.5.10), so that it becomes

$$z_{n+1} = z_n \exp\left[a_n(1 - z_n)\right] , \tag{6.5.11}$$

where the values a_n form a periodic sequence of period τ, obtained by the repetition of a finite subsequence $\{a_1, a_2, \ldots, a_\tau\}$.

Suppose that all members a_i of a subsequence belong to the set A_c, i.e., they yield chaotic dynamics if the parameter a remains constant. It can then be proved [6.69] that a subsequence $\{a_1^d, a_2^d, \ldots, a_\tau^d\}$ of certain length τ can always be found, such that the perturbed map (6.5.11) has stable periodic orbits with the period T representing a multiple of τ.

Actually, the one-dimensional map (6.5.10) is closely related to the Poincaré map of the Rössler system (6.3.1) in the cross-section $y = 0$. Therefore, its analysis provides further evidence that the dynamical chaos in this system can be suppressed by periodic variations of its parameters.

We have described in this section only a few methods that are employed to control chaotic dynamics. Further examples and applications can be found in [6.70–77].

7. Spatiotemporal Chaos

Though we have considered only dynamical systems with a finite (and small) number of variables, the results of this analysis are also applicable to a wide class of distributed active systems with spatial coherence. The actual number of degrees of freedom for such systems is very large, but the majority of them are enslaved and adiabatically follow variations of only a few variables that control the dynamics (regular or chaotic) of a coherent spatial pattern. Numerical techniques are available which allow us to determine, proceeding from experimental data, the minimal number of independent variables needed to reproduce the observed chaotic dynamics of a coherent pattern.

A fundamentally new situation, described as developed turbulence, is encountered if the correlation radius is much shorter than the spatial dimensions of the medium, which thus breaks into a great number of uncorrelated spatial domains. The transition to developed turbulence will be illustrated using examples of coupled chaotic maps and the complex Ginzburg–Landau equation.

7.1 Analysis of Time Series

The state of a distributed system is described by spatial fields of its variables, such as temperature, pressure, density, etc. All possible spatial distributions form a functional space of infinite dimensionality. The evolution of a distributed dynamical system can therefore be viewed as motion inside an infinite-dimensional phase space which is governed by partial differential equations. However, it turns out that in many cases the dynamics of a distributed system can be well approximated by a few ordinary differential equations [7.1–7]. Figure 7.1 shows a characteristic example of convection in the Hele–Shaw cell [7.8, 9]. This cell consists of a flat rectangular vessel filled by a fluid that is steadily heated from below. Provided the heating is weak enough, the fluid remains still. For stronger heat flows, however, motion starts and the fluid performs first periodic and then, at higher heating intensity, aperiodic oscillations. The fluid in the cell is obviously a distributed system and its motion can be completely described only in terms of partial differential equations. Nonetheless, the established motion of such a system can be very well approximated using a simple set of eight ordinary differential equations.

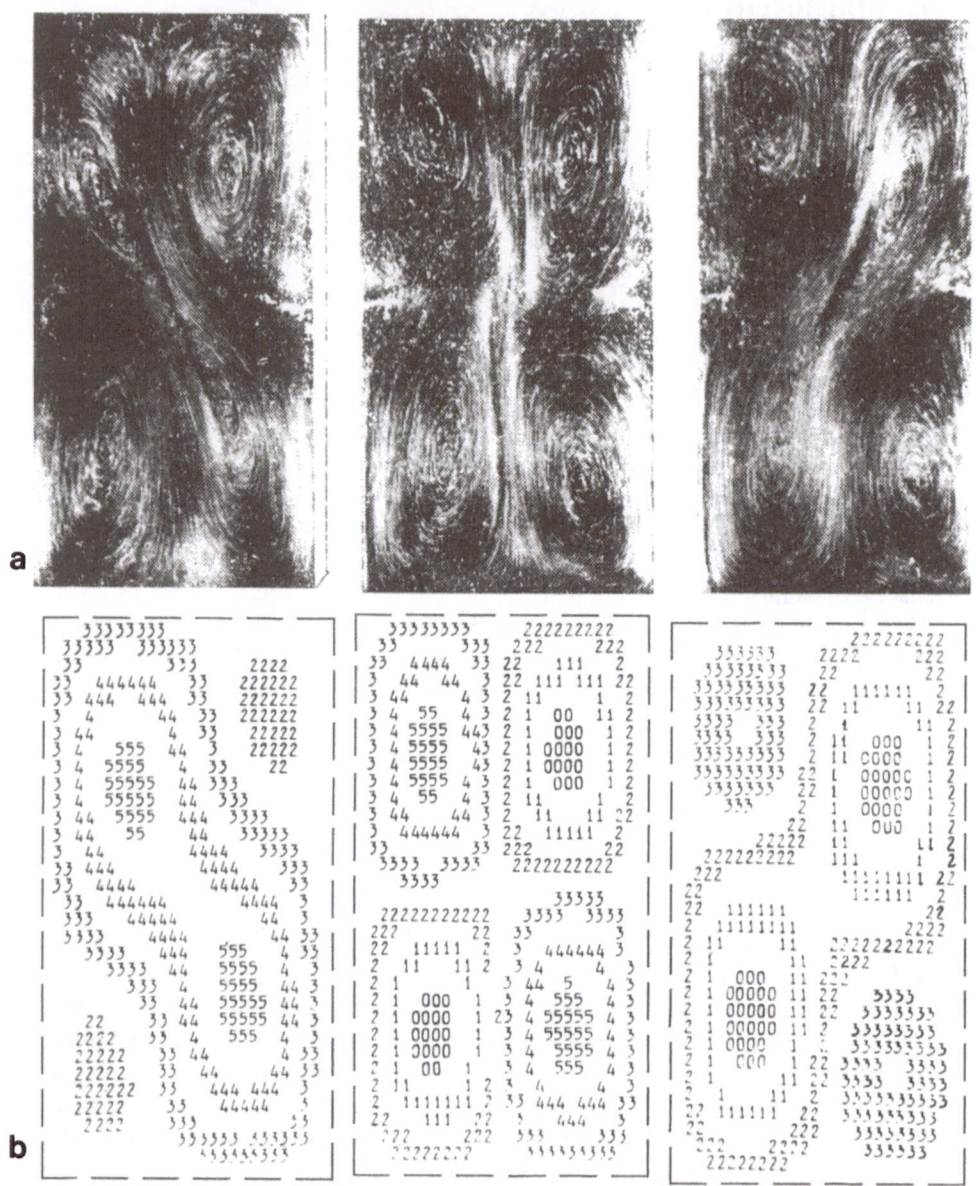

Fig. 7.1. Convection in the Hele–Shaw cell (from [7.8, 9]); (**a**) the experimental flow pattern, (**b**) its approximation obtained by numerical integration of a system of eight differential equations

The physical explanation of why a finite-dimensional approximation becomes applicable is, of course, that the great majority of spatial modes are damped and adjust adiabatically to variations in the amplitudes of a few principal modes.

Following *Haken* [7.10], this situation can be described as *enslaving* of damped modes. The mathematical equivalent of such arguments is that, since evolution of dissipative distributed dynamical systems is accompanied by compression of the phase-space volume, at long times their trajectories may asymptotically approach a low-dimensional subset of the system's infinite-dimensional functional phase space. Such attractive subsets may represent simple attractors, i.e., fixed points or limit cycles. However, they can also be strange attractors – then the dynamics of a distributed dynamical system would be chaotic though it is well described by only a few dynamical variables.

The smallest number of independent variables which uniquely determine the established motion of a dissipative distributed dynamical system is called its (minimal) *embedding dimension* d_e. Generally, the embedding dimension of an attractor represents the minimal dimension of a phase subspace in which one can embed, without self-intersections, a smooth manifold that completely contains this attractor [7.11, 12]. The dimensionality of such a subspace containing the attractor will, as a rule, be larger than the proper dimensionality of this attractor.

Indeed, a toroidal surface can be embedded into a three-dimensional space. A plane curve can be placed without self-intersection onto a plane. However, a non-planar curve, such as a complicated cycle on a torus, can be immersed without self-intersections only into a three-dimensional space. There is a theorem [7.13, 14] which states that any smooth manifold of m dimensions can *always* be embedded (i.e., placed without self-intersections) into a space of $2m+1$ dimensions. It means, for instance, that any curve ($m = 1$) can be embedded into a three-dimensional space.

A strange attractor, characterized by fractal dimension D_F, can always be embedded into a space of integer dimensionality k that satisfies the condition $k \geq 2D_F + 1$ [7.12, 15]. However, this is only a sufficient condition. In some cases the embedding dimension can actually be smaller, yet it never goes below $[D_F] + 1$ where $[D_F]$ is the integer part of the number D_F.

To determine the embedding dimension of a distributed system from experimental data, a procedure developed by *Packard* et al. [7.11] and *Takens* [7.12] (see also [7.15]) can be used. According to this procedure, it is sufficient to know the time dependence of a *single* variable at a certain spatial point (for instance, to monitor the temperature or fluid velocity in the Rayleigh–Bénard convection experiment).

Suppose we investigate a distributed system in the established regime, after all the transients have died out. By repeatedly measuring the values of one of the variables (we denote it as x) at equal time intervals τ, the sequence

$$x(t) = y_1, x(t + \tau) = y_2, \dots, x(t + (n - 1)\tau) = y_n, \dots \tag{7.1.1}$$

can be obtained.

If the dynamics of our system is effectively described by just one first-order differential equation, this sequence is uniquely determined by the initial value y_1 while all other values y_n are functionally dependent on it. This means that a function $f(y)$ should exist, such that $y_2 = f(y_1)$, $y_3 = f(f(y_1))$ and so on. A consequence of this is that all points formed by the pairs (y_{n+1}, y_n) from the sequence (7.1.1) should lie *on the same curve* $y_{n+1} = f(y_n)$ in the plane (y_{n+1}, y_n).

If the asymptotic dynamics of our system is actually specified by two first-order differential equations, the points of (7.1.1) would no longer belong to the same curve. Instead, they will be scattered in a complex fashion on the plane (y_{n+1}, y_n). To uniquely determine the entire sequence (7.1.1), now we must fix two initial values y_1 and y_2. In this case any two successive values y_{n-1} and y_n would uniquely determine the next value y_{n+1} and hence a function f must exist such that $y_{n+1} = f(y_{n-1}, y_n)$.

Generally, if a dynamical system is described by k independent variables, the value of y_{n+1} is uniquely determined by k previous values $y_n, y_{n-1}, \ldots, y_{n-k+1}$ in the sequence. This means that y_{n+1} is functionally dependent on $y_n, y_{n-1}, \ldots$ $\ldots, y_{n-k+1}$, i.e., a function f exists, such that $y_{n+1} = f(y_n, y_{n-1}, \ldots, y_{n-k+1})$.

The method, which was proposed in [7.11, 12], begins with the construction of the sequence $y_1, y_2, \ldots, y_n, y_{n-1}, \ldots$ formed by the values of some observed variable taken at time intervals τ. After that, starting from $k = 1$, one checks whether at any n the value of y_{n+1} in the sequence is functionally dependent on the k previous values $y_n, y_{n-1}, \ldots, y_{n-k+1}$. If the functional dependence is lacking, k is increased by one and the test is repeated. The value of k for which the functional dependence is first found yields the minimal embedding dimension d_e.

The question, of course, is how the existence of a functional dependence can be tested without an actual construction of the mapping function f, which would be too difficult a task. A simple geometric criterion of the functional dependence has been suggested in [7.16, 17]:

Let us pick from the sequence of measurements (7.1.1) various subsequences of length k starting from different nth elements and consider vectors $w(n) = \{y_{n-k+1}, y_{n-k+2}, \ldots, y_{n-1}, y_n\}$ belonging to a certain k-dimensional space. Choosing a subsequence which begins at some element n_0 and denoting the respective vector as $w(0)$, we define the distance $\varrho_k(n, n_0)$ between any vector $w(n)$ and the vector $w(0)$,

$$\varrho_k(n, n_0) = |w^{(n)} - w^{(0)}| = \left(\sum_{i=1}^{k} (y_{n-k+i} - y_{n_0-k+i})^2 \right)^{1/2} \tag{7.1.2}$$

and consider the quantity

$$r(n, n_0) = |y_{n+1} - y_{n_0+1}| . \tag{7.1.3}$$

If the functional dependence between y_{n+1} and the previous k values $y_n, y_{n-1}, \ldots, y_{n-k+1}$ exists and therefore $y_{n+1} = f(w(n))$, we have

$$r(n, n_0) = |f(w(n)) - f(w(0))| . \tag{7.1.4}$$

Hence, because f is a continuous function, the quantity $r(n, n_0)$ should in this case go to zero as the distance $\varrho_k(n, n_0)$ between the two vectors $w(n)$ and $w(0)$

is decreased. On the other hand, this quantity remains arbitrary when no functional dependence is present and the vector $w(n)$ does not fix the value of y_{n+1}.

To approximately test the functional dependence, we plot a set of points (r, ϱ) for various moments n (for convenience, the subsequent points in this sequence can be connected to form a curve). By looking at the plot, we can examine the behavior of this curve for small values of ϱ. If the functional dependence is present, all parts of the curve in this region must lie close to the origin of coordinates.

By repeatedly testing the functional dependence at increasing values of k, the embedding dimension can be found. Figure 7.2 shows an example of the application of such an algorithm in the experimental investigation of spin-wave turbulence in an antiferromagnetic crystal [7.18].

Though the above technique provides a simple procedure for determination of the embedding dimension, it is less popular than the method of *Grassberger* and *Procaccia* [7.19, 20] which yields in addition the (fractal) correlation dimension of a strange attractor.

Suppose that an experiment has produced a sequence (7.1.1) of values of a certain variable taken at fixed time interval τ. We can again choose various

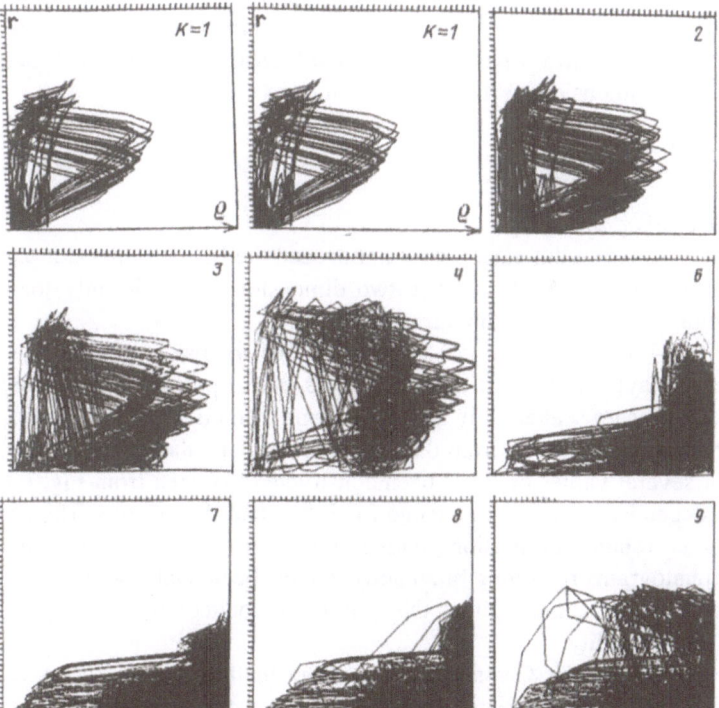

Fig. 7.2. Experimental determination of the embedding dimension (from [7.18]). The measured dependence of r on ϱ is plotted for different values of k, ranging from 1 to 9. The functional dependence is first achieved at $k = 5$

subsequences of length k and construct from them vectors $w(n)$. Using (7.1.2), the distance $\varrho_k(n, n')$ between any pair $w(n)$ and $w(n')$ of such vectors can be determined. Let us take some number l and calculate the *correlation integral* $C_k(l)$ defined as

$$C_k(l) = \lim_{N \to \infty} N^{-2} \sum_{n,n'=1}^{N} H(l - \varrho_k(n, n')) . \tag{7.1.5}$$

Here $H(z)$ is the step function, such that $H(z) = 0$ for $z < 0$ and $H(z) = 1$ for $z > 0$, while $N+k$ is the total number of the elements in the sequence (7.1.1). Note that in an actual calculation, instead of taking the limit, one can simply consider the longest available recorded sequence of measurements.

As follows from (7.1.5), contributions to the correlation integral $C_k(l)$ are given only by those pairs of vectors $w(n)$ and $w(n')$ that are separated by a distance not exceeding l. For small values of l the correlation integral should go to zero as some power of l so that

$$C_k(l) \sim l^{\nu_k}. \tag{7.1.6}$$

The exponent ν_k for different k's can be found by plotting the dependence of $\ln C_k(l)$ on $\ln l$ and taking the slope of this dependence. Starting from some k, the exponent ν_k must cease to grow when k is increased [7.20]. This value of k then yields the embedding dimension of the attractor, whereas

$$D_2 = \lim_{l \to \infty} \frac{\ln C_k(l)}{\ln l} \tag{7.1.7}$$

represents its *correlation dimension*.

Generally, the correlation dimension does not exceed the fractal dimension D_F of an attractor, introduced in Sect. 4.2. The two dimensions coincide only for a uniform distribution of points over the attractor [7.20].

The Grassberger–Procaccia method has been applied to many different systems. For example, Rayleigh–Bénard convection has been optically investigated in [7.21]. A thin light beam scattered by a small volume of fluid bears information about the fluid velocity. Proceeding from the experimental data, the dependence of $\ln C_k(l)$ on $\ln l$ for several values of k has been constructed. As seen from Fig. 7.3, the slope of this dependence ceases to change for $k > 4$ and thus $d_e = 4$. The determination of the correlation dimension yields in this case $D_2 = 2.8$. In a similar way, electroencephalograms recording brain activity have been analyzed in [7.22], with the conclusion that in some regimes they can be reproduced by a dynamical system of low dimensionality.

It should be noted that the Grassberger–Procaccia method becomes less reliable for high embedding dimensions because of a fast increase in the required number of experimental points in the time series. The method is most effective when an optimal time step τ is chosen (see [7.23–25]). Modifications improving the convergence rate of this method have been proposed in [7.26–30]. Similar

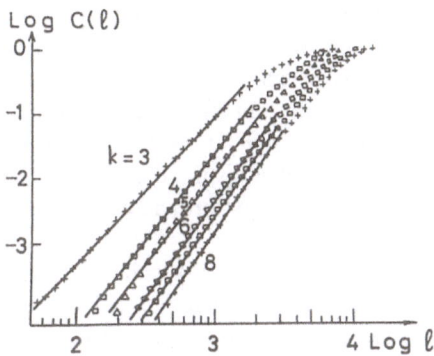

Log C(ℓ)

Fig. 7.3. The experimental dependence of the correlation integral C_k on l on a logarithmic scale for several different values of k. From [7.21]

techniques for determination of other fractal dimensions D_q, of the entropy and the Lyapunov exponents on the basis of experimental time series have also been developed [7.24, 25, 31, 32].

It is important to realize that the application of any method of time series analysis is based on the physical assumption that the dynamical process is statistically stationary, i.e., the considered system has already passed a transient state and reached the asymptotic regime where its properties are independent of the initial conditions. This also implies that the system's parameters must remain constant with time. When a recorded time series is produced by a complex system whose behavior is not completely controlled in the experiment, it can be very difficult to verify this assumption. Any experimental time series contains also a certain amount of noise which has a very high dimensionality and special procedures have been proposed to distinguish dynamical chaos from environmental noise [7.33]. A discussion of the problems related to the analysis of time series obtained in neurophysiological experiments is given in [7.34].

It has been tacitly assumed above that a time series recorded at an arbitrary location in a distributed dynamical system is sufficient to specify its global dynamics. This implies a spatial coherence which could be expected only for relatively small volumes. As the system size grows, it becomes larger than the spatial correlation length of the process. Then the system breaks down into spatial domains where the local processes are statistically independent. When this occurs, a series of observations performed at one spatial point does not generally allow any conclusions to be made about the dynamics in its distant parts.

Suppose that we study a turbulent fluid flow in a channel by recording the time evolution of the local velocity, which could be performed, for example, by optical methods. When the probe is taken in the middle of the channel, large-amplitude chaotic pulsations with a high effective dimensionality would be observed. However, if we place the probe near the channel boundary, a fairly regular flow would be found. Indeed, the viscosity effects dominate inside the boundary layer making the flow laminar there. Hence, by varying the probe position one can observe here a wide range of different local dynamical regimes.

In a large distributed chaotic system, the dynamics remains coherent within domains whose size is about the correlation radius. We can view this system as an ensemble of individual elements representing such spatial domains. The local dynamics of an element can be regular or intrinsically chaotic. The absence of statistical correlations between different elements does not, of course, mean that the dynamics of a given element is not influenced by other spatial elements in the ensemble. What happens is that all other elements collectively form an *environment* of a given element and generate *noise* which acts on it.

If the dynamics of a distributed dynamical system is spatially coherent (even though chaotic in time), it can effectively be described by a finite number of variables representing the amplitudes of a few active modes. In this case, all methods and results of the analysis of finite-dimensional dynamical systems described in the previous chapters remain valid. A new situation, requiring special analysis, is encountered however when the spatial coherence is destroyed and the system is found in a *turbulent* state characterized by spatial correlations with a radius that is much shorter than the size of the system considered.

7.2 Turbulence in Distributed Active Systems

The classical examples of turbulence are provided by fluid flows. However, there is one property of hydrodynamic turbulence which makes it very special. It is related to the presence of convective terms in the Navier–Stokes equations. Because of such terms, small turbulent fluid elements are transported over large distances by large-scale flows which could also be chaotic. This leads to a complex hierarchical structure spanning a wide range of characteristic eddy sizes.

The situation is simpler and more amenable to theoretical analysis when such global mixing is suppressed. This can be achieved, for instance, by taking shallow fluid layers where viscosity effects make lateral flows difficult. Under these conditions, the transition to developed fluid turbulence has, as we show below, much similarity to the respective phenomena in reaction–diffusion systems.

In the experiments with Rayleigh–Bénard convection, a thin horizontal layer of fluid is uniformly heated from below. Because the warmer fluid at the bottom is lighter, it tends to rise to the surface and replace the colder and heavier fluid there. But, obviously, this replacement cannot be realized in a uniform way: the local upwards flow of the warmer fluid must alternate in space with the downwards flow of colder fluid. If the layer is sufficiently thin, the flows remain fairly regular and there is no strong lateral mixing. Near the onset of convection and depending on the system's parameters, the flow pattern corresponds either to a set of rolls or to a lattice arrangement of hexagonal cells.

When the system is not too large, the asymptotic pattern is coherent and can be described by just a few modes. The spatial coherence then persists even when the pattern is no longer stationary and chaotic pulsations begin. In a fact, the Lorenz model (3.2.1) gives a rough description of this convection regime.

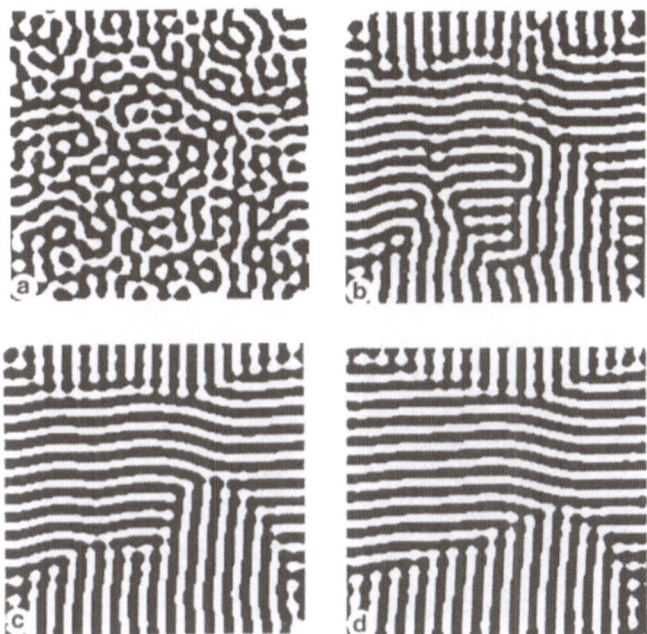

Fig. 7.4a–d. Development of a roll pattern in simulations of fluid convection. From [7.35]

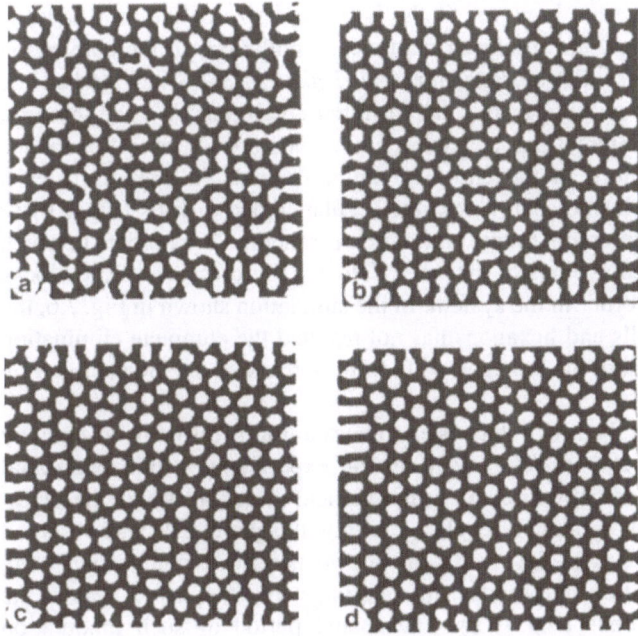

Fig. 7.5a–d. Development of a hexagon pattern in fluid convection. From [7.35]

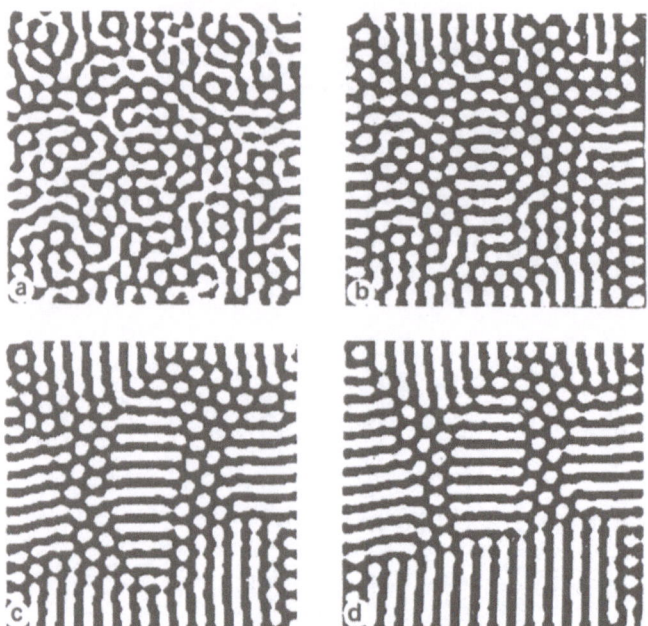

Fig. 7.6a–d. Formation of a mixed pattern in fluid convection. From [7.35]

The situation is different, however, if the horizontal size of the fluid layer is larger. Figures 7.4–6 show the results of numerical simulations of the convection instability in fluids, performed by *Bestehorn* and *Haken* [7.35]. They studied the evolution of the system starting from random initial conditions. We see that in a large system the local interactions between its elements are not able to produce a coherent periodic pattern in the entire medium. Instead, the medium breaks up into several blocks. Inside each block, the regular lattice of rolls (Fig. 7.4) or hexagons (Fig. 7.5) is preserved, but its orientation is random. The borders between blocks can be considered as lines of extended defects. The presence of the blocks destroys the long-range order in the system. In the simulation shown in Fig. 7.6, the competition between rolls and hexagons has not resulted the complete elimination of one sort of pattern: Here the hexagon cells form borders of the blocks filled by the roll patterns.

A strikingly similar behavior has been reported for a different system by *Gaponov-Grekhov* and *Rabinovich* [7.36]. In the experiment, a thin horizontal layer of liquid was periodically vibrated in the vertical direction. When the vibration amplitude was small enough, a regular grating with square cells was observed (Fig. 7.7a); it corresponded to the excitation of two pairs of counterpropagating waves. For larger vibration amplitudes, modulation waves appeared against the background of square cells (Fig. 7.7b). The spatial period of such modulation gradually decreased and, beginning at a certain vibration intensity, the pattern lost its long-range order. It now consisted of large blocks with periodically spaced cells

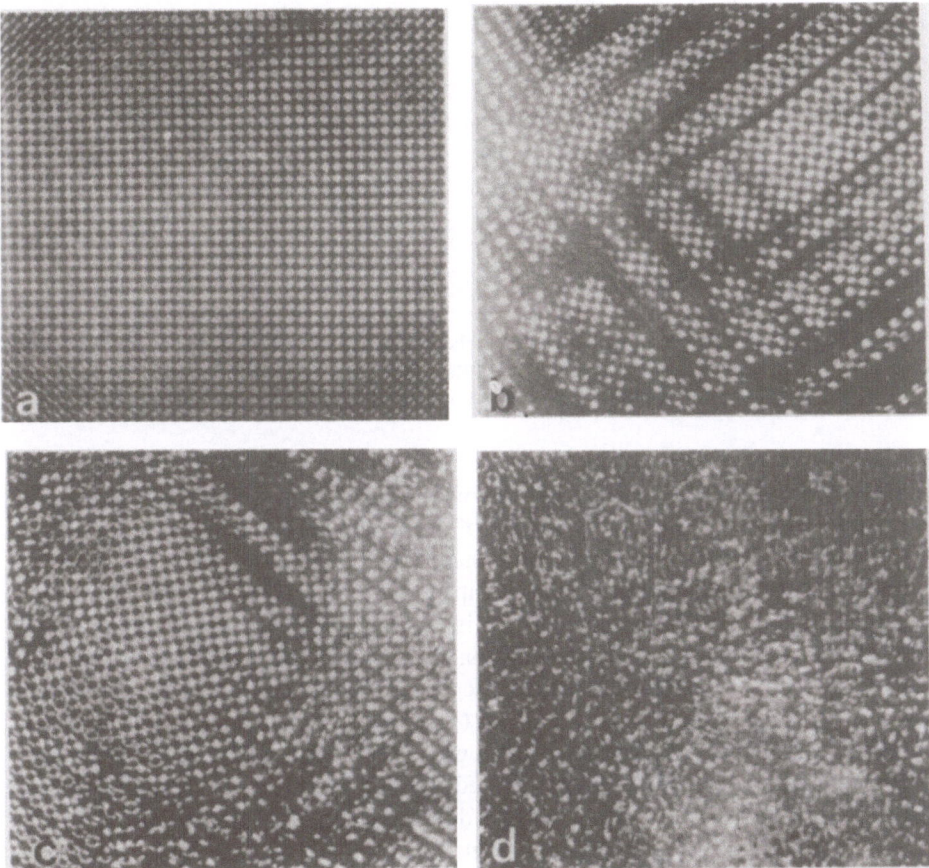

Fig. 7.7a–d. Transition to turbulence in a horizontal layer of vibrating fluid. From [7.36]

separated by interfaces where chaotically located cells were found (Fig. 7.7c). For still stronger vibrations, the typical size of the regular blocks became smaller while the chaotic interfaces widened. Eventually, a state of developed turbulence was observed (Fig. 7.7d).

We see that this transition to turbulence resembles the melting of a crystal. The block boundaries here correspond to borders between small crystallites. As the temperature of a crystal is increased, lattice defects, such as vacancies and dislocations, start to appear near such borders and their number gradually grows. Accumulation of these defects leads to a situation where rigid crystal blocks are seen floating inside the liquid which is formed by the melted material. The sizes of the blocks decrease and eventually, at a high temperature, the entire system is found in the liquid state.

Recently, a transition to turbulence from Turing patterns has been experimentally investigated in a chemical reaction–diffusion system by *Ouyang* and *Swinney*

(a)　　　　　　　　　(b)　　　　　　　　　(c)

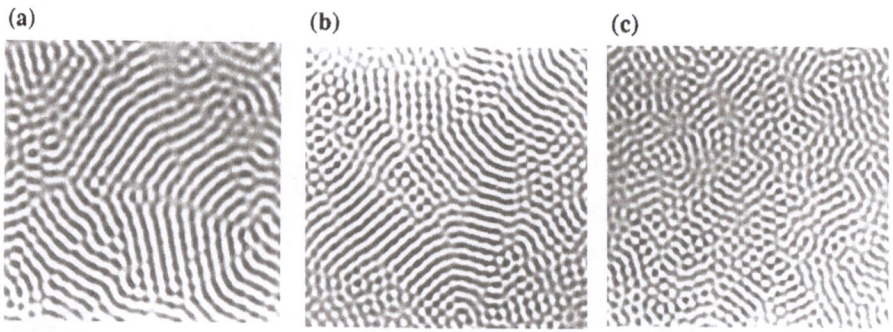

Fig. 7.8a–c. Transition from stripes to turbulence in a chemical reaction as the concentration of a certain chemical species is gradually increased: (**a**) stripes, (**b**) stripes near the limit of stability, and (**c**) turbulence. The region shown is 5.8 mm × 5.8 mm. From [7.37]

[7.37, 38]. As explained in Chap. 5 of our first volume [7.39], elementary stationary patterns of chemical concentrations formed by the Turing systems slightly above the instability onset represent lattices of stripes or hexagons ('dots'), similar to convection patterns in the Rayleigh–Bénard problem. Figure 7.8 shows the experimentally observed transition from a striped to a turbulent pattern under a gradual change of the control parameter which is the malonic acid concentration.

The reaction proceeds in a thin, non-mixed layer and the patterns are effectively two-dimensional. Immediately after the instability onset, domains filled with stripes of various orientations are found (Fig. 7.8a). The borders between these domains remain almost stationary in this regime. If the control parameter is increased above a certain critical value, the striped state becomes unstable (Fig. 7.8b). Defects spontaneously arise and cut large domains of stripes into smaller ones. The turbulent pattern found beyond the transition consists of small domains of dots and stripes randomly mixed together and continuously evolving as a function of time (Fig. 7.8c). Each domain is typically only about two wavelengths wide, much smaller than the domains in the stationary hexagonal or striped states.

In the turbulent state, both the pattern within the domains and the size of the domains fluctuate on a time scale which is much shorter than the characteristic times for motion of the domain boundaries in the ordered state. The amplitude and the speed of the fluctuations were found to increase abruptly above the transition point. To demonstrate that this behavior does not simply represent a long transient, the pattern was brought into a stable stationary state by applying a strong perturbation. When the perturbation was lifted, the system returned to the turbulent regime.

As we noted previously in Chap. 5 of the first volume, reaction–diffusion systems with a Turing instability are closely related to oscillatory systems and can be transformed into them by a continuous change of their parameters. *De Kepper* et al. [7.40, 41] have studied chemical turbulence near the Turing–Hopf bifurcation. Because in this parameter region the system is very sensitive to local variations of

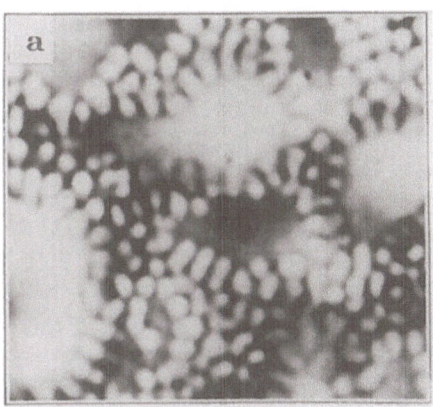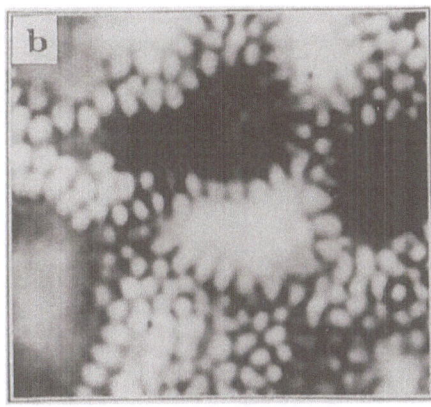

Fig. 7.9a,b. The turbulent Turing–Hopf mode in a chemical reaction. Snapshots (**a**) and (**b**) are separated by 15 s, the window size is 4.2 mm × 4.2 mm. From [7.41]

its properties, an intermittent turbulent state was found where domains filled with hexagons randomly alternated with oscillatory regions (Fig. 7.9).

Instabilities of localized structures in activator–inhibitor systems with long-range inhibition (see Sect. 5.5 in the first volume) can also lead to very complex and irregular spatio-temporal regimes, as indicated by numerical simulations [7.42] and experimental data [7.43]. *Krischer* and *Mikhailov* [7.44] have found a bifurcation leading to steadily traveling spots in two-dimensional media. Near the bifurcation threshold, binary collisions between the spots (Fig. 7.10) and their reflections from the medium's borders are elastic. Hence, their collective dynamics resembles essentially that of a hard-sphere gas and one might expect the chaotic behavior typical for billiards.

In contrast to Turing turbulence, studies of turbulence in excitable media (see Chap. 3 of the first volume) have a long history. Electrical waves in cardiac tissue, representing a classical example of excitation waves, were already observed at the beginning of this century (see, e.g., the article "On circulating excitations on heart muscles and their possible relation to tachicardia and fibrillation" published by *Mines* [7.45] in 1914!). A model for the propagation of excitation and the formation of spiral waves was proposed in 1946 by *Wiener* and *Rosenblueth* [7.46]. Under certain conditions, regular wave propagation in the heart is replaced by chaotic fibrillation. Its appearance means termination of the heart beat, i.e., clinical death. Therefore, investigations into the conditions leading to the onset of fibrillation and its prevention are of great importance.

It has been generally believed that fibrillation is related to the presence of some inhomogeneities in the medium, whose effect might be negligible under normal physiological conditions. Indeed, in 1964 computer simulations by *Moe* et al. [7.47] showed that propagation of excitation waves through a medium with a random spatial distribution of the duration of the refractory phase results in a

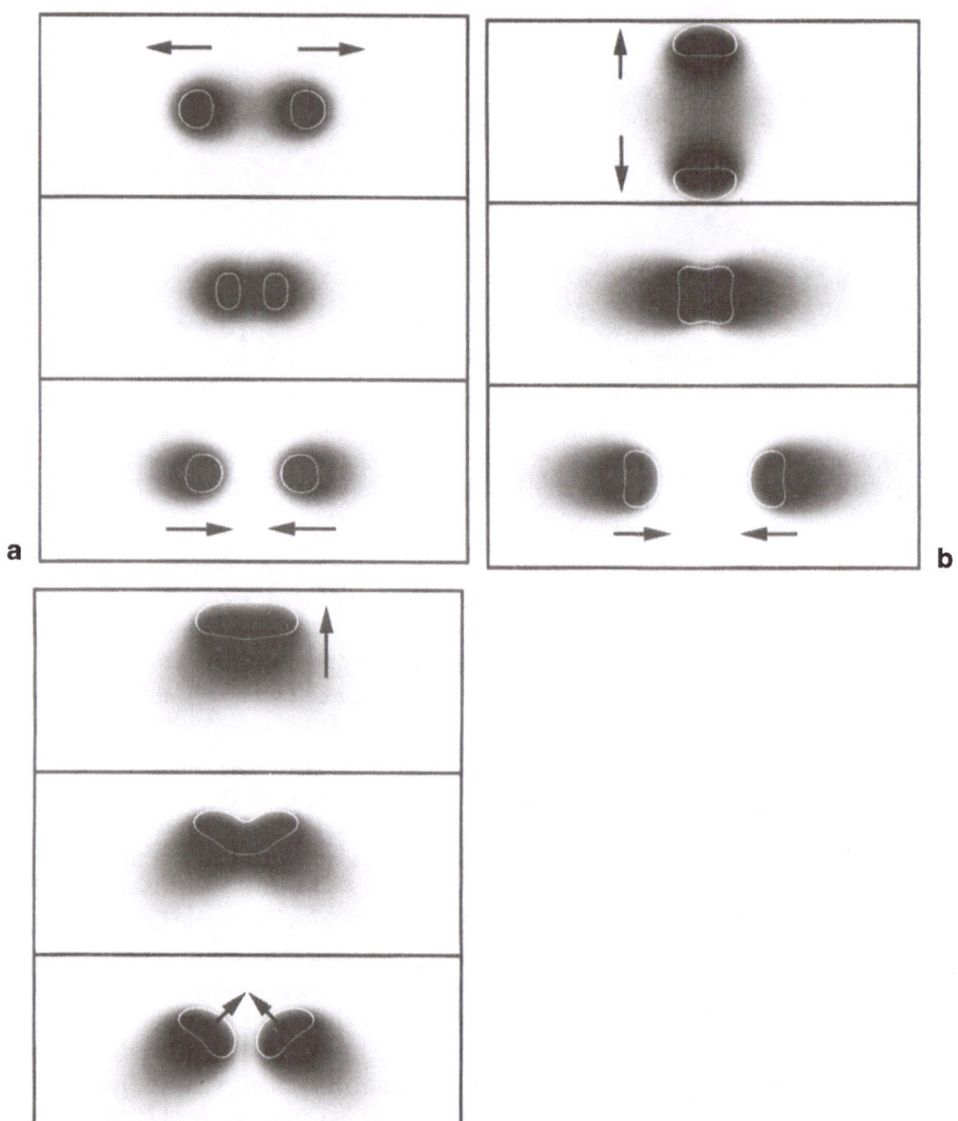

Fig. 7.10a–c. Collisions between moving spots in a medium with long-range inhibition: (**a**) close to the onset of translational motion, colliding spots are elastically scattered; (**b**) frontal collision of fast moving spots leads to a 90° change in the direction of motion; (**c**) oblique collision of fast-moving spots results in a single moving spot of a larger size. From [7.44]

fibrillation-like pattern of excitation. Using the Wiener–Rosenblueth model, *Krinsky* [7.48] demonstrated in 1968 that a wavefront can break on a boundary between two regions with a jump in the duration of the refractory phase and this leads to the formation of a spiral wave in the medium. The entire experimental and theoretical research on pattern formation in the Belousov–Zhabotinskii reaction in Russia has been significantly influenced by the preceding analysis of the respective phenomena in cardiac tissue.

In their book [7.49] published in 1984, *Krinsky* and *Mikhailov* discussed wave instabilities and chaos in active media. They wrote:[1]

"Spiral waves are created by breakups of plane waves. In a homogeneous active medium, such a breakup can be achieved by means of an external perturbation. For instance, in a liquid solution with the Belousov–Zhabotinskii reaction a front may easily be broken mechanically, e.g., by touching it with a thin wire.

When properties of a medium are not uniform, however, generation of breakups and creation of spiral waves can occur without any external perturbations. Multiple creation of spiral waves eventually leads to chaotization of a wave pattern, i.e., to the onset of a 'turbulent' regime in this active medium.

If a medium is not uniform, the recovery time (i.e., the duration of the refractory state) is not the same in different spatial regions. As seen from Fig. 7.11, when two excitation waves, separated by a sufficiently short time interval, propagate through a medium with varying refractoriness, the front of the second wave is ruptured.

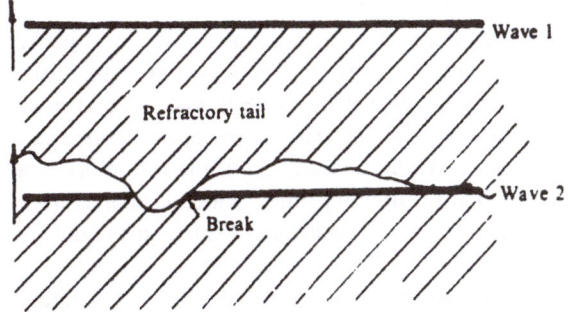

Fig. 7.11. Local destruction of the wavefront in a medium with nonuniform refractoriness. The interval t between the waves satisfies the condition $R_{max} > t > R_{min}$, where R_{max} and R_{min} are the maximal and the minimal durations of the refractory state in the medium. From [7.49, 50]

Creation of a spiral wave on an inhomogeneity is illustrated in Fig. 7.12. The initial positions of two waves are shown in Fig. 7.12a. Excitation cannot penetrate into region D with an increased duration of the refractory state. As a result, the front of the second wave becomes broken (Fig. 7.12b).

Subsequently, the broken wave turns, due to diffraction, over the region D (Fig. 7.12c). As the elements inside the region D go back from the refractory state to the state of rest, the excitation wave starts to penetrate inside the region D and curls into a spiral (Fig. 7.12d).

---

[1] The following text is translated from the original Russian of [7.49]. The figure numbers, however, have been altered to correspond to the figures in the present volume.

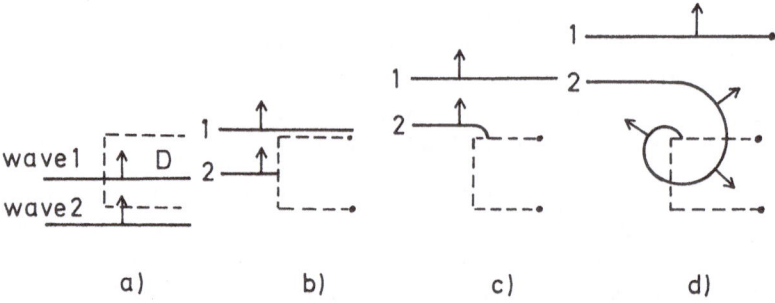

Fig. 7.12a–d. Creation of a spiral wave. The dashed line marks the region D with an increased refractory time. The solid lines show the wave fronts. (**a**) Initial positions of two waves, arrows indicate the direction of motion; (**b**) the second wave does not enter the region D, and the front breaks up; (**c**) the right end of the broken wave starts to curl into a spiral; (**d**) a spiral wave is created.

The instability of self-propagating waves, related to creation of a spiral wave, provides an explanation for the dangerous cardiac arrhythmia known as paroxysmal tachicardia. With this arrhythmia, the rate of heart contractions suddenly increases by several times but then returns to its initial level.

The cardiac tissue is not homogeneous with respect to many of its properties – and particularly the duration of the refractory state. The variations are further enhanced when the normal nutrition of cardiac cells is destroyed as a result of a heart infarct. Under normal conditions, the interval between two subsequent excitation waves in the heart is about 1s, while the duration of the refractory state is only about 0.1s. Hence, the dispersion of refractory times, which is characteristic for the cardiac tissue, does not, under normal conditions, lead to a breakup of excitation waves.

In certain heart disorders, an additional extrasystolic wave, which is separated only by a very small time interval from the preceding wave, is generated in the cardiac tissue. If this time interval is short enough (i.e., less than the maximal duration of the refractory state), the front of such an extrasystolic wave breaks up on the boundary of the region with increased refractoriness, and a spiral wave is thus produced. [...] The spiral wave that appears is a more rapid wave source and therefore it suppresses the waves emitted by the sinus node. The normal operation of the heart is destroyed.

The lifetime of a spiral wave in an inhomogeneous active medium is finite. Due to the inhomogeneity of the medium, the center of the spiral wave slowly drifts with time. As a result of this drift, the core of the spiral wave eventually reaches the boundary of the medium and then the spiral wave disappears. With the disappearance of the spiral wave, the dangerous attack of rapid heart contractions comes to an end.

Another outcome, however, is also possible. Secondary breakups of the spiral may start to develop when its excitation front (i.e., the arm of the spiral wave) passes through regions with an increased refractory time. As a result, *reproduction* of spiral waves occurs.

If the reproduction rate of spiral waves exceeds the rate of their disappearance at the boundaries, the number of spiral waves in the medium grows. The growth continues until the whole medium is filled by short 'fragments' of spiral waves and its elements perform irregular, or chaotic, oscillations. This chaotic wave regime can be interpreted as turbulence in an active medium.

In order that turbulence develops, the size of an active medium must exceed a certain threshold. Indeed, if we assume that the refractory time varies randomly over the medium, the probability of creation of a new spiral wave would be approximately the same for every element. The total number of new spiral waves created in the whole medium per unit time is then proportional to the area occupied by the medium. Hence, this reproduction rate is $n_+ = \alpha N L^2$ where N is the number [per unit area] of spiral waves already present, α is a proportionality factor, and L is the linear size of the active medium (we assume that it is two-dimensional). The total number of spiral waves per unit time reaching boundaries of the medium due to random drift, and hence disappearing, is $n_- = \beta N L$. The disappearance rate is thus proportional to the boundary perimeter (β is a constant coefficient). The chaotic regime is established when the reproduction rate of spiral waves exceeds the rate of their disappearance, i.e., if $n_+ > n_-$ As seen from the above expressions, this is possible only if the linear size $L_c = \alpha/\beta$. In media of smaller sizes, reproduction of spiral waves is dominated by the process of their exit through the medium's boundaries.

Quite remarkably, the same arguments are used in the calculation of the critical mass for nuclear chain reactions which involve the reproduction of neutrons and their departure through the boundaries of the system.

For a biological active medium – the tissue of the cardiac muscle – a chaotic wave regime is known. It represents fibrillation, in which contractions of the heart as a whole disappear and only chaotic activity in individual elements continues, so that the heart ceases to pump blood. As found in experiments with small pieces of the cardiac tissue, development of fibrillation indeed requires a certain 'critical mass': up to a certain critical size of a piece of tissue, it remains impossible to induce self-supported fibrillation in this tissue".

The transition to chaos in a non-stirred Belousov–Zhabotinskii reaction has been investigated by *Agladze* et al. [7.50]; it was attributed to breakups of excitation waves on the boundaries of Bénard cells under the conditions of thermal convection in the solution layer (Fig. 7.13). *Maselko* and *Showalter* [7.51] have systematically studied chaotic wave patterns in randomly heterogeneous excitable media. *Mikhailov* and *Zykov* [7.52] showed that the breakup of *curved* wavefronts is facilitated because it requires weaker variations of the recovery time over the medium.

We emphasize that the role of inhomogeneities in the above examples consists only in triggering a transition to the turbulent state. The asymptotic statistical properties of the established turbulent state can be almost independent of the in-

Fig. 7.13. Transition to chaos in the non-stirred Belousov–Zhabotinskii reaction. Interaction of propagating waves with stationary convection cells leads to reproduction of spiral waves. The Petri dish was tilted, and the emerging spiral waves are seen only at the top, where the depth of the solution layer exceeded the threshold for the onset of convection, The time interval of 5 min between snapshots (**a**) and (**b**). From [7.50]

homogeneities that have led to its emergence. Recently, spontaneous breaking of spiral waves in uniform excitable media has been observed in numerical simulations by *Panfilov* and *Holden* [7.53] (see also [7.54]), *Courtemanche* and *Winfree* [7.55], *Karma* [7.56] and *Bär* and *Eiswirth* [7.57]. Figure 7.14 shows the development of turbulence after the breakup of a spiral wave in a model of an excitable medium describing a catalytic surface chemical reaction.

Various mechanisms could be responsible for the spontaneous breakup of spirals. In some cases, such breakup is preceded by meandering of spiral waves which can create sufficient fluctuations of the residiual inhibitor concentration in the medium to rupture a wave front. From a general point of view, any dynamic instability in an excitable medium that gives rise to small-amplitude chaotic variations may lead to this effect, since it is equivalent to the action of a noise source that creates spatial inhomogeneities in the medium.

Turbulence can also spontaneously develop in oscillatory active media – in fact, they provide the best studied examples of such phenomena. The respective mathematical models will be presented in Sect. 7.4. However, we first want to consider the emergence of turbulence in coupled map lattices.

Fig. 7.14. Spontaneous breakup of a spiral wave in a model of a catalytic surface reaction; simulations with the grid of 256×256, the gray level is proportional to the local inhibitor concentration. From [7.57]

7.3 Coupled Chaotic Maps

As noted in Chap. 5, iterative maps

$$x_{n+1} = \phi(x_n, \mu) \tag{7.3.1}$$

are intimately related to differential equations. When the control parameter μ is varied, they can display a sequence of bifurcations leading to chaotic dynamics. A simple example of an iterative map is the logistic map (5.2.2) exhibiting an infinite sequence of period-doubling bifurcations that results in chaos. In the present section, we will write the mapping function in a slightly different form, namely

$$\phi(x_n, \mu) = 1 - \mu x_n^2 . \tag{7.3.2}$$

When the control parameter μ lies between 0 and 2, the map (7.3.1) with the function (7.3.2) projects the segment $[-1, 1]$ onto itself. It is reduced to (5.2.2) by a shift of the variable. Here, the transition to chaos occurs at $\mu = 1.40155 \ldots$.

Our aim is to consider lattices of coupled chaotic maps. Therefore we take a lattice of sites i occupied by the elements whose individual dynamics is described by some chaotic map (7.3.1). Only the simplest case of a one-dimensional lattice, i.e., of a linear chain, is discussed below. The state of an element i in the chain at a time moment n is denoted as $x_n(i)$.

There are different ways of introducing coupling between neighboring elements (see, e.g., [7.58]). However, most numerical simulations have been performed for a special form of coupling given by

$$x_{n+1}(i) = (1 - \varepsilon)\phi(x_n(i), \mu)$$
$$+ \frac{\varepsilon}{2}[\phi(x_n(i + 1), \mu) + \phi(x_n(i - 1), \mu)] , \tag{7.3.3}$$

where ε is a small parameter that specifies the strength of coupling. Note that this coupling is symmetric and has no effect when the states of all elements are identical.

Kaneko [7.59] has carried out a detailed numerical study of pattern formation and turbulence in the linear chain of logistic maps (7.3.2) coupled according to (7.3.3). Random initial conditions were used in these simulations. If the coupling coefficient ε was small ($\varepsilon = 0.1$), the following sequence of spatiotemporal regimes (Fig. 7.15) was observed as the control parameter μ was increased: At sufficiently small values of μ (yet exceeding the critical value $\mu = 1.40155 \ldots$ when the chaotic dynamics of an individual map begins), the system splits into a collection of coexisting elementary patterns of various sizes. In the largest elementary patterns the motion is chaotic, while it is almost period-8 in smaller patterns, period-4 in much smaller patterns and period-2 in the smallest ones. The boundaries and positions of all such domains remain fixed; they are determined by the initial conditions. The frozen domains are clearly seen in Fig. 7.15a, which shows the amplitudes $x_n(i)$ overlayed for 50 time steps (to guarantee that the transient stage is over, this is done after 1000 iterations from the initial state).

As the control parameter μ is increased, competition among the elementary patterns begins. In the transient stage some domains start to grow at the expense of others. If the coupling constant is sufficiently small ($\varepsilon = 0.1$), competition results in survival of the simplest period-2 "zigzag" pattern, which dominates in the greater part of the medium (Fig. 7.15b). Blocks bearing the "zigzag" pattern are separated by defects which represent narrow domains with higher-period motion. The defects remain quiescent. Note that the amplitudes of oscillations within any "zigzag" block are not perfectly constant. Small periodic and, at higher μ, chaotic modulation is observed in these regions.

At approximately $\mu = 1.74$, the dynamical regime undergoes a qualitative change. The defects begin to wander chaotically within the system. At the same time, they are widened and chaotic motion sets in inside any individual defect

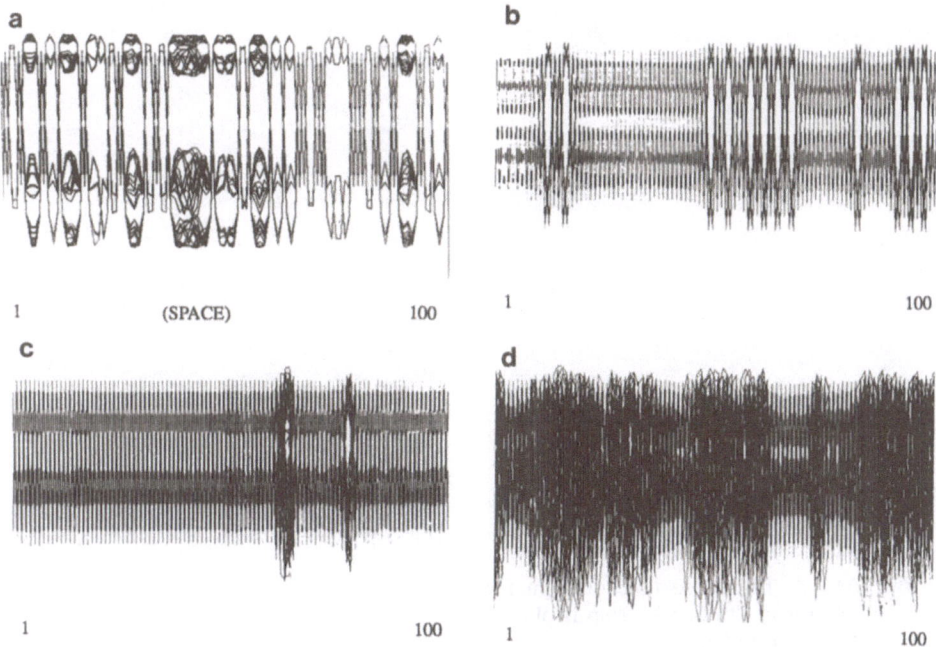

Fig. 7.15a–d. The space-amplitude plot for the coupled logistic map. Amplitudes $x_n(i)$ are overlayed for 50 time steps after 1000 iterations of the transients for the model with $\varepsilon = 0.1$, $N = 100$, and random initial conditions. (a) $\mu = 1.44$, (b) $\mu = 1.72$, (c) $\mu = 1.80$, (d) $\mu = 1.90$. From [7.59]

(Fig. 7.15c). Investigations show that the random motion of a defect can be described as diffusion. It turns out that the average square of the defect displacement grows linearly with time, as should occur for a diffusion process. The diffusion constant (i.e. the mobility of defects) increases rapidly when the control parameter μ is increased.

In contrast to the usual diffusion of Brownian particles in a liquid, the diffusive motion of a defect is the consequence of processes taking place inside it, rather then the effect of random impulses from the environment. This allows us to classify it as *diffusive self-propagation*.

The dynamics of defects is clearly seen in Fig. 7.16. If $x_n(i)$ is greater than the unstable fixed point of the logistic map

$$x^* = (2\mu)^{-1}\left[(1 + 4\mu)^{1/2} - 1\right] , \tag{7.3.4}$$

the corresponding space-time pixel in this diagram is painted black, and is otherwise left blank. The values of $x_n(i)$ are thus plotted at every 8th or 64th time moment. If some element performs a period-2 motion, its state will be the same at each of these moments. Depending on the phase of oscillations, it will be constantly found either in the state with $x_n(i) > x^*$ (black pixel) or in the state with

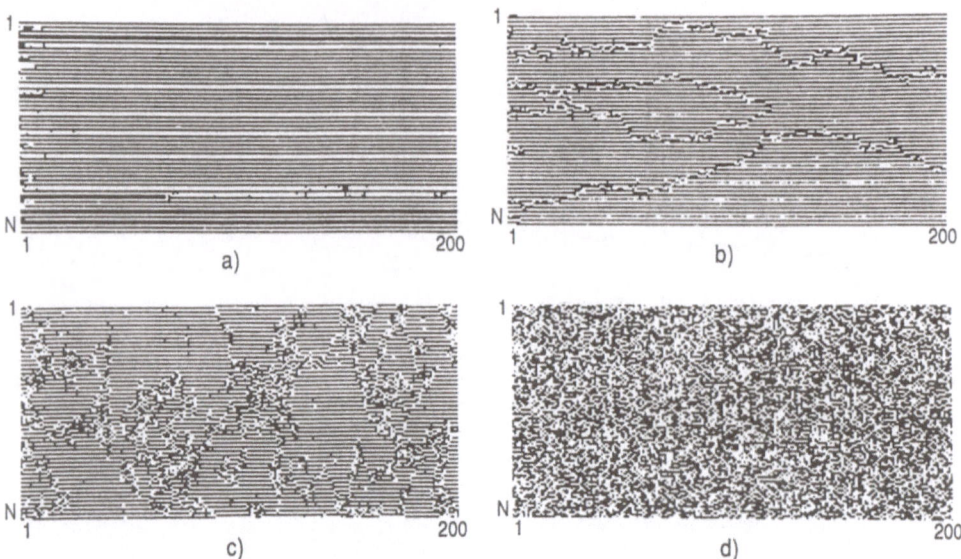

Fig. 7.16a–d. The space–time diagram for the coupled logistic map with $\varepsilon = 0.1$ and $N = 100$. Every 8th or 64th step is plotted. (**a**) $\mu = 1.74$, (**b**) $\mu = 1.80$, (**c**) $\mu = 1.89$, (**d**) $\mu = 1.94$. From [7.59]

$x_n(i) < x^{ast}$ (blank pixel). Since in a perfect "zigzag" pattern the phases of oscillations of any two adjacent elements are opposite, it will be portrayed as a grid of horizontal lines. In the background of this grid, the defects are clearly discernable. When any two defects collide, they are annihilated (Fig. 7.16b). Therefore, if we wait sufficiently long, all defects present in the initial state will eventually disappear and the perfect regularity of the "zigzag" pattern will be established in the entire system.

However, at still larger values of the control parameter μ, a new mechanism comes into play which induces pair-production and multiplication of defects. These effects counteract the extinction of defects due to pair-annihilation, and result in the persistence of defects in the system at all times (Fig. 7.16c). Since the defects continue to perform random diffusive motion, the state of "defect turbulence" is established, characterized by dynamical intermittency of regular and chaotic regions (Fig. 7.15d).

Notably, similar behavior is observed even in a system of two coupled logistic maps. At $\mu = 1.92$ the attractor which corresponds to stable period-2 motion in such a system undergoes a crisis that leads to the formation of a strange attractor. In systems consisting of larger numbers of elements, such a transition is found at a lower value of the control parameter. It also becomes more complicated because the background "zigzag" pattern already possesses a small chaotic inhomogeneous modulation prior to the transition. As a result, the defects, which represent bursts of strongly chaotic activity, are locally and spontaneously produced in a distributed system.

As the control parameter is further increased, the ordered structure in the space-time diagram becomes completely smeared out (Fig. 7.16d) and the stage of developed turbulence is reached. This is characterized by the exponential decay of spatial correlations.

We can define the spatial power spectrum as

$$S(k) = \left\langle \left| \frac{1}{N} \sum_{j=1}^{N} x_n(j) \exp\left(2\pi i k j\right) \right|^2 \right\rangle , \tag{7.3.5}$$

where $\langle \dots \rangle$ denotes the long-time average.

Some examples of spatial power spectra observed in numerical simulations at different values of the control parameter μ are shown in Fig. 7.17. In the state of defect turbulence (Fig. 7.17a), the spectrum consists of a sharp peak at $k = 1/2$, corresponding to the background "zigzag" pattern, and the broad noise band around $k = 0$. When the control parameter is increased, the contribution of broadband noise increases (Fig. 7.17b,c). When the stage of developed turbulence is reached, the sharp peaks disappear completely (Fig. 7.17d) and the spatial spectrum decays monotonically with k. It can then be roughly approximated by

$$S(k) \sim \exp\left[-\left(r_c k\right)^2\right] , \tag{7.3.6}$$

where r_c is the correlation radius.

Kaneko [7.59] performed a detailed investigation of developed turbulence for the system of coupled logistic maps, using different techniques of statistical analysis. He came to the conclusion that, at this stage, the behavior of the system can be described as a simple random Markov process (Chap. 8). Namely, it turns out that in the stage of developed turbulence, the action of all other elements on any given element in the chain is effectively reduced to the action of a random "environmental" noise $\xi_n(i)$ with a short memory in space and time, so that

$$x_{n+1}(i) = \phi\big(x_n(i)\big) + \xi_n(i) . \tag{7.3.7}$$

This result, conjectured from numerical simulations, is quite remarkable. It supports the intuitive understanding that the influence of a turbulent environment can often be modelled by a statistically independent random noise.

Bunimovich and *Sinai* [7.60] carried out a rigorous analytical investigation of space-time chaos in a chain of coupled "saw" maps (Fig. 7.18). In this case, each individual map is given by

$$\phi(x) = \{f(x)\} , \tag{7.3.8}$$

where $\{ \dots \}$ denotes taking the fractional part of the argument and $f(x)$ is any monotonically increasing function, such that $f(0) = 0$ and $f(1) = m$ where m is an integer greater than one. The map is defined on a segment $[0, 1]$; for technical

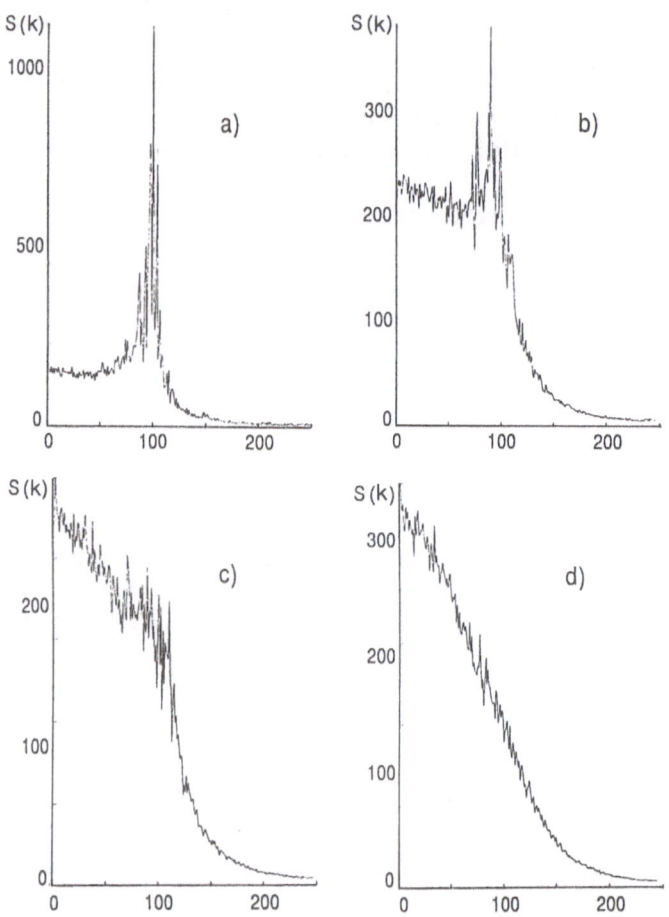

Fig. 7.17a–d. The spatial power spectrum $S(k)$ for the coupled logistic map with $\varepsilon = 0.1$, $N = 512$, and random initial conditions. Calculated from 1000 time step averages discarding 10 000 transients. (**a**) $\mu = 1.75$, (**b**) $\mu = 1.8$, (**c**) $\mu = 1.85$, (**d**) $\mu = 1.95$. From [7.59]

reasons it was convenient to assume that the coupling parameter ε in (7.3.3) depends on $x_n(i)$, vanishing near the borders of this segment and remaining almost constant in the interior.

For a system of maps (7.3.8) coupled according to (7.3.3), *Bunimovich* and *Sinai* [7.60] proved that there is mixing and generation of a continuous "invariant measure" (i.e., roughly speaking, a steady probability distribution). It follows from their analysis that spatial and temporal correlation functions decay exponentially fast in this case, so that the behavior of the coupled maps system can indeed be described as a random process with a short-term memory and a small correlation radius.

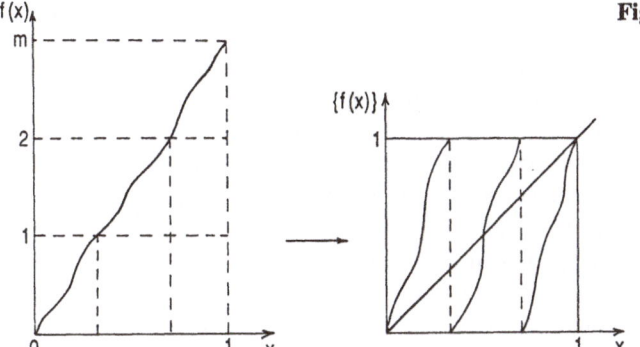

Fig. 7.18. The "saw" map

7.4 The Complex Ginzburg–Landau Equation

To extend the theory of coupled chaotic maps to distributed dynamical systems, it would have been natural to consider next media in which small local elements have chaotic dynamics and processes in neighboring elements are coupled by diffusion. Since the chaotic dynamics in a single element is possible only when it is described by at least three continuous variables (as in the Lorenz model, for example) nonlinear reaction–diffusion systems with at least three interacting species should then be considered. However, research on such systems remains very scarce. Numerical simulations and analytical studies are dominated by investigations of reaction–diffusion systems that have only two species (or can be easily reduced to such a description). An isolated small element of such a system may have, of course, only very simple dynamics: its attractors may be either fixed points or limit cycles. Therefore, whenever spatiotemporal chaos is found in these models, it results entirely from the spatial *coupling* between the elements.

One particular model has attracted much attention in recent years. This is the complex Ginzburg–Landau equation which describes a population of harmonic limit-cycle oscillators locally coupled by diffusion. The importance of this equation lies in its universality: it represents a normal form of the distributed dynamical system near an Andronov–Hopf bifurcation, i.e., at the onset of uniform oscillations.

After rescaling the variables, the complex Ginzburg–Landau equation for the complex oscillation amplitude η can be written in the dimensionless form

$$\dot{\eta} = (1 - \mathrm{i}\omega)\eta - (1 + \mathrm{i}\beta)|\eta|^2\eta + (1 + \mathrm{i}\varepsilon)\nabla^2\eta \,, \tag{7.4.1}$$

where β and ε are the parameters of this system. The value of the (linear) oscillation frequency ω plays no significant role here because, by going to a rotating coordinate frame $\eta \to \eta \exp(\mathrm{i}\omega t)$, the term $-\mathrm{i}\omega\eta$ can be eliminated (and therefore it is often simply omitted).

The dynamics of an individual element of this system is described by (7.4.1) without the last term. Introducing the modulus $\varrho = |\eta|$ of the complex oscillation amplitude and noting that $d(\varrho^2)/dt = \eta^*(d\eta/dt) + \eta(d\eta^*/dt)$ we obtain for this variable the simple equation

$$\dot{\varrho} = (1 - \varrho)\varrho \ . \tag{7.4.2}$$

Hence, it has an attractive fixed point $\varrho = 1$ which corresponds to stable limit-cycle oscillations $\eta = \exp(-i\omega_0 t)$; the oscillation frequency is $\omega_0 = \omega + \beta\varrho^2 = \omega + \beta$. The meaning of the parameter β in the complex Ginzburg–Landau equation is thus clear: it specifies the nonlinear shift of the oscillation frequency of individual oscillators (note that this parameter may be either positive or negative).

The last term in (7.4.1) takes into account coupling between neighboring elements. In reaction–diffusion systems it arises due to diffusion of reacting species. A frequently asked question is why the coupling coefficient can then be complex, i.e. is given by $1 + i\varepsilon$. The answer is that, in order to have oscillations, at least two reacting species are needed. Generally, they would have two different diffusion constants. After a transformation to complex oscillation amplitudes, their combinations give rise to the real and the imaginary parts of the coupling coefficient (the real part is later made equal to unity by rescaling the coordinates).

We have already used the complex Ginzburg–Landau equation as an example of a distributed oscillatory system in Chap. 5 of the first volume. There, however, we considered its long-wavelength properties described by the phase dynamics equation and also constructed an approximate solution for spiral waves. Below, our aim is to show how turbulence develops in a certain parameter region in this system. To make the discussion more instructive, we first present the known exact solutions of the one-dimensional Ginzburg–Landau equation.

One family of nonlinear steady solutions is given by *traveling plane waves*,

$$\eta(x, t) = \varrho_k \exp(ikx - i\omega_k t) \ . \tag{7.4.3}$$

In contrast to plane waves in linear systems, not only the wave frequency ω_k but also the wave amplitude ϱ_k are determined by the wavenumber k, i.e., we have

$$\varrho_k = 1 - k^2 \ , \tag{7.4.4}$$

$$\omega_k = \omega_0 + (\varepsilon - \beta)k^2 \ . \tag{7.4.5}$$

When $\varepsilon > \beta$, the wave frequency increases with the wave number, so that the waves are characterized by a positive dispersion. The wave amplitude ϱ_k becomes smaller for higher wave numbers; it vanishes when $k = 1$. No plane wave solutions with $k > 1$ are possible. Note also that as $k \to 0$ such solutions transform into uniform oscillations.

The stability analysis of uniform oscillations and plane waves with respect to small perturbations yields the followinq results:

Uniform oscillations are unstable with respect to phase modulation if the *Benjamin–Feir condition*

$$1 + \varepsilon\beta < 0 \tag{7.4.6}$$

is realized (see Chap. 4 of the first volume [7.39]).

When the uniform oscillations are stable, i.e., the condition opposite to (7.4.6) holds, traveling plane waves undergo the *Eckhaus instability* [7.61, 62] if their wave numbers exceed the critical value given by

$$k_E^2 = \frac{1 + \varepsilon\beta}{2(1 + \beta^2) + 1 + \varepsilon\beta} . \tag{7.4.7}$$

Note that $k_E < 1$ and hence the waves always become unstable before the boundary of their existence ($k = 1$) is reached. Moreover, the boundary of the Eckhaus instability already lies in the region of small wave numbers k if the system is close to the Benjamin–Feir instability.

The Eckhaus instability leads to the modulation of a traveling plane wave. Because it develops in the propagating wave, i.e., in the coordinate frame moving with the wave velocity, it has a *convective* nature. This means that if, for instance, we have a source emitting a plane wave, the instability would first appear at a certain distance 'along the flow' from the source.

Another family of exact solutions has been obtained by *Bekki* and *Nozaki* [7.63, 64] by applying the Hirota method, previously used in investigations of solitons in the nonlinear Schrödinger equation. These solutions describe steadily moving or standing localized patterns ('holes') that asymptotically approach the traveling plane waves at a large distance from the center of the pattern. Below we present the original Bekki–Nozaki solutions expressed in slightly improved notation.

Using the coordinate $\xi = x - ct$ in the frame moving with the velocity c of the pattern, a Bekki–Nozaki hole is described by the following equation for the local complex oscillation amplitude:

$$\eta(\xi, t) = \frac{\varrho_+ e^{\kappa\xi + i\delta} - \varrho_- e^{-\kappa\xi}}{e^{\kappa\xi} + e^{-\kappa\xi}}$$
$$\times \exp\left\{ i\left[\frac{1}{2}(k_+ + k_-)\xi - \Omega t + r\ln(e^{\kappa\xi} + e^{-\kappa\xi}) \right] \right\} . \tag{7.4.8}$$

Here we have

$$\varrho_+ = (1 - k_+^2)^{1/2}, \varrho_- = (1 - k_-^2)^{1/2} , \tag{7.4.9}$$

$$c = (\varepsilon - \beta)(k_+ + k_-) , \tag{7.4.10}$$

$$\Omega = \omega_0 + (\varepsilon - \beta)(k_-^2 - k_+^2) , \tag{7.4.11}$$

$$k_\pm = k_{\max}\sin\theta \pm k_0\cos\theta , \tag{7.4.12}$$

$$\kappa = (k_0/r)\cos\theta , \tag{7.4.13}$$

$$\tan\delta = \frac{r[r^2(\varepsilon - \beta) - (1 + \varepsilon\beta)]k_0 k_{\max}\sin\theta\cos\theta}{r^2(1 + \beta^2)k_{\max}^2(\sin\theta)^2 - (1 + r^4)(1 + \varepsilon^2)k_0^2(\cos\theta)^2} , \tag{7.4.14}$$

$$k_0 = \left[1 + \frac{3(1 + \varepsilon^2)}{r(\varepsilon - \beta)} \right]^{-1/2} , \tag{7.4.15}$$

$$k_{max} = \left[1 + \frac{3r(1 + \beta^2)}{(1 + r^2)(\varepsilon - \beta)}\right]^{-1/2} \tag{7.4.16}$$

$$r = \begin{cases} b + (1 + b^2)^{1/2}, & \text{if } \varepsilon > \beta, \\ b - (1 + b^2)^{1/2}, & \text{if } \varepsilon > \beta, \end{cases} \tag{7.4.17}$$

$$b = -\frac{3(1 + \varepsilon\beta)}{2(\varepsilon - \beta)}. \tag{7.4.18}$$

This family of special solutions has a free parameter θ that can be varied between $-\pi/2$ and $\pi/2$. When $|\xi| \to \infty$, the solutions behave asymptotically as plane waves,

$$\eta(\xi, t) \approx \begin{cases} \varrho_+ \exp[i(k_+\xi - \Omega t + \delta)], & \text{if } \xi \to \infty \\ -\varrho_- \exp[i(-k_-\xi - \Omega t)], & \text{if } \xi \to -\infty \end{cases} \tag{7.4.19}$$

Thus, k_+ and k_- are the wave numbers in the regions $\xi \to +\infty$ and $\xi \to -\infty$, respectively.

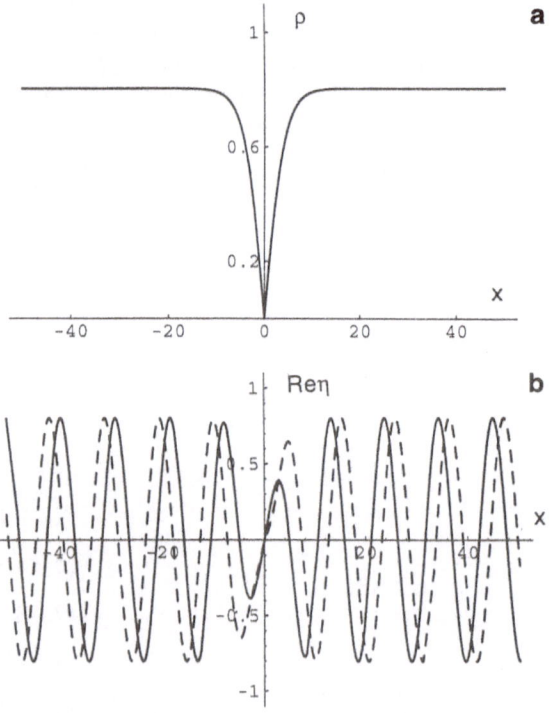

Fig. 7.19a,b. A standing Bekki–Nozaki hole. The profiles (**a**) of the modulus $\varrho = |\eta|$ and (**b**) of the real part $\mathrm{Re}\,\eta$ of the complex oscillation amplitude are shown. The solid and dashed curves in (**b**) correspond to successive time moments $t = 0$ and $t = 0.7$; the parameters are $\varepsilon = 2.0$, $\beta = -1.4$, $\omega = 2.0$ and $\theta = 0$

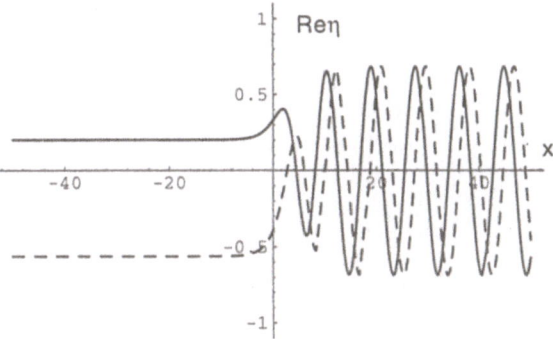

Fig. 7.20. The exact Bekki–Nozaki solution for $\varepsilon = 2.0$, $\beta = -0.4$, $\omega = 2.0$ and $\theta = 0.495$. The solid and dashed curves show spatial profiles of Re η at two subsequent time moments $t = 2.0$ and $t = 2.5$

As follows from (7.4.12), we have $k_+ = k_0$ and $k_- = -k_0$ if the value $\theta = 0$ of the free parameter is taken. In this case $c = 0$ and hence the pattern represents a standing hole (this solution has been constructed in [7.65]). The standing hole is a source of waves with a wave number k_0 satisfying equation (7.4.15). For such a pattern, $\delta = 0$ and, according to (7.4.19), $\eta(x,t) \approx -\eta(-x,t)$ for $x \to \infty$ (this means that the emitted plane waves are shifted in phase by π). Because $\varrho_+ = \varrho_-$ and $\delta = 0$, the complex oscillation amplitude η vanishes at the center $x = 0$ of this pattern. Figure 7.19 shows computed profiles of $\varrho = |\eta|$ and Re η for a standing hole.

If $\theta \neq 0$ the holes are moving. Indeed, equation (7.4.10) for the velocity c can also be written as $c = 2(\varepsilon - \beta)k_{max} \sin\theta$. The waves emitted by a moving source have different wavelengths (and hence differrent frequencies) on the left and the right sides of it (this is a manifestation of the Doppler effect). Note that the holes always move towards the region with the shorter wavelength (here and below we assume that the condition $\varepsilon > \beta$ is satisfied). The oscillation amplitude ϱ no longer vanishes in the center of a moving hole and, moreover, we have now $\varrho_+ > \varrho_-$.

At $\theta = \arctan(k_0/k_{max})$, the wavenumber of the wave left behind by a propagating pattern vanishes (i.e., $k_- = 0$) and the pattern represents instead a front separating the region with uniform oscillations and the region filled by a plane wave (Fig. 7.20). This front moves into the wave region, emitting the wave in its forward direction.

When the parameter θ is further increased, both wave numbers k_+ and k_- become *positive* and therefore the waves in front of and behind the moving pattern propagate in the same direction, coinciding with the direction of motion of the pattern. If θ is near $\pi/2$, the difference in the wave numbers of these two waves is small and the suppression of the oscillation amplitude in its center is not very pronounced (Fig. 7.21). At $\theta = \pi/2$, the pattern transforms continuously into a traveling plane wave with the wave number $k = k_{max}$.

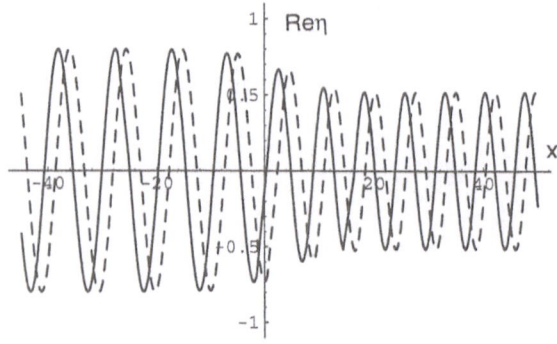

Fig. 7.21. The exact Bekki–Nozaki solution for $\varepsilon = 2.0$, $\beta = -0.4$, $\omega = 2.0$ and $\theta = 1.26$. The solid and dashed curves show spatial profiles of $\mathrm{Re}\,\eta$ at two successive time moments $t = 2.0$ and $t = 2.5$

The stability of the Bekki–Nozaki solutions has been investigated in [7.66, 67]; it has also been demonstrated that the introduction of additional small quintic terms in the complex Ginzburg–Landau equation may significantly influence the observed behavior, giving rise to evolution along the family of solutions specified by the free parameter θ [7.68].

The two-dimensional systems described by the complex Ginzburg–Landau equation have, as their exact solutions, traveling plane waves and Bekki–Nozaki strips which are obtained simply by extending the one-dimensional solution along the second spatial dimension. Because of the isotropy of the medium, the orientation of these patterns is arbitrary. The stability of the Bekki–Nozaki strips with respect to their spatial deformations has been discussed in [7.69].

In addition to the patterns that already exist in the one-dimensional case, the two-dimensional complex Ginzburg–Landau equation has a characteristic new solution, i.e., a rotating *spiral wave*. If we introduce the polar coordinate system whose origin coincides with the center of the spiral wave, this solution is given by

$$\eta(r, \phi, t) = \varrho(r) \exp\{-i\Omega^* t - i[\phi \pm \chi(r)]\} \,, \qquad (7.4.20)$$

where $\varrho(r)$ and $\chi(r)$ are functions of the polar radius r and ϕ is the polar angle; the minus (plus) sign corresponds to the wave rotating in the clockwise (counterclockwise) direction. The exact form of the functions ϱ and χ is not known. However, an approximate solution, using the method of matched asymptotic expansions, has been constructed in the limit

$$1 + \varepsilon\beta \gg |\varepsilon - \beta| \qquad (7.4.21)$$

by *Hagan* [7.70] (see also Sect. 4.4. in the first volume [7.39]). At a point located far from the center, a rotating spiral wave is seen as a source of (asymptotically) plane waves with a certain wave number k^*. In the limit (7.4.21) this wave number is

$$k^* = a^* \frac{(1 + \varepsilon\beta)^{1/2}}{|\varepsilon - \beta|} \exp\left(-\frac{\pi(1 + \varepsilon\beta)}{2|\varepsilon - \beta|}\right) , \tag{7.4.22}$$

where the numerical coefficient is $a^* \cong 1.018\ldots$. The rotation frequency of the spiral wave is $\Omega^* = \omega_0 + (\varepsilon - \beta)(k^*)^2$.

The exact and approximate analytical solutions for wave patterns in the complex Ginzburg–Landau equation are helpful in the analysis of the very complex spatiotemporal regimes that are displayed by this system.

Below in this section we consider the nonlinear dynamics of this system in the vicinity of a limit cycle corresponding to uniform oscillations. It is then convenient to write the complex oscillation amplitude in the form

$$\eta(x, t) = \varrho(x, t) \exp\left[-i\omega_0 t - i\phi(x, t)\right] , \tag{7.4.23}$$

thus introducing the oscillation phase ϕ. The uniform oscillations correspond to $\varrho(x, t) = 1$ and $\phi(x, t) = \text{const}$. Since the phase of the uniform oscillations is arbitrary, slow spatial variations of the phase variable can be expected to evolve with characteristic times that are much larger than that of the amplitude ϱ. This allows one to derive a separate *phase dynamics equation*. When the uniform oscillations are stable with respect to small perturbations, this equation is [7.71–73] (see also Chap. 4 of our first volume [7.39])

$$\frac{\partial\phi}{\partial t} = (\varepsilon - \beta)\left(\frac{\partial\phi}{\partial x}\right)^2 + (1 + \varepsilon\beta)\frac{\partial^2\phi}{\partial x^2} . \tag{7.4.24}$$

The local oscillation amplitude is then given approximately by

$$\varrho = 1 + \frac{\varepsilon}{2}\frac{\partial^2\phi}{\partial x^2} - \frac{1}{2}\left(\frac{\partial\phi}{\partial x}\right)^2 . \tag{7.4.25}$$

The phase dynamics description remains applicable so long as the corrections due to the last two terms in (7.4.25) are much smaller than unity, i.e. the amplitude variations are weak.

When $1 + \varepsilon\beta < 0$, so that the Benjamin–Feir instability of the uniform oscillations is realized, the coefficient in the last diffusion-like term in (7.4.24) becomes negative. In this case, initially smooth phase variations grow with time and the rate of growth is *greater* for higher spatial harmonics. This means that spatial phase modulation develops spontaneously in the system.

Such spontaneous spatial modulation would generally push the system far from the vicinity of the (now unstable) limit cycle corresponding to uniform oscillations. However, if the coefficient $\nu = -(1 + \varepsilon\beta)$ is small, the development of the instability should be slow and therefore the system may spend a long time in the vicinity of uniform oscillations before leaving this region in its infinite-dimensional phase space.

Since phase gradients still remain relatively small at this stage, we can try to extend the expansion to higher orders of the spatial derivatives in equation (7.4.24). The cubic terms, such as $(\partial\phi/\partial x)^3$, $(\partial\phi/\partial x)(\partial^2\phi/\partial x^2)$ or $\partial^3\phi/\partial x^3$ should be absent in this expansion because they are not invariant under reflection $x \to -x$. Hence, one would get an equation of the form

$$
\frac{\partial \phi}{\partial t} = -\nu \frac{\partial^2 \phi}{\partial x^2} + (\varepsilon - \beta) \left(\frac{\partial \phi}{\partial x} \right)^2 + c_0 \frac{\partial^4 \phi}{\partial x^4} + c_1 \frac{\partial \phi}{\partial x} \frac{\partial^3 \phi}{\partial x^3}
$$
$$
+ c_2 \left(\frac{\partial^2 \phi}{\partial x^2} \right)^2 + c_3 \left(\frac{\partial \phi}{\partial x} \right)^2 \frac{\partial^2 \phi}{\partial x^2} + c_4 \left(\frac{\partial \phi}{\partial x} \right)^4 . \tag{7.4.26}
$$

The coefficient c_0 can easily be obtained from the linearized equations, it is given by [7.74]

$$
c_0 = -\tfrac{1}{2} \varepsilon^2 (1 + \beta^2) . \tag{7.4.27}
$$

Other coefficients c_i in (7.4.26) are combinations of the parameters ε and β; their explicit form is irrelevant for the subsequent analysis.

The expansion in (7.4.26) is based only on the assumption that the spatial variations are smooth. Our next step is to take into account the presence of a small parameter ν in this equation, which reflects the fact that we are close to the Benjamin–Feir boundary. We introduce new variables of the phase φ, time τ and the spatial coordinate s, defined by

$$
\phi = \nu^{1/2} (\varepsilon - \beta)^{-1} \varphi , \quad t = \nu^{-2} \varepsilon^2 (1 + \beta^2) \tau,
$$
$$
x = \nu^{-1/2} \varepsilon (1 + \beta^2) s . \tag{7.4.28}
$$

Applying this transformation of variables to (7.4.26), we obtain

$$
\frac{\partial \varphi}{\partial \tau} = -\frac{\partial^2 \varphi}{\partial s^2} + \left(\frac{\partial \varphi}{\partial s} \right)^2 - \frac{1}{2} \frac{\partial^4 \varphi}{\partial s^4} + \tilde{c}_1 \frac{\partial \varphi}{\partial s} \frac{\partial^3 \varphi}{\partial s^3}
$$
$$
+ \tilde{c}_2 \left(\frac{\partial^2 \varphi}{\partial s^2} \right)^2 + \tilde{c}_3 \left(\frac{\partial \varphi}{\partial s} \right)^2 \frac{\partial^2 \varphi}{\partial s^2} + \tilde{c}_3 \left(\frac{\partial \varphi}{\partial s} \right)^4 . \tag{7.4.29}
$$

All coefficients $\tilde{c}$ in this equation are small, i.e. we have $\tilde{c}_1 \sim \tilde{c}_2 \sim \nu^{1/2}$, $c_3 \sim \nu$, $\tilde{c}_4 \sim \nu^{3/2}$ and therefore the terms with these coefficients can be neglected.

Thus we arrive at the *Kuramoto–Sivashinsky equation*

$$
\dot{\varphi} = -\Delta \varphi - \tfrac{1}{2} \Delta^2 \varphi + (\nabla \varphi)^2 . \tag{7.4.30}
$$

This equation has been independently derived by *Kuramoto* and *Tsuzuki* [7.75] in their study of oscillatory reaction–diffusion systems and, two years later, by *Sivashinsky* [7.76] who investigated instabilities of flame fronts. We note that, although only the one-dimensional case has been considered above, equation (7.4.30) is also valid for two-dimensional systems.

The numerical simulations [7.77] of the Kuramoto–Sivashinsky equation show that it gives rise to chaotic dynamics, characterized by the presence of positive Lyapunov exponents and decay of temporal correlations. A remarkable property of this equation is that, in an infinite system, it has no intrinsic control parameters. Indeed, even if we put some coefficients in front of the various derivative terms in

(7.4.30), all of them can be eliminated by rescaling the variables. The consequence of such universality is that all possible probability distributions and correlation functions for the Kuramoto–Sivashinsky equation can in principle be tabulated once and forever.

The spatiotemporal chaos generated by the Kuramoto–Sivashinsky equation in the limit of large system sizes ($L \rightarrow \infty$) represents developed turbulence, where the system breaks into a large number of statistically uncorrelated subsystems. It is therefore expected that the collective action of all these subsystems on a given element can be modelled by the introduction of noise. Indeed, it was shown analytically in 1981 by *Yakhot* [7.78] that the properties of the Kuramoto–Sivashinsky equation can be reproduced by a certain *stochastic* differential equation.

The dynamical variable φ in the Kuramoto–Sivashinsky equation can be written as $\varphi(s,\tau) = \bar{\varphi}(s,\tau) + \delta\varphi(s,\tau)$, where $\bar{\varphi}$ contains only contributions from Fourier modes with sufficiently small wave numbers $0 < k < \Lambda$. Then, as shown in [7.78], the slowly varying component approximately obeys the equation

$$\frac{\partial \bar{\varphi}}{\partial \tau} = \mu \frac{\partial^2 \bar{\varphi}}{\partial s^2} + \frac{\lambda}{2} \left(\frac{\partial \bar{\varphi}}{\partial s} \right)^2 + \zeta(s,\tau) . \tag{7.4.31}$$

Here $\zeta(s,\tau)$ is expressed solely in terms of the fastly varying component $\delta\bar{\varphi}(s,\tau)$. An important result is that, under an appropriate choice of the cut-off Λ, the statistical properties of $\zeta(s,\tau)$ are the same as those of 'white noise' with correlations described by

$$\langle \zeta(s,\tau)\zeta(s',\tau) \rangle = 2S\delta(s - s')\delta(\tau - \tau') . \tag{7.4.32}$$

This correspondence has been confirmed by numerical simulations [7.79, 81]. For a cut-off of $\Lambda = 0.5$, the following approximate values of the coefficients have been found: $\mu = 7.5$, $\lambda = 1$ and $S = 17.9$ [7.81].

The stochastic equation (7.4.31) was used in 1986 by *Kardar, Parisi* and *Zhang* [7.82] to describe the growth of solid interfaces; it is closely related to the noisy Burgers equation that was investigated in [7.83].

Above, the Kuramoto–Sivashinsky equation was derived as an approximation for the initial stage of the system's evolution from the unstable state of uniform oscillations. We see now that even this first stage, during which phase gradients remain relatively weak, is chaotic in large systems. Such a dynamical regime can therefore be characterized as *phase turbulence* [7.73].

Since the Kuramoto–Sivashinsky equation has no free parameters, it generates a set of *universal* probability distributions that can be numerically determined with any required precision. Such a universal distribution for the gradient $v = \partial\varphi/\partial s$ is given by a certain function $p = F(v)$. For small gradients, this distribution should be Gaussian, i.e., it has the form

$$F(v) = z_0 \exp(-\gamma_0 v^2) , \tag{7.4.33}$$

where γ_0 and z_0 are numerical coefficients.

Note that the variables φ, τ and in the Kuramoto–Sivashinsky equation are related by the scaling transformation (7.4.28) to the original variables of the Ginzburg–Landau equation. By reversing this transformation, we find that $\nu = a(\partial\phi/\partial x)$ where the proportionality factor is

$$a = v^{-1}\varepsilon(\varepsilon - \beta)(1 + \beta^2)^{1/2} . \tag{7.4.34}$$

Hence, the actual probability distribution of phase gradients in the state of phase turbulence for the one-dimensional complex Ginzburg–Landau equation is given by

$$P = aF\left(a\frac{\partial\phi}{\partial x}\right) . \tag{7.4.35}$$

In the limit of weak gradients, it can be approximated as

$$p = z_0 a \exp\left[-\gamma_0 a^2 \left(\frac{\partial\phi}{\partial x}\right)^2\right] . \tag{7.4.36}$$

Typical phase gradients in the state of phase turbulence have the magnitude $\partial\phi/\partial x \sim 1/a$. Near the Benjamin–Feir boundary, the parameter $\nu = -(1 + \varepsilon\beta)$ is small. Since $a \sim 1/\nu$, this means that the typical phase gradient behaves as $\partial\phi/\partial x \sim \nu$ in this limit and, therefore, the phase fluctuations become increasingly smooth as the Benjamin–Feir boundary is approached.

In essence, the internal dynamics of this system gives rise to a noise whose probability distribution in the region of small gradients is given by (7.4.36). Though the typical phase fluctuations are weak, strong fluctuations with $\partial\phi/\partial x \sim 1$ can also occur. Once a strong fluctuation takes place, the reduced phase decription in terms of the Kuramoto–Sivashinsky equation is no longer valid. Indeed, according to (7.4.25) the oscillation amplitude ϱ is significantly changed within such a fluctuation. A strong phase fluctuation can produce amplitude defects; these are discussed in the next section.

7.5 Statistics of Defects

The breakdown of order in a distributed system occurs through the formation of defects. This is well known from the equilibrium statistical physics of condensed matter. The defects in a crystal lattice are vacancies and various atomic dislocations. When the temperature is increased and the crystal begins to melt, the number of defects rapidly grows, destroying the long-range order in the system. Similar phenomena take place in magnetic materials such as ferromagnets. In the completely ordered ferromagnetic state, the magnetic moments of all atoms have the same orientation. However, in the absence of strong external fields which tend to align the magnetic moments, a ferromagnet often has a domain structure.

Then, the orientation of magnetic moments is uniform only within a single spatial domain; it varies from one domain to another. The walls separating different magnetic domains are extended defects: the magnetic moment vector undergoes a rotation when a domain wall is crossed.

Some magnetic materials have such strong anisotropy that their magnetic moments can lie only in a certain plane. The orientation of a magnetic moment vector can then be specified by indicating its angle, ψ, measured from a reference direction in the plane. Inside a domain wall, this angle quickly changes and thus the orientations of magnetic moments on the two sides of the wall are different.

The localized defects in magnetic materials are *vortices*. By taking a contour surrounding the center of a vortex and moving around it, so that a full turn is made, we would find that the vector of the magnetic moment has performed a complete rotation, i.e., its angle ψ has changed by 2π. Hence, the angle variable must have a singularity at the vortex center. Vortices represent a fairly typical form of defects: they are also found in many other condensed systems, such as liquid crystals, superconductors and superfluids.

Defects can be created by applying an appropriate external perturbation even in perfectly ordered systems, i.e., at zero temperature. Thermal fluctuations, which are observed at finite temperatures, represent an internal source of such perturbations. Because of these, defects are spontaneously produced by the system itself.

In order to creat a defect, the perturbation should exceed a certain threshold. Therefore, at low temperatures, when typical thermal fluctuations are small, the production of a defect is an exponentially rare event and the number of defects is extremely small. As the temperature is increased and the mean magnitude of thermal fluctuations approaches the threshold, and the defect production rate is rapidly enhanced. According to *Kosterlitz* and *Thouless* [7.84], this causes the phase transition which destroys the spatial order in the system.

Today, we have many indications that the breakdown of spatial order in distributed-active systems proceeds in a similar manner to such equilibrium phase transitions. The difference is simply that the defects in such systems can also be *active*, i.e., they can periodically emit waves or move themselves through the medium.

The analogies are especially close for an oscillatory medium near the Hopf bifurcation, described by the complex Ginzburg–Landau equation. In this case, the local phase ϕ of the oscillations is directly related to the planar angle ψ of magnetic moments in the above example. The walls of magnetic domains then correspond to phase flips, whereas the vortices in two-dimensional oscillatory media are rotating spiral waves.

The analogies between distributed oscillatory media and equilibrium condensed systems with complex order parameters have been known for a long time; see, e.g. [7.85, 86]. *Kuramoto* [7.73] has noted that spiral waves give rise to phase singularities and has suggested that the development of amplitude turbulence proceeds via a reproduction cascade of these localized defects. *Coullet* et al. [7.74, 87] have performed computer studies of the transition to amplitude turbulence in the two-dimensional complex Ginzburg–Landau equation.

In the previous section we showed that as the Benjamin–Feir boundary is crossed, phase turbulence appears. In a sense, such phase turbulence plays the role of thermal fluctuations in the considered system. It represents an intrinsic noise source that supplies the perturbations necessary to produce defects. We have seen that the typical magnitude of fluctuations in the phase turbulence regime increases with the distance from the Benjamin–Feir boundary, and this makes the production of defects more probable.

We first discuss the phenomena related to the generation of defects for the one-dimensional complex Ginzburg–Landau equation. This case is special because *topological defects*, such as vortices, do not exist in one-dimensional systems. Indeed, any one-dimensional phase distribution can be continuously transformed into the uniform state, whereas a continuous transformation eliminating a vortex is not possible. However, one-dimensional systems are more amenable to large-scale numerical simulations and therefore they have been investigated numerically in great detal [7.88, 91]. Despite the above-mentioned difference, the development of amplitude turbulence in such systems has much in common with the two-dimensional case.

In one dimension, all defects essentially represent phase flips. Inside a defect, the oscillation phase changes significantly over a narrow spatial interval. An effect of the rapid phase variation is that, where it occurs, the oscillation amplitude is reduced. Hence, the defect is also associated with a local suppression of the oscillation amplitude. Moreover, it generally separates regions filled with plane waves of different wave numbers. The defects are not stationary and move through the medium.

The defects are apparently related to the family of exact one-dimensional solutions representing the Bekki–Nozaki holes. *Chate* [7.91] has found that the velocity c of a single moving defect is related to the wave numbers k_+ and k_- of the plane waves on either side of it by the same equation (7.4.10) as for the Bekki–Nozaki hole.

Near the Benjamin–Feir boundary, only individual defects are generated in the medium; they have finite lifetimes and do not interact significantly with one another (Fig. 7.22) At a larger distance from this boundary, the number of defects is very large and they form a dense, strongly interacting subsystem (Fig. 7.23).

As noted by *Shraiman* et al. [7.88], there is also a parameter region where amplitude turbulence, characterized by the presence of multiple defects, can be observed *before* the Benjamin–Feir boundary is crossed, i.e., where uniform oscillations still remain stable with respect to small perturbations. If we 'infect' the state of uniform oscillations by creating a single defect in this parameter region, it gives rise to a reproduction cascade (Fig. 7.24) that completely destroys uniform oscillations and establishes amplitude turbulence in the system.

Note that, in the latter case, the dynamical system has two principal attractors: one corresponding to stable uniform oscillations and the other corresponding to

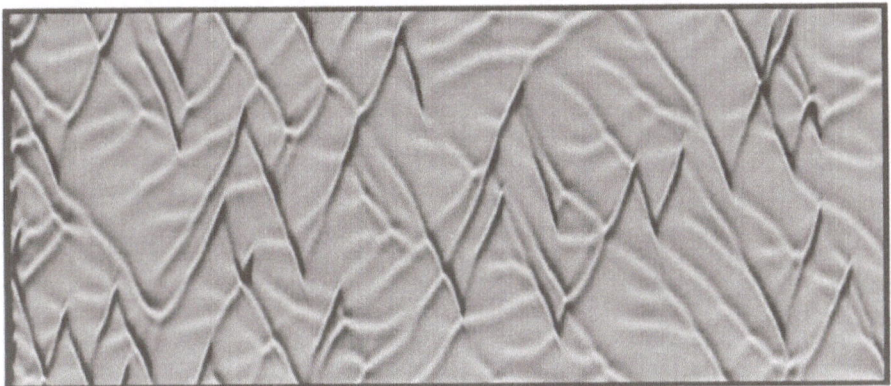

Fig. 7.22. Spontaneous formation of defects on the background of phase turbulence for $\beta = -1.0$ and $\varepsilon = 2.0$. Time is increasing from left to right in the horizontal direction, the vertical direction is used to display the coordinate x; the total length of the system is $L = 256$, no-flux boundary conditions are used. Local values of the modulus ϱ of the oscillation amplitude are shown in gray scale; the lighter areas correspond to larger values of ϱ, and the defects are seen as moving black spots. (Reproduced by D. Battogtokh after [7.92]

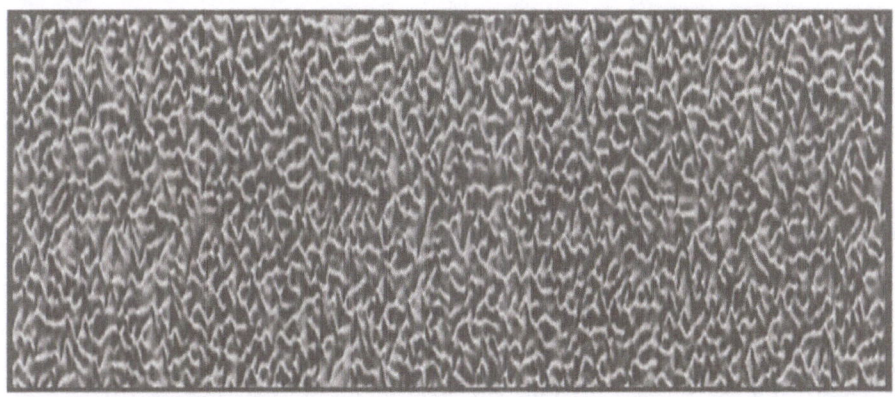

Fig. 7.23. Developed amplitude turbulence at $\beta = -2.0$ and $\varepsilon = 2.0$. The same explanations as in Fig. 7.22. (Reproduced by D. Battogtokh after [7.92])

developed amplitude turbulence. When a suitable perturbation brings the system out of the basin of attraction of the first attractor, it evolves towards the second. *Pomeau* [7.93] suggested that this process should be similar to *directed percolation*, which describes, for example, the spread of epidemics. When the initial perturbation is local, the transition to developed amplitude turbulence occurs by the propagation of a front that leaves behind it the state with a high density of defects.

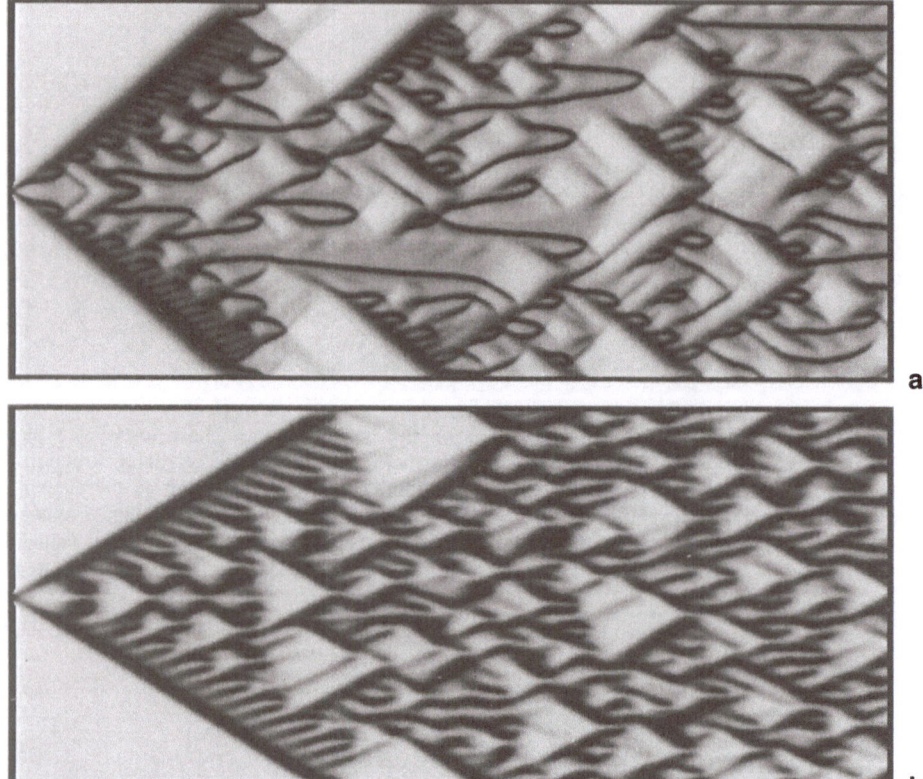

Fig. 7.24a,b. The reproduction cascades of defects initiated by a local perturbation of the state with uniform oscillations; the parameters are (**a**) $\beta = -5.555$ and $\varepsilon = -0.75$ and (**b**) $\beta = -0.33$ and $\varepsilon = -0.1$, other details are the same as in Fig. 7.22. (D. Battogtokh, unpublished)

Another situation is found when hysteresis is absent and turbulence is observed only below the Benjamin–Feir boundary. The uniform oscillations are unstable here but, while losing their stability, they first give rise to phase turbulence. In this dynamical regime, trajectories lie in the vicinity of uniform oscillations and the amplitude variations remain small. However, the behavior of the system is chaotic and, albeit rarely, the trajectories can visit more distant regions in the space of the system's states corresponding to strong amplitude fluctuations. Such fluctuations produce amplitude defects.

When this has occurred, the subsequent behavior of the system is determined by the properties of defects. Since we assume that the global amplitude-turbulence attractor is absent in the considered case above the Benjamin–Feir instability and by continuity at least in some region below it, the defects cannot start to indefinitely reproduce here. Instead, only individual defects or local bursts of defects could appear. This state with a mixture of defects and phase turbulence can be called *intermittent*. Because the defects have finite life-times, a certain mean concentration of defects is then maintained in the system. This concentration decreases as the Benjamin–Feir boundary is approached.

Turbulence in two-dimensional systems described by the complex Ginzburg–Landau equation has a number of special features. Its genesis is determined fundamentally by phase noise, Bekki–Nozaki strips, and spiral waves. The phase noise is again effectively described by the Kuramoto–Sivashinsky equation and its long–wavelength statistical properties seem to be well reproduced by the Kardar–Parisi–Zhang equation [7.82].

The Bekki–Nozaki strips are obtained simply by continuing the respective one-dimensional solutions along the second coordinate axis. They represent *extended defects* [7.93]. By bending a strip, defects with complex shapes may also be produced. Generally, the properties of a strip (such as its width and local velocity) can vary along it. Note that, unless the strip is standing, the oscillation amplitude in its filament does not vanish.

Spiral waves represent true topological defects in two–dimensional oscillatory media. The properties of individual spirals have been studied for a long time [7.85, 7.94–98] and the approximate analytical solution for a spiral wave has been constructed by *Hagan* [7.70]. These wave patterns were discussed in Chap. 4 of our first volume. We noted there that a spiral wave is associated with a phase singularity: When a closed contour surrounding the center of a spiral wave is traversed, the phase is changed by $\phi = +2\pi$ or $\phi = -2\pi$, depending on the rotation direction of the spiral. Hence, the oscillation phase is not defined in the center, which is physically admissible only if the oscillation amplitude vanishes in this point.

A spiral wave represents a topological defect of the phase distribution field – this structure cannot be eliminated by any continuous deformation of the solution. The *topological charge* of the defect is defined as $Q = \Delta\phi/2\pi$ [7.99, 100]. Thus, spiral waves rotating in opposite directions have topological charges $+1$ and -1. Higher values of the topological charge correspond to unstable multi-armed spirals. Topological defects interact according to the signs of their topological charges. They can form bound pair states (see Chap. 4 of the first volume). Defects with opposite charges *annihilate* when they come sufficiently close together.

When uniform oscillations are unstable with respect to modulation, i.e., the Benjamin–Feir boundary is crossed, pairs of spiral waves may spontaneously appear in the two-dimensional system. Once the defects are created, they can move apart and behave as relatively independent particles.

Kuramoto [7.73] suggested that the development of phase defects gives rise to a transition to amplitude turbulence. This conjecture has been confirmed by *Coullet* et al. [7.74, 101] who performed numerical simulations and statistical analysis of the turbulence associated with defects. They have shown that the transition from phase to amplitude turbulence is characterized by a qualitative change of the spatial correlation function for the complex oscillation amplitude. The correlation radius decreases strongly as defects appear in the system and becomes very close to the mean distance between the defects.

Another important aspect of two-dimensional amplitude turbulence found in [7.74, 101] is that it is associated with hysteresis, i.e., this form of turbulence persisted even on the other side of the Benjamin–Feir boundary. This resembles the behavior of two-dimensional Turing patterns (see Chap. 5 of the first volume) and is probably also explained by mutual enhancement of spatial modes. A systematic exploration of the parameter space for the two-dimensional complex Ginzburg–Landau equation has not yet been performed.

Gil et al. [7.102] have considered the statistics of topological defects in the two-dimensional complex Ginzburg–Landau equation. They counted the total number of defects at different time moments. An attempt has been made to explain the statistics by means of a simple stochastic model. The model assumes that the defects are generated at a constant rate but disappear only in pair collisions, by the process of binary annihilation. But it was only shown in [7.102] that, for large numbers of defects, the standard deviation $\Sigma = (\langle N^2 \rangle - \langle N \rangle^2)^{1/2}$ is close to the mean number $\langle N \rangle$ of the topological defects in the system. This, however, is a characteristic property of the Poissonian process corresponding to individual independent generation and decay of particles.

As noted above, topological defects are always created in pairs with opposite topological charges and, when they annihilate, the topological charges of the two particles must be opposite. The full statistical description should also include reproduction of the defects. Moreover, interactions between the defects can modify their statistics.

In an experimental study of defects in a nonlinear active optical system, *Ramazza* et al. [7.103] have observed deviations from the Poissonian statistics in time series obtained by recording the time moments of a defect's appearance inside a given spatial volume. This set of data was mainly sensitive to *motion* of defects. Statistical analysis revealed the presence of a 'refractory time' between succeccive appearance of defects, consistent with repulsive interactions between these particle-like structures.

Hildebrand et al. [7.104] have investigated the statistical properties of spatial defect distributions for a model of an excitable medium describing a catalytic surface chemical reaction. They have found that, as the control parameter is increased, defects first arrange themselves in such a way as to maintain a certain minimum distance between neighboring defects. This resembles the kind of spatial pattern found in liquids. For higher values of the control parameter, diffusive motion of defects sets in (Fig. 7.25). The evolution of the pair correlation function of defects in this transition is similar to that of individual molecules in the liquid–vapor phase transition.

With the appearance of defects, a new level of structural hierarchy is reached. The behavior of a system can now, to a large extent, be rationalized in terms of interactions and reactions between these elementary structures. The complex nature of a single defect is not fully expressed in such processes and therefore

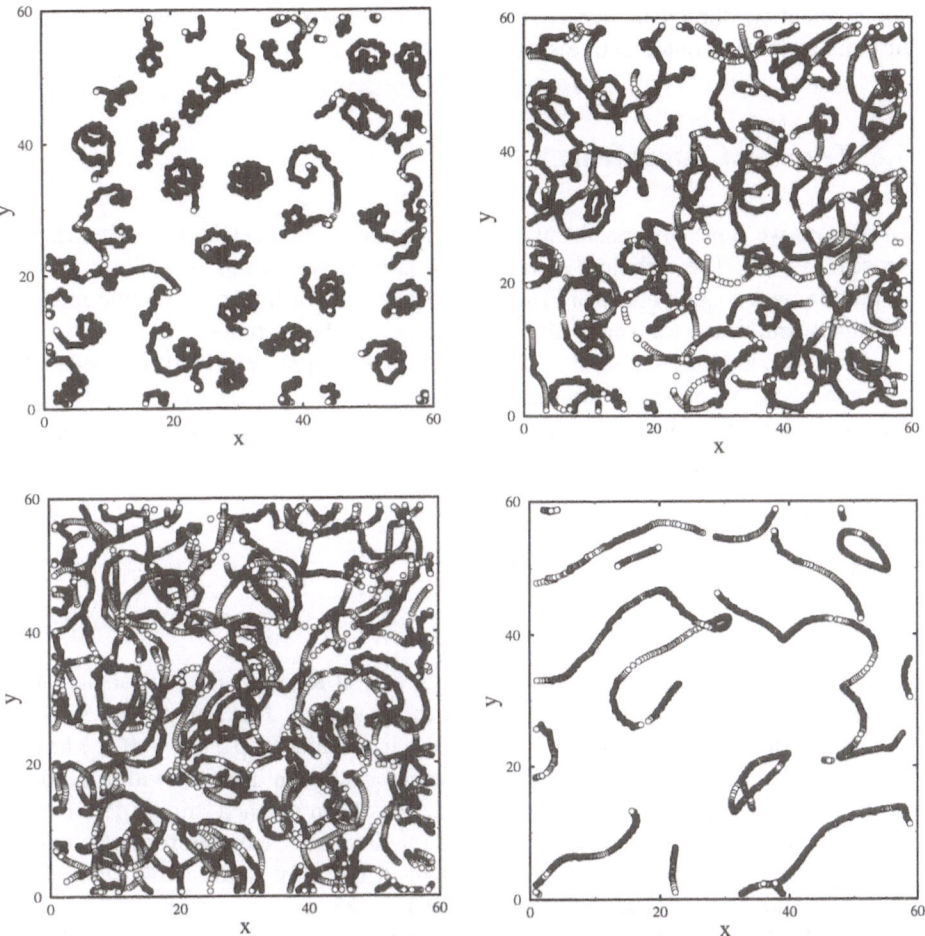

Fig. 7.25. Trajectories of topological defects in a model of a catalytic surface reaction for four different increasing values of a control parameter (from top left to bottom right). From [7.104, 105]

simple models, formulated in terms of reaction rates and coordinate-dependent physical interactions, may be used for their description.

The role of an active medium which has produced defects consists further in providing an environment for their collective evolution. Small-amplitude chaotic dynamics in this environment (cf. phase turbulence in the complex Ginzburg–Landau equation) gives rise to *noise*, which act on the defects and induces, for instance, their diffusive motion.

Thus, our analysis of spatiotemporal chaos has naturally brought us to what constitutes the main theme of the following chapters: a general study of noise effects in reaction–diffusion models. However, before we proceed to such a study, some attention should addressed to a different, and quite surprising, aspect of turbulence in distributed-active systems.

7.6 Transient Turbulence

Until now we have discussed only steady regimes which set in after a transient process is over. The turbulent behavior that is observed in this limit corresponds to motion on a chaotic attractor. The time required to reach this asymptotic regime is usually small and transients do not play any significant role.

However, another possibility exists: *Crutchfield* and *Kaneko* [7.106] noted that, in some cases, the observed turbulence might correspond instead to extremely long quasi-stationary transients. As an illustration, they considered a system of coupled maps described by

$$x_{n+1}(i) = \tfrac{1}{3}[\phi(x_n(i+1)) + \phi(x_n(i)) + \phi(x_n(i-1))] , \tag{7.6.1}$$

where

$$\phi(x) = \{Ax + B\} . \tag{7.6.2}$$

This map is defined on the segment $[0, 1]$; A and B are numerical constants, and $\{\dots\}$ denotes taking the fractional part of the argument; for example, $\{1.333\} = 0.333$. When $0 < A < 1$ and $0 < B < 1$, the individual map $x_{n+1} = \phi(x_n)$ has a dynamics that consists in raising x_n with each successive iteration with a sudden drop of x_n after a threshold $x_c = (1 - B)/A$ has been reached. Since $A < 1$, the Lyapunov exponent of this map is negative (cf. Sect. 5.2) and therefore no chaotic behavior is displayed by the individual map – it has only an attractive cycle. For $A = 0.91$ and $B = 0.1$, the period of this cycle is 25.

Simulations of (7.6.1) with these parameter values have shown that, after a transient, stable periodic motion is always established. However, the average duration $T(N)$ of the transient was found to increase rapidly with the number N of elements in a chain. The numerical data in [7.106] for the chains with lengths up to $N = 47$ are consistent with a hyperexponential dependence of the transition time on N, but subsequent simulations [7.107] of this model for longer chains up to $N = 62$ yield an exponential law (Fig. 7.26).

The same behavior of transients has been found in [7.107] for another system of coupled maps described by (7.6.1) with a different function

$$\phi(x) = \begin{cases} bx, & 0 < x < 1/b, \\ a + c(x - 1)/b, & 1/b \le x < 1 \end{cases} \tag{7.6.3}$$

for the parameters $a = 0.07$, $b = 2.7$ and $c = 0.1$. Each individual map has in this case an attractive period-3 cycle whose Lyapunov exponent is $\lambda = \log(b^2 c) = -0.105 \dots$. For a chain of these maps, stable periodic oscillations have always

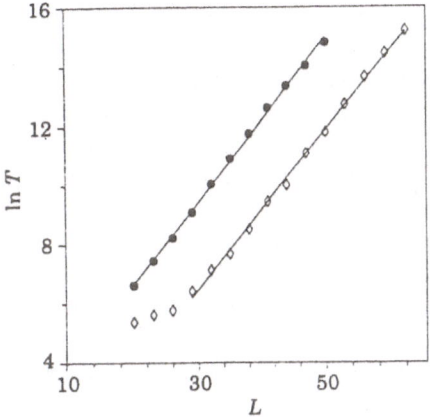

Fig. 7.26. The dependence of the average transient time T on the length L of the chain for functions $\phi(x)$ given by equation (7.6.2) (diamonds) and by equation (7.6.3) (full circles). The averages are taken over 300 different realizations of the initial conditions. From [7.107]

been observed after a long transient. The duration of the transient increases exponentially with the length of the chain (Fig. 7.26).

Such a strong dependence of the transient time means that, even for relatively short chains with length about 100, the duration of transients is extremely large. It increases to astronomically large numbers when still longer chains are considered. Hence, the only behavior observable for such large systems in a numerical simulation or in an experiment must be a *transient*.

Lattices of coupled maps are closely related to distributed dynamical systems with continuous time. Therefore, similar behavior can also be expected for distributed systems. Indeed, the exponential dependence of the transient time on the length of a system was found in 1986 by *Shraiman* [7.108] and confirmed later by *Hyman* et al. [7.77] for the Kuramoto–Sivashinsky equation (7.4.30). Recently, transient spatiotemporal chaos has been extensively studied by *Wacker* et al. [7.109] in numerical simulations of a reaction–diffusion system that describes charge transport in layered semiconductors. The model is given by two equations:

$$\dot{u} = \frac{v - u}{(v - u)^2 + 1} - \gamma u + \nabla^2 u,$$
$$\dot{v} = \alpha(j + u - v) + D\nabla^2 v . \tag{7.6.4}$$

Numerical simulations of this system have been performed in the one-dimensional case for varying total length L of the medium. The parameter values were $\gamma = 0.05$, $\alpha = 0.02$, $D = 8$ and $j = 1.21$; no-flux boundary conditions for both components were taken at $x = 0$ and $x = L$. Note that the variable u corresponds here to an activator, and the variable v describes an inhibitor.

With this choice of parameters, stable dynamics of the system represents periodic spatiotemporal spiking, where a spatially periodic pattern forms and vanishes periodically in time. However, this asymptotic periodic regime is preceded by a long transient with irregular behavior. While the periodic pattern is characterized by spikes which arise periodically at two different sets of positions separated by half a period, the transient behavior is dominated by similar spikes occurring at irregular positions.

The principal result of the numerical investigation [7.109] is that the average duration $\langle T(L) \rangle$ of the transient is found to grow approximately *exponentially* from $\langle T \rangle = 12787$ at $L = 400$ to $\langle T \rangle = 345674$ at $L = 1400$. The observed statistical variations in the transient time T are well reproduced by the law of exponential decay $P(T) \sim \exp[-T/\langle T(L) \rangle]$.

Intermediate Lyapunov exponents in the transient state were computed by taking the gliding mean over an interval of 5000 time units. It is found that the largest exponent is positive during the transient period. It does not depend significantly on L and has a value of about 2.5×10^{-3}. At the end of the transient the largest Lyapunov exponent drops to zero and the second largest exponent becomes negative, as expected for stable periodic behavior.

While Lyapunov exponents characterize mainly the temporal behavior of perturbations, spatiotemporal properties of the system can be examined by considering the spreading of an initially localized perturbation. Suppose that we add at the initial time moment a small perturbation localized near a certain spatial point. It can be expected that the influence of this perturbation will spread over the system. To visualize deviations resulting from the spreading of the perturbation, the logarithm of the local difference between the unperturbed and perturbed trajectories can be plotted as a function of time within the system.

If such a numerical experiment is performed for a distributed system characterized by local relaxation to the turbulent state, one would find that, after some initial spreading, the deviations fade away and soon cease to be discernible. Surprisingly, the influence of local perturbations on the state of the system (7.6.4) in its transient regime is completely different [7.109]. As seen from Fig. 7.27, the perturbation spreads over the whole system. This spreading proceeds at a constant velocity that is apparently independent of the size L of the system. Similar behavior was found earlier by *Politi* et al. [7.108], who studied a system of coupled maps exhibiting transient turbulence.

Such spreading of perturbations closely resembles the propagation of trigger waves in bistable systems (cf. Chap. 2 of the first volume). In the latter case, a propagating wave represents a front of a transition between two stable uniform states of a distributed-active system. To initiate it, a sufficiently strong local perturbation must be applied. The wave propagates at a constant velocity through the medium until it reaches its boundary. After such a wave has passed through the entire medium, a new steady uniform state is established.

The observed response of transient turbulence to local perturbations suggests that, in this regime, the dynamical system may have a large number of (almost) attractive global states. However, in contrast to the case of simple bistable media, all these states represent various spatiotemporal patterns. Each of them is characterized by chaotic spatial and temporal behavior, so that the correlation time and the correlation radius for a given pattern are finite and small. However, these patterns are still 'rigid'. When a perturbation is applied, it does not lead to a

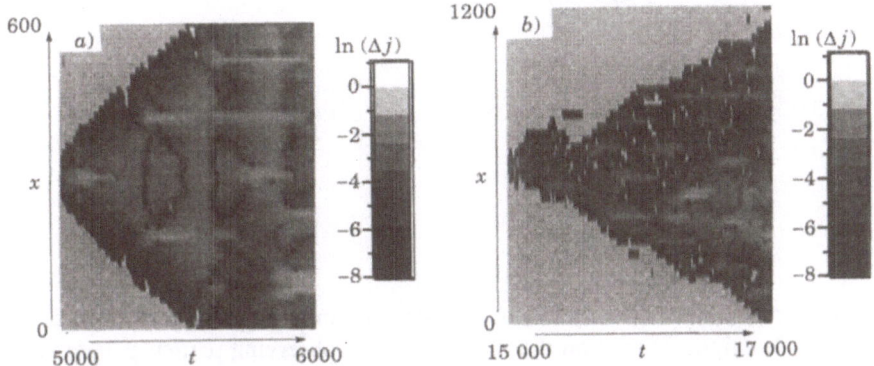

Fig. 7.27. Temporal evolution of a localized perturbation in the transient phase of the system (7.6.4) for (**a**) $L = 600$ and (**b**) $L = 1200$. The grey scale corresponds to the difference $v - u$ for the perturbed and the unperturbed system. From [7.109]

spatial mixture of patterns but rather initiates a transition to another pattern that subsequently spreads over the entire medium.

The transient turbulence, which becomes the only observable regime in the limit of large systems, represents a kind of spatiotemporal chaos which is different from the usual turbulence characterized by local statistical relaxation. The existence of two classes of spatiotemporal chaos was first pointed out by *Wolfram* [7.110] in his extensive study of cellular automata.

8. Random Processes

When a deterministic description of a system is impossible, one can still use a statistical description based on the *probabilities* of observing particular outcomes. In the simplest case, only the probabilities of single random events are considered. However, in studies of dynamical phenomena it is often necessary to know the probabilities of entire sequences of random events that are not independent of one another. This description can be constructed within the mathematical theory of random processes.

We begin this chapter with an analysis of random processes in very simple systems which have only a few distinct states. Their dynamics consists of random transitions between these separate states. Today, such systems are commonly called probabilistic automata. Continuous random processes are much more complex mathematical objects. Their description is based on probability distributions over sets of random trajectories.

8.1 Probabilistic Automata

The general concept of an automaton was introduced in Chap. 1 of the first volume. Usually, an automaton has several distinct states which can be conveniently enumerated by some integer variable a. The state of an automton is changed at discrete time moments t_n. Its evolution can be specified by the sequence of states $a_1, a_2, \ldots, a_n, a_{n+1}, \ldots$ taken at these discrete time moments. If an automaton is deterministic, its next state is always uniquely determined by the state at the present moment. Hence, there should be some regular map $a_{n+1} = f(a_n)$ which generates the observed sequence.

In probabilistic automata the next state of an element cannot be unambiguously predicted from its state at a given moment[1]. Since the evolution of an individual probabilistic automaton is random, it can be described only in statistical terms for an ensemble of identical automata.

---

[1] This can happen, for instance, if an actual element modelled by the automaton has a complex internal dynamics that is very sensitive to small perturbations. It might occur that such a dynamics, which involves some internal variables, is not directly manifested in interactions between the elements and is not seen by an "outside observer". Nonetheless, the chaotic internal dynamics triggers transitions between distinct states of the element, making them random.

A probabilistic automaton is defined if we specify the probabilities of transitions $w(a|a')$ from any state a' to any other state a. If we know the probability $p_n(a)$ of finding the automaton in state a at moment t_n, the probability distribution $p_{n+1}(a)$ at the next moment t_{n+1} will be given by

$$p_{n+1}(a) = \sum_{a'} w(a|a')p_n(a') . \tag{8.1.1}$$

This equation simply takes into account that the automaton can enter state a from all possible states a' with transition probabilities $w(a|a')$; it can also remain in the same state so that $a' = a$.

Note that at any moment t_n the probability distribution should satisfy the obvious normalization condition

$$\sum_a p_n(a) = 1 . \tag{8.1.2}$$

We also have the condition for the transition probabilities

$$\sum_a w(a|a') = 1 , \tag{8.1.3}$$

since the probability of transition from state a' to *any* other (or the same) state is equal to one.

According to the above definition of a probabilistic automaton, its transition probabilities are independent of its states at previous moments. This means that such an automaton has no memory.

It is convenient to introduce a matrix W with elements $w(a|a')$ and to construct formal vectors p_n with components $p_n(a)$. Then (8.1.1) can be written as a linear map, i.e. as

$$p_{n+1} = W p_n . \tag{8.1.4}$$

Subsequent iterations of this map yield the probability distribution at any moment t_n in terms of the initial distribution:

$$p_n = W^n p_0 . \tag{8.1.5}$$

The properties of the linear maps (8.1.4) and (8.1.5) are determined by the eigenvalues of matrix W. When the initial probability distribution vector p_0 is decomposed in terms of the eigenfunctions Φ_α of W, (8.1.5) yields

$$p_n = \sum_\alpha C_\alpha \lambda_\alpha^n \Phi_\alpha , \tag{8.1.6}$$

where λ_α are the eigenvalues of W and C_α are the decomposition coefficients.

As a simple example of a probabilistic automaton, we consider the *stochastic oscillator*. This automaton has a quiescent state $a=0$ and an active state $a=1$.

Transitions from the quiescent to the active state occur with the probability $w(1|0) = w^+$. The reverse transitions have the probability $w(0|1) = w^-$.

One can easily verify that matrix W of such an automaton has two eigenvalues $\lambda_0 = 1$ and $\lambda_1 = 1 - w^- - w^+$. Because the sum of probabilities $p_n(0)$ and $p_n(1)$ is always equal to one, it is sufficient to consider only the behavior of $p_n(1)$. Application of (8.1.6) then yields

$$p_n(1) = w^+(w^+ + w^-)^{-1}$$
$$+ \left[p_0(1) - w^+(w^+ + w^-)^{-1}\right](1 - w^+ - w^-)^n . \tag{8.1.7}$$

We see that, since $|\lambda_1| < 1$, the last term vanishes in the limit $n \to \infty$. Therefore, (8.1.7) describes relaxation to the equilibrium probability distribution, given by the first term in this equation. When λ_1 is negative, this relaxation is accompanied by damped oscillations of the probability $p_n(1)$.

Note that $(1 - w^-)^k$ is the probability that the stochastic oscillator remains in the active state for k consecutive time steps. Hence the *mean* duration of stay in the active state is obtained by calculation of the weighted sum

$$T_1 = \sum_{k=0}^{\infty} k(1 - w^-)^k , \tag{8.1.8}$$

which yields $T_1 = (1 - w^-)/(w^-)^2$. Similarly, one can find that the mean duration of stay in the quiescent state is $T_0 = (1 - w^+)/(w^+)^2$. The mean oscillation period is thus $T = T_0 + T_1$.

We now discuss the general properties of networks or lattices of interacting automata. The instantaneous *activity pattern* of a network is specified by indicating the states $\{a_j\}$ of all individual automata. Any such activity pattern is realized with its respective probability $P_n(\{a_j\})$. Since the entire network can be formally considered as a single multi-state automaton, these probabilities at two successive time moments will be related by

$$P_{n+1}(\{a_j\}) = \sum_{\{a'_j\}} W(\{a_j\}|\{a'_j\}) P_n(\{a'_j\}) , \tag{8.1.9}$$

which corresponds to (8.1.1) for an individual isolated automaton. Note that here the summation is carried out over all possible activity patterns $\{a'_j\}$.

Various forms of cooperative behavior are possible in networks of probabilistic automata. For instance, automata may perform transitions strictly in pairs or triplets. Then one should specify the probabilities of simultaneous transitions for such groups of elements. Usually, however, transitions are individual, i.e. only the final state of a single element is changed as the result of a transition.

Even when transitions are individual, the probability of transition from state a'_j into some other state a_j for automaton j can still depend on the instantaneous states $\{a'_k\}$ of other automata, i.e.

$$w = w(a_j|a'_j , \{a'_k\}) . \tag{8.1.10}$$

The global probabilities of transitions between activity patterns of the entire network then factorize into the product of individual transition probabilities:

$$W(\{a_j\}|\{a'_j\}) = \prod_j w(a_j|a'_j, \{a'_k\}) . \tag{8.1.11}$$

If the network represents a regular lattice and the transition probabilities $w = w(a_j|a'_j, \{a'_k\})$ depend only on the states $\{a'_k\}$ of other elements in some neighborhood of element j, this is a *probabilistic cellular automaton*. Different examples of the application of such automata for modeling physical and biological phenomena can be found in [8.1–7].

It was previously assumed that automata could perform transitions only at prescribed moments of time (such probabilistic automata are called *synchronous*). However, in certain applications we also find *asynchronous* probabilistic automata which can change their states at any arbitrary moment in time.

For asynchronous automata we specify the transition probability $w(a, t|a', t')$ which gives the probability of finding the automaton in some state a at time t, provided it was in state a' at a previous moment t'. Note that such transition probabilities can be introduced only if the automaton has no memory – otherwise they would also depend on earlier states of this automaton, prior to moment t'. This requirement is rather restrictive because it must hold for *any* two consecutive time moments t and t', however close they are.

Real physical processes usually retain some memory at least for a short time. Mathematical models of "forgetful" automata are thus an idealization which can be used on time-scales that are large in comparison to the intrinsic correlation time of a process.

For asynchronous probabilistic automata, (8.1.1) is replaced by

$$p(a, t) = \sum_{a'} w(a, t|a', t')p(a', t') , \tag{8.1.12}$$

where t' is *any* previous time moment $(t' < t)$.

Consider the probabilities of transitions within a very small time interval $\Delta t = t - t'$. In a real system, the probability of transition to a different state should vanish in the limit $\Delta t \to 0$. We can introduce the *transition rates* $\nu(a|a')$ by

$$w(a, t + \Delta t|a', t) = \nu(a|a')\Delta t + \dots , \quad a \neq a' . \tag{8.1.13}$$

Note that the set of transition probabilities formally includes the probability $w(a', t|a', t')$ that an automaton remains in its initial state. When the transition interval Δt vanishes, this probability should approach unity. Therefore, we have

$$w(a', t + \Delta t|a', t) = 1 + \nu(a'|a')\Delta t + \dots . \tag{8.1.14}$$

Since within any interval Δt, the automaton either goes into some other state or remains in the original state, the condition

$$\sum_a w(a, t|a', t') = 1 \tag{8.1.15}$$

holds and we obtain

$$w(a', t|a', t') = 1 - \sum_{a \neq a'} w(a, t|a', t') . \tag{8.1.16}$$

Substitution of (8.1.13) and (8.1.14) into (8.1.16) then yields

$$\nu(a'|a') = - \sum_{a \neq a'} \nu(a|a') . \tag{8.1.17}$$

In a typical problem with probabilistic asynchronous automata, one knows the rates of all transitions $\nu(a|a')$ and the problem consists in finding the evolution of the probability distribution. This evolution is described by the *master equation* which is derived from (8.1.12) in a straightforward manner.

Consider (8.1.12) for two close time moments $t + \Delta t$ and t, i.e.

$$p(a, t + \Delta t) = \sum_{a'} w(a, t + \Delta t|a', t)p(a', t) . \tag{8.1.18}$$

Using (8.1.13) and (8.1.14), we find from (8.1.18) that

$$p(a, t + \Delta t) - p(a, t) = \sum_{a'} \nu(a|a')p(a', t)\Delta t + \dots , \tag{8.1.19}$$

where the terms neglected have higher order in Δt. If we divide both sides of (8.1.19) by Δt and take the limit $\Delta t \to 0$, this yields

$$\frac{\partial p(a, t)}{\partial t} = \sum_{a'} \nu(a|a')p(a', t) . \tag{8.1.20}$$

The right-hand side of this equation can be put into a slightly different form if we single out the term with $a' = a$ and use the identity (8.1.17). This transformation results in

$$\frac{\partial p(a, t)}{\partial t} = \sum_{a' \neq a} [\nu(a|a')p(a', t) - \nu(a'|a)p(a, t)] . \tag{8.1.21}$$

The master equation (8.1.21) can be given a simple interpretation. Its right-hand side represents the rate of change of the probability of finding the automaton in a particular state a. This rate consists of two contributions with opposite signs. The first (positive) contribution describes the increase of this probability due to transitions *into* state a from all other states a'. The second (negative) contribution corresponds to the decrease of $p(a, t)$ due to transitions *from* state a into all other states a'.

Although we derived the master equation only for an individual probabilistic automaton, it can also be derived, using the same arguments, for an arbitrary network of asynchronous probabilistic automata. This is evident from the fact that we can consider any such network as a single automaton with a very large number of different states (i.e. activity patterns).

Note that in the standard mathematical classification, the synchronous probabilistic automata analyzed in the first part of this section are called *Markov chains*. Asynchronous probabilistic automata are known as *discrete Markov processes*. A detailed discussion of Markov chains and discrete Markov processes can be found in many standard texts on probability theory (e.g. [8.8–13]). A good introduction to these problems, including a discussion of the properties of the master equation, is given by *Haken* [8.14].

8.2 Continuous Random Processes

There are many situations in which the state of a system is specified by one or several continuous properties which vary randomly with time. The description of these systems is constructed in terms of the mathematical theory of continuous random processes.

Consider some random process $x(t)$. If we plot in the plane (x, t) the values of $x(t)$ at all consecutive time moments t, this would give us a trajectory. When one says that the process $x(t)$ is random, it is implicitly assumed that there is some probability distribution which determines "how probable" is any given trajectory representing a particular realization of this random process. Obviously, the exact probability of a single trajectory is zero, because the set of all allowed trajectories is infinitely large. Therefore, we are able to speak only about a probability *density* in the space formed by all possible trajectories.

Note that, for a single random quantity x, the probability density $p(x)$ is defined by requiring that the product $p(x)$ dx gives the probability of finding this quantity within the interval of length dx around the value x. If we turn to random trajectories, we can try to define the probability functional $P[x(t)]$ by assuming that $P[x(t)] \mathcal{D}x(t)$ would give us the probability of finding a trajectory within a *tube* characterized by some element $\mathcal{D}x(t)$ (Fig. 8.1).

Let us fix in the plane (x, t) a system of "gates" $[x_i, x_i + dx_i]$ at discrete time moments t_i where $i = 0, 1, 2, \ldots, N$ (Fig. 8.2). We are then justified in asking what is the probability that the trajectory passes through all these gates. We can write this probability as

$$P(\{x_i\}) \prod_i dx_i , \tag{8.2.1}$$

where P is some probability distribution.

A random process is defined by specifying *all* probability distributions $P(\{x_i\})$ for *arbitrary* placements of the gates and for *any* total number N. Note that this definition does not as yet imply that a random process should have only continuous realizations. It allows a trajectory to perform abrupt jumps (or even to be discontinuous at any time moment!).

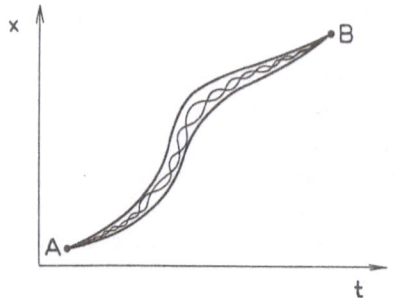

Fig. 8.1. Random trajectories inside a tube

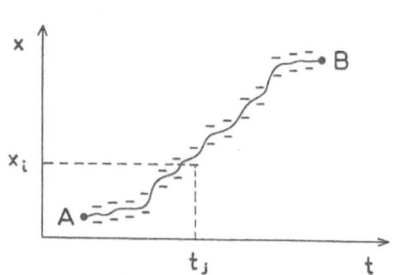

Fig. 8.2. A trajectory passing through the system of gates

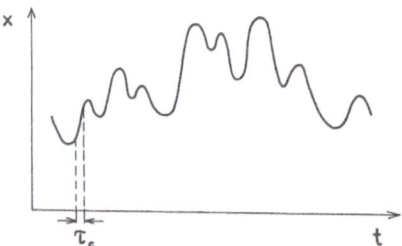

Fig. 8.3. A typical trajectory of a random process with finite correlation time

However, in actual physical problems there is almost always some ultimate time resolution at which the trajectory eventually becomes smooth. We can then determine a fine time scale such that within any smaller time intervals, the process $x(t)$ does not vary significantly (Fig. 8.3). This characteristic time-scale corresponds to the *correlation time* τ_c of a random process.

If we limit our discussion to continuous random processes with finite (but possibly very small) correlation times, the mathematical theory becomes greatly simplified. Indeed, in this case we can always choose such a fine time grading $\Delta t \ll \tau_c$ that any trajectory is well approximated by a piecewise-linear curve that links values $x(t_i)$ at successive moments $t_i = i\Delta t$. Therefore, it is sufficient to specify a joint probability distribution (8.2.1) for this particular time grading.

In principle, it is possible to perform, within such a discrete time grading, all operations required to determine statistical averages, correlation functions, etc. However, it would make the notation rather cumbersome. A better description is obtained if we use the formal concept of *path integration*.

Looking at the probability distribution (8.2.1) for a given continuous random process, one can usually split it into a product of the two terms. The first term can be interpreted as a value of a certain functional $P[x(t)]$ taken on a piecewise-linear trajectory $x_\Gamma(t)$ that links together all points $x(t_i)$. The second term should then include all the remaining factors, such as the product of dx_i and the normalization constants corresponding to a particular time grading. Denoting this last term as $\mathcal{D}x(t)$, we can write (8.2.1) as

$$P(\{x_i\}) \prod_i dx_i = P[x_\Gamma(t)] \mathcal{D}x(t) . \tag{8.2.2}$$

When very fine grading is used (so that $\Delta t \ll \tau_c$), any trajectory $x(t)$ of the random process can be well approximated by some piecewise-linear trajectory $x_\Gamma(t)$. Therefore the functional $P[(x(t)]$ for any such trajectory can be easily determined.

Note that, if the trajectories are not allowed to have large variations between the gates set at time moments $t_i = i\Delta t$, the system of such gates effectively defines a certain smooth tube. Hence, at sufficiently fine grading, the product $P[x(t)]\mathcal{D}x(t)$ would give the probability of finding a trajectory within such a tube. Because of this, $P[x(t)]$ can indeed be interpreted as the *probability functional* for a particular continuous random process. In the limit of very fine time grading, the term $\mathcal{D}x(t)$ is interpreted as the *"path differential"*.

It should be emphasized that the precise form of the terms included in the "path differential" $\mathcal{D}x(t)$ depends on the particular random process (in contrast to the usual differential dx that enters into the expression $p(x)dx$ in the case of a single random variable). Furthermore, the splitting of the probability into the product of $P[x(t)]$ and $\mathcal{D}x(t)$ is not unique. One must remember that it is only the combination of these terms that is meaningful.

When a probability distribution is known, it can be used to estimate various mean values. For instance, if we have a random quantity x with some probability distribution $p(x)$, we can find the mean value $\langle f(x) \rangle$ of any function $f(x)$ simply as

$$\langle f(x) \rangle = \int f(x)p(x)\,dx . \tag{8.2.3}$$

Likewise, for a random process $x(t)$ we can ask what would be the mean value $\langle F[x(t)] \rangle$ of some functional $F[x(t)]$. This mean value should be understood in the following sense. Let us take a narrow tube. Since the trajectories which belong to the tube are very close, they will correspond to practically the same value of the functional F. On the other hand, we know the probability that a trajectory will fall into this tube. Hence, to determine the mean value of the functional F we should simply sum its values for all possible tubes, multiplying each value by the probability assigned to the individual tube.

Now we can try to make this definition more rigorous. As noted above, for continuous random processes with finite correlation times, a tube can be specified by a set of gates $[x_i, x_i + dx_i]$ for a sufficiently fine time grading $t_i = i\Delta t$. The value of the functional F for the trajectories inside such a tube can be well approximated by its value $F[x_\Gamma(t)]$ for the piecewise-linear trajectory $x_\Gamma(t)$ that links points $x(t_i) = x_i$. "Summation" over all possible positions of the gates is equivalent to estimation of the multiple integral and thus the mean value of F is given by

$$\langle F[x(t)] \rangle = \int \cdots \int F[x_\Gamma(t)] P(\{x_i\}) \prod_i dx_i . \tag{8.2.4}$$

Varying the position of each gate independently, we scan all possible tubes. Hence, the right-hand side of (8.2.4) can be interpreted as an integral taken over all paths. It has a special notation

$$\int F[x(t)]P[x(t)]\mathcal{D}x(t) = \int \ldots \int F[x_\Gamma(t)]P(\{x_i\}\prod_i dx_i \ . \tag{8.2.5}$$

Consequently, the mean value of any functional can be estimated by calculating the path integral

$$\langle F[x(t)]\rangle = \int F[x(t)]P[x(t)]\mathcal{D}x(t) \ . \tag{8.2.6}$$

The probability functional $P[x(t)]$ (together with the path differential $\mathcal{D}x(t)$!) bears the complete information about the random process $x(t)$. However, it is often enough to have only the information contained in some of the *multi-moment* probability distributions.

We may ask what is the probability density that, at moment t_1, the trajectory of a random process passes through the point (x_1, t_1), so that $x(t_1) = x_1$. In order to find this probability density $p_1(x_1, t_1)$, we should sum the contributions from all trajectories which pass through the point (x_1, t_1), i.e. estimate the path integral

$$p_1(x_1, t_1) = \int \delta\big(x(t_1) - x_1\big)P[x(t)]\mathcal{D}x(t) \ . \tag{8.2.7}$$

Furthermore, one can define the two-moment probability distribution $p_2(x_1, t_1; x_2, t_2)$, which specifies the joint probability density of reaching the value x_1 at moment t_1 and the value x_2 at t_2. Now we should integrate $P[x(t)]$ over the bundle of trajectories that are fixed at the two points (x_1, t_1) and (x_2, t_2):

$$p_2(x_1, t_1; x_2, t_2) = \int \delta\big(x(t_1) - x_1\big)\delta\big(x(t_2) - x_2\big)P[x(t)]\mathcal{D}x(t) \ . \tag{8.2.8}$$

In a similar way one can determine other multi-moment distributions $p_M = p_M(x_1, t_1; x_2, t_2; \ldots; x_M, t_M)$.

Using multi-moment probability distributions, we can define *correlation functions* $\langle x(t_1)x(t_2) \ldots x(t_M)\rangle$ as

$$\langle x(t_1)x(t_2) \ldots x(t_M)\rangle = \int x_1 x_2 \ldots x_M$$
$$\times\, p_M(x_1, t_1; x_2, t_2; \ldots; x_M, t_M)dx_1 dx_2 \ldots dx_M \ . \tag{8.2.9}$$

A special role is played by the *mean value*

$$\langle x(t)\rangle = \int x p_1(x, t)dx \tag{8.2.10}$$

and the *pair correlation function*

$$\langle x(t_1)x(t_2)\rangle = \int x_1 x_2 p_2(x_1, t_1;\ x_2, t_2)dx_1 dx_2 \ . \tag{8.2.11}$$

Note that every random process $x(t)$ can be related to another process, $y(t) = x(t) - \langle x(t)\rangle$, which has a zero mean value. It can be proved that a random process is uniquely determined if all its correlation functions are known.

The variety of possible random processes is enormous. However, there is one particular class which is very common in practical applications. This is the class of *Gaussian random processes*. A Gaussian random process is completely determined by its first two correlation functions, i.e. by the mean value $\langle x(t)\rangle$ and the pair correlation function

$$\langle x(t_1)x(t_2)\rangle = \sigma(t_1, t_2) \ . \tag{8.2.12}$$

All correlation functions of higher orders can be expressed in terms of products of the first two correlators.

If a Gaussian random process has a zero mean value, its probability functional is given by

$$P[x(t)] = \exp\left[-\frac{1}{2}\int\int B(t, t')x(t)x(t')dt\, dt'\right] \tag{8.2.13}$$

where $B(t, t')$ is the inverse of the pair correlation function $\sigma(t, t')$, i.e.

$$\int \sigma(t, \tau)B(\tau, t')d\tau = \delta(t - t') \ . \tag{8.2.14}$$

A Gaussian distribution for a random quantity x is usually found when this quantity represents the sum of a large number of small independent random contributions (as stated in the "central limit theorem" of probability theory, e.g. [8.9]). A similar property holds for Gaussian random processes.

Consider an example of a random process which consists of a sequence of identical pulses appearing at independent random time moments t_j,

$$x(t) = \sum_j g(t - t_j) \ , \tag{8.2.15}$$

where the function $g = g(\tau)$ describes the form of an individual pulse. Suppose that the mean number of pulses per unit time is μ, whereas the characteristic width of a single pulse is Δ. In the limit of very frequent and strongly overlapping weak pulses ($\mu\Delta \gg 1$) this random process becomes Gaussian. Its mean value is

$$\langle x(t)\rangle = \mu S \tag{8.2.16}$$

and its pair correlation function $\langle x(t_1)x(t_2)\rangle = \sigma(t_1 - t_2)$ is given by

$$\sigma(t_1 - t_2) = \mu\chi(t_1 - t_2) + \mu^2 S^2 \ . \tag{8.2.17}$$

Here S is the total intensity of a single pulse,

$$S = \int g(\tau)d\tau ,$$

(8.2.18)

and $\chi(t_1 - t_2)$ represents the overlap between two pulses located at time moments t_1 and t_2, i.e.

$$\chi(t_1 - t_2) = \int g(t_1 - \tau)g(\tau - t_2)d\tau .$$

(8.2.19)

Note that, since this Gaussian random process is composed of pulses with finite width, it cannot undergo sharp variations within the time width of a single pulse. This means that the characteristic correlation time τ_c coincides for such process with the pulse width Δ. On the other hand, since the overlap (8.2.19) vanishes when two pulses are separated by a time much larger than Δ, this correlation time τ_c also determines the characteristic time scale for the pair correlation function (8.2.17).

Convergence to the Gaussian process is also observed for other random sequences of independent pulses. If the form and polarity of each pulse in the sequence are chosen at random, i.e.

$$x(t) = \sum_{j}(-1)^{\kappa_j} g_j(t - t_j) ,$$

(8.2.20)

where κ_j are independent random integers, the resulting Gaussian process has zero mean value. Its pair correlation function is

$$\sigma(t_1 - t_2) = \mu \left\langle \int g_j(t_1 - \tau)g_j(\tau - t_2)d\tau \right\rangle ,$$

(8.2.21)

where averaging over all possible choices of the shape function $g_j(\tau)$ is implied.

When pulses are very narrow (but still significantly overlapping), we can write approximately $g_j(\tau) = S_j \delta(\tau)$, where $\delta(\tau)$ is the delta function. In this case, (8.2.21) is also given by a delta function because

$$\sigma(t_1 - t_2) = \mu \langle S_j^2 \rangle \int \delta(t_1 - \tau)\delta(\tau - t_2)d\tau$$
$$= \mu \langle S_j^2 \rangle \delta(t_1 - t_2) .$$

(8.2.22)

The delta-correlated Gaussian random process with zero mean and the pair correlation function

$$\langle x(t_1)x(t_2)\rangle = \sigma \delta(t_1 - t_2)$$

(8.2.23)

is known as *white noise*. It has the probability functional

$$P[x(t)] = \exp\left[-\frac{1}{2\sigma}\int x^2(t)dt\right] .$$

(8.2.24)

When dealing with white noise, one should keep certain precautions in mind. From a formal point of view, its correlation time is zero and the trajectories of such a process are discontinuous at almost every point. However, we cannot then define correctly the usual integrals over such a process. The situation resembles to some extent the mathematical difficulties which arise when dealing with the Dirac delta function $\delta(x)$. Its singular properties forced mathematicians to develop a special theory of generalized functions. However, physisists usually treat the delta function on the same footing as ordinary functions, by assuming that it has some vanishingly small yet finite width.

The same kind of approach can be taken in operations involving white noise. Namely, we can consider it as an idealization of an actual random process with some finite but very small correlation time. If this correlation time is smaller than the available temporal resolution, on a macro-scale such a process would exhibit the same properties as ideal white noise. However, at the ultimate "micro" level it would have smooth trajectories, a fact which enables us to use the standard rules of differential calculus.

White noise is not the only example of a random process with vanishing correlation time. In effect, such processes constitute an entire class of Markov random processes; these will be discussed in the next section.

Investigations of continuous random processes represent an important branch of modern probability theory. A detailed introduction can be found in [8.13–20].

8.3 The Fokker-Planck Equation

If a random process has zero correlation time it cannot remember its previous history. This means that the probability $w(x, t|x', t')$ of transition from the state x' at t' to another state x at a later moment t, cannot depend on the values taken by the random process $x(t)$ before the time moment t'. A process with this property is known as a *Markov process*.

In close analogy with (8.1.12), the probability distribution for a Markov process satisfies

$$p(x, t) = \int w(x, t|x', t')p(x', t')dx' \tag{8.3.1}$$

for any two time moments t and t'. Proceeding from (8.3.1) we can try to obtain a differential equation that describes the time evolution of the probability distribution $p(x, t)$.

Let us take in (8.3.1) two close time moments $t + \Delta t$ and t, and define the characteristic function ϕ as

$$\phi(u, x') = \int_{-\infty}^{\infty} \exp[iu(x - x')]w(x, t + \Delta t|x', t)dx . \tag{8.3.2}$$

We can expand ϕ as a Taylor series

$$\phi = \sum_{s=0}^{\infty} \left(\frac{d^s\phi(u)}{du^s}\right)_0 \frac{u^s}{s!} , \tag{8.3.3}$$

where we assume $0! = 1$. The derivatives of ϕ can be estimated using the definition (8.3.2), i.e.

$$\frac{d^s\phi(u)}{du^s} = \int_{-\infty}^{\infty} i^s(x - x')^s \exp[iu(x - x')]$$
$$\times w(x, t + \Delta t|x', t)dx .$$ (8.3.4)

Therefore we find

$$\left(\frac{d^s\phi(u)}{du^s}\right)_0 = i^s\langle(x - x')^s\rangle ,$$ (8.3.5)

where $\langle(x - x')^s\rangle$ are the different statistical moments of the random process increment $x(t + \Delta t) - x(t)$ within the time interval Δt; they are given by the expressions

$$\langle(x - x')^s\rangle = \int_{-\infty}^{\infty} (x - x')^s w(x, t + \Delta t|x', t)dx .$$ (8.3.6)

Introducing for these moments the special notation

$$M_s(x', t) = \langle(x - x')^s\rangle ,$$ (8.3.7)

we can write (8.3.3) as

$$\phi(u, x') = 1 + \sum_{s=1}^{\infty} \frac{i^s}{s!} u^s M_s(x', t) .$$ (8.3.8)

This allows us to express the transition probability $w(x, t + \Delta t|x', t)$ in terms of the moments $M_s(x', t)$ by inverting (8.32), i.e.

$$w(x, t + \Delta t|x', t) = \frac{1}{2\pi} \int_{-\infty}^{\infty} \exp[-iu(x - x')]\phi(u, x')du .$$ (8.3.9)

Substitution of expansion (8.3.8) into (8.3.9) yields

$$w(x, t + \Delta t|x', t) = \delta(x - x') + \sum_{s=1}^{\infty} \frac{i^s}{s!} M_s(x', t)$$
$$\times \frac{1}{2\pi} \int_{-\infty}^{\infty} \exp[-iu(x - x')]u^s \, du .$$ (8.3.10)

The integrals in (8.3.10) can be transformed into the formal derivatives of the delta function,

$$\frac{1}{2\pi} \int_{-\infty}^{\infty} \exp[-iu(x - x')]u^s \, du = i^s \left(\frac{\partial}{\partial x}\right)^s \delta(x - x') .$$ (8.3.11)

We thereby obtain the following *Kramers-Moyal expansion* for the transition probability of a Markov random process

$$w(x, t + \Delta t | x', t) = \delta(x - x')$$

$$+ \sum_{s=1}^{\infty} \frac{(-1)^s}{s!} M_s(x', t) \left(\frac{\partial}{\partial x} \right)^s \delta(x - x') . \tag{8.3.12}$$

Substitution of this expansion into (8.3.1) gives

$$p(x, t + \Delta t) = p(x, t) + \sum_{s=1}^{\infty} \frac{(-1)^s}{s!} \frac{\partial^s}{\partial x^s} \left[M_s(x, t) p(x, t) \right] . \tag{8.3.13}$$

If we move the term $p(x, t)$ to the left-hand side of this equation, divide both sides by Δt and take the limit $\Delta t \to 0$, this results in the following equation for the probability distribution

$$\frac{\partial}{\partial t} p(x, t) = \sum_{s=1}^{\infty} \frac{(-1)^s}{s!} \frac{\partial^s}{\partial x^s} \left[K_s(x, t) p(x, t) \right] , \tag{8.3.14}$$

where we use the notation

$$K_s(x, t) = \lim_{\Delta t \to 0} \Delta t^{-1} M_s(x, t)$$

$$= \lim_{\Delta t \to 0} \Delta t^{-1} \langle [x(t + \Delta t) - x(t)]^s \rangle . \tag{8.3.15}$$

Thus we have derived a general differential equation that describes the temporal evolution of the probability distribution for an arbitrary Markov process. Equation (8.3.4) includes derivatives with respect to x of infinitely high orders. It seems natural to inquire whether all the terms with very high-order derivatives are really essential or whether we can truncate this expansion at some finite order.

The answer is given by the theorem of *Pawula* [8.21] (see also [8.18]) which states that this equation permits[2] only two possibilities: either the expansion is truncated at $s = 2$ (so that all further terms with $s \geqslant 3$ are absent) or it must include terms of indefinitely high order.

It turns out that the terms of all orders should be kept in (8.3.14) only if the random process $x(t)$ is discontinuous, i.e. if it has finite jumps. When continuous Markov random processes are considered, the evolution of the probability distribution is always described by the *Fokker-Planck equation*

$$\frac{\partial}{\partial t} p(x, t) = - \frac{\partial}{\partial x} \left[K_1(x, t) p(x, t) \right]$$

$$+ \frac{1}{2} \frac{\partial^2}{\partial x^2} \left[K_2(x, t) p(x, t) \right] , \tag{8.3.16}$$

which includes only first and second derivatives.

If a continuous random process has no memory, its trajectory looks locally like the trajectory of Brownian motion of a diffusing particle (Fig. 8.4). Indeed, the velocity dx/dt at any time moment then depends only on the position x at the previous time moment – which contains no information about the velocity at that

---

[2] Otherwise the condition that the probability distribution $p(x, t)$ is positive is violated.

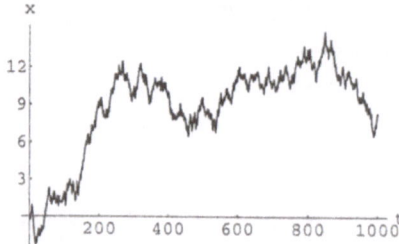

Fig. 8.4. A typical trajectory of a continuous random process without memory

time moment. Hence, the time dependence of the velocity for a typical realization of such a random process cannot retain continuity.

A complete survey of the properties of the Fokker-Planck equation and a discussion of related problems is given by *Risken* [8.18]. These questions are also discussed in many other textbooks, such as [8.13–17, 19]. In our derivation of the Fokker-Planck equation, we have followed *Stratonovich* [8.16].

9. Active Systems with Noise

The application of external noise causes a dynamical system to wander in its phase space. Such random wandering is superimposed on the steady drift produced by the deterministic dynamics. The resulting process is similar to the Brownian motion of a particle in the presence of a permanent driving force.

In this chapter we describe the mathematical framework for the analysis of noise-induced behavior in distributed active systems. It will be used further in the following chapters to describe the emergence of complex patterns through selective amplification of fluctuations induced by noise.

9.1 Generalized Brownian Motion

Since standard Brownian motion serves as a good illustration of general noise-induced behavior, we begin this section with a brief outline of the theory of this phenomenon (a detailed discussion can be found in e.g. [9.1–3]).

Brownian motion is performed by a heavy particle immersed in a liquid. The particle experiences frequent collisions with the molecules of the liquid. Each collision results in a weak impulse acting briefly on the particle. The total random force $F(t)$ consists of the sum of such weak impulses; it is this force that brings about stochastic wandering of the particle.

Because the particle is relatively large, it will experience many collisions with different molecules at the same time and the pulses which form the force $F(t)$ will be strongly overlapping. All these pulses are independent because collisions are not correlated. Taking these into account, we expect that $F(t)$ will be a Gaussian random process.

In addition to the random force F, any particle moving through the liquid also feels the force $F_v = -\gamma v$ of viscous friction, which is proportional to its velocity v. If the viscosity coefficient γ is very large, the force F is almost compensated by the frictional force F_v, so that we have approximately $v = F/\gamma$. The Brownian motion of the particle is then described by

$$\dot{x} = f(t) \,, \tag{9.1.1}$$

where $f(t) = F(t)/\gamma$.

The correlation time τ_c of the Gaussian random process $f(t)$ is determined by the characteristic time of individual collisions. If we do not wish to resolve time moments closer than τ_c, this random process can be approximated by a Gaussian process with vanishing correlation time, i.e. by white noise:

$$\langle f(t) \rangle = 0, \quad \langle f(t)f(t') \rangle = 2\sigma\delta(t - t') . \tag{9.1.2}$$

It is well known that the Brownian motion described by (9.1.1) gives rise to diffusion of the particle. Proceeding from (9.1.1), we can derive the equation governing the evolution of the probability density $p(x, t)$.

In the limit of vanishing correlation times τ_c, the random process $x(t)$ has no memory, i.e. it approaches some Markov random process. Indeed, we have

$$x(t_2) = x(t_1) + \int_{t_1}^{t_2} f(t') \, dt' . \tag{9.1.3}$$

It follows from (9.1.3) that $x(t_2)$ does not explicitly depend on the values of $x(t)$ at moments $t < t_1$ earlier than t_1. Moreover, in the limit considered $x(t_2)$ cannot be statistically correlated with $x(t)$ at $t < t_1$ through correlations in the values of f, as would have been the case for a finite correlation time τ_c of the random force $f(t)$.

Since $x(t)$ is a Markov random process, its probability density $p(x, t)$ should obey (8.3.14). We can determine the coefficients K_s in this equation using their definition (8.3.15) directly.

As follows from (9.1.3),

$$\langle [x(t + \Delta t) - x(t)]^s \rangle = \left\langle \left[\int_t^{t+\Delta t} f(t') dt' \right]^s \right\rangle . \tag{9.1.4}$$

Since all odd-order correlation functions of the white noise $f(t)$ are zero, we see that the moments (9.1.4) are non-vanishing only for even integers s. Moreover, we can use the Gaussian property of white noise and express any of its even-order correlation functions as a sum over all possible decompositions into a product of its pair correlation functions $\delta(t_i - t_k)$. Taking this into account, one finds from (9.1.4) and (9.1.2):

$$\langle [x(t + \Delta t) - x(t)]^s \rangle = \frac{s(s - 1)}{2}(2\sigma\Delta t)^{s/2} . \tag{9.1.5}$$

We see that the sth moment is proportional to $\Delta t^{s/2}$. Definition (8.3.15) of the coefficients K_s then implies that all such coefficients with $s > 2$ should vanish and that $K_2 = 2\sigma$. Thus the probability density $p(x, t)$ obeys the Fokker-Planck equation

$$\frac{\partial p}{\partial t} = \sigma\frac{\partial^2 p}{\partial x^2} , \tag{9.1.6}$$

which is simply the diffusion equation.

In a more general situation, the particle may experience not only the random force $f(t)$, but also some permanent driving force q,

$$\dot{x} = q + f(t) \ . \tag{9.1.7}$$

By similar arguments one then derives the equation

$$\frac{\partial p}{\partial t} = -\frac{\partial(qp)}{\partial x} + \sigma \frac{\partial^2 p}{\partial x^2} \ . \tag{9.1.8}$$

This differs from (9.1.6) by the presence of the additional *drift* term, which describes the permanent driven component of motion. Thus the motion consists of the steady drift in the direction of q superimposed on the random wandering produced by the white noise $f(t)$.

The same diffusion equation (9.1.8) is valid if the regular driving force depends on the position x of the particle, i.e. if $q = q(x)$. This is because, in deriving (9.1.8), we need only the local value of the driving force q near a given point x.

We conclude that the stochastic differential equation

$$\dot{x} = q(x) + f(t) \tag{9.1.9}$$

determines a diffusive process whose probability density obeys (9.1.8). Obviously, this conclusion is not directly related to the particular mechanical interpretation that we have given to (9.1.9). Actually, this equation may describe the action of white noise on any dynamical system, regardless of its origin. The only special assumption is that this noise enters *additively* into the differential equation, i.e. $f(t)$ is not multiplied by a function of x.

Next we discuss the situation with *multiplicative* noise. We now assume that the stochastic differential equation is

$$\dot{x} = q(x) + g(x)f(t) \ , \tag{9.1.10}$$

where $f(t)$ is the white noise with correlators (9.1.2) and $q(x)$ and $g(x)$ are known functions of x. Such stochastic differential equations are called *Langevin equations* (this class also includes equations with additive noise). A detailed discussion of the properties of Langevin equations is given in [9.4–9]. Below we merely derive a few results that will be used in the subsequent analysis.

The essential new feature of the general case (9.1.10) is that the multiplicative noise can itself contribute to the drift term in the Fokker-Planck equation. To illustrate the origins of this contribution, we consider first an example of the differential equation

$$\dot{x} = \pm x f_1(t) \ , \tag{9.1.11}$$

where $f_1(t)$ represents a single short pulse centered at $t = t_1$. If this pulse of width $\tau_{\rm c}$ lies completely inside the interval from t to $t + \Delta t$, the solution to (9.1.11) is

$$x(t + \Delta t) = C_{\pm} x(t) \ , \quad \text{where} \tag{9.1.12}$$

$$C_\pm = \exp\left[\pm \int_t^{t+\Delta t} f_1(t')dt'\right] \tag{9.1.13}$$

are some constant factors. Therefore, when $\Delta t \gg \tau_c$ and $t + \Delta t > t_1 > t$, the increment of x would be

$$x(t + \Delta t) - x(t) = (C_\pm - 1)x(t) . \tag{9.1.14}$$

Consider now two identical pulses differing only in their polarity, i.e. in the sign in (9.1.11). The positive pulse increases x, while the negative pulse decreases it, as seen from (9.1.14). However, (9.1.14) in combination with (9.1.13) shows that the absolute value of the difference $|x(t + \Delta t) - x(t)|$ is not the same in the two cases. The positive pulse leads to an increase in x that is larger than the decrease caused by the same pulse with a negative polarity. Thus, if we have a random sequence of independent pulses with alternating polarities, the asymmetry of their effects would result in the statistical domination of drift towards the greater positive values of x.

Similar arguments can be applied for a general stochastic differential equation (9.1.10). Below, following *Stratonovich* [9.4], we determine the drift and diffusion coefficients in the Fokker-Planck equation corresponding to (9.1.10). Note that (9.1.10) is equivalent to the integral equation

$$x(t) = x(t_0) + \int_{t_0}^t q(x(\tau))d\tau + \int_{t_0}^t g(x(\tau))f(\tau)d\tau , \tag{9.1.15}$$

which can be used to calculate iteratively $x(t_0 + \Delta t)$ at sufficiently small values of Δt.

The zeroth iteration of (9.1.15) is $x^{(0)}(t) = x(t_0)$. To obtain the first iteration $x^{(1)}(t)$, we substitute $x^{(0)}$ into the right-hand side of (9.1.15) for $x(\tau)$:

$$x^{(1)}(t) = q(x(t_0))(t - t_0) + g(x(t_0))\int_{t_0}^t f(\tau)d\tau . \tag{9.1.16}$$

The second iteration is obtained if we substitute $x^{(0)} + x^{(1)}$ for $x(\tau)$:

$$\begin{aligned}
x^{(2)}(t) &= q'(x(t_0))\int_{t_0}^t x^{(1)}(\tau)\,d\tau + g'(x(t_0))\int_{t_0}^t x^{(1)}(\tau)f(\tau)\,d\tau \\
&= q'(x(t_0))q(x(t_0))\tfrac{1}{2}(t - t_0)^2 \\
&\quad + q'(x(t_0))g(x(t_0))\int_{t_0}^t \int_{t_0}^\tau f(\tau_1)d\tau_1\,d\tau \\
&\quad + g'(x(t_0))q(x(t_0))\int_{t_0}^t (\tau - t_0)f(\tau)d\tau \\
&\quad + g'(x(t_0))g(x(t_0))\int_{t_0}^t \int_{t_0}^\tau f(\tau)f(\tau')d\tau\,d\tau' .
\end{aligned} \tag{9.1.17}$$

Here $q' = dq/dx$ and $g' = dg/dx$. Note that, since $f(t)$ may vary rapidly, we leave it under the sign of the integral.

According to (8.3.15), the drift coefficient K_1 is given by

$$K_1 = \lim_{\Delta t \to 0} \frac{1}{\Delta t} \langle (x(t + \Delta t) - x(t) \rangle . \tag{9.1.18}$$

On the other hand, combining the results of the first two iterations, we have

$$x(t + \Delta t) = x(t) + x^{(1)}(t + \Delta t) + x^{(2)}(t + \Delta t) . \tag{9.1.19}$$

Now we must substitute (9.1.16) and (9.1.17) into (9.1.19), and then use it to estimate K_1 with the help of (9.1.18). Although the resulting expression has many different terms, only a few yield a nonvanishing contribution into K_1. All the terms that include only one factor f should vanish after averaging because $\langle f \rangle = 0$. Furthermore, in the remaining terms we should keep only the contributions which do not vanish in the limit $\Delta t \to 0$. Thereby we find

$$K_1(x) = q(x) + g'(x)g(x) \lim_{\Delta t \to 0} \Delta t^{-1} \int_t^{t+\Delta t} \int_t^{\tau} \langle f(\tau)f(\tau') \rangle d\tau \, d\tau' . \tag{9.1.20}$$

The last term in (9.1.20) describes drift induced by the multiplicative noise. Using (9.1.2), we can write the integral in this terms as

$$I = \int_t^{t+\Delta t} \int_t^{\tau} \langle f(\tau)f(\tau') \rangle d\tau \, d\tau' = 2\sigma \int_t^{t+\Delta t} \int_t^{\tau} \delta(\tau - \tau') d\tau \, d\tau' . \tag{9.1.21}$$

The evaluation of such an integral requires some care. We should calculate the integral of the delta function $\delta(\tau - \tau')$ over τ' with the upper limit equal to τ. However, the delta function is singular and it diverges at precisely $\tau' = \tau$! Therefore, some additional convention must be brought in to make this integral meaningful.

Recall now that our white noise is actually an idealization of a Gaussian random process with some vanishingly small, but still finite, correlation time τ_c. Therefore, its correlation function is proportional to a "smoothed" delta function that has a width of about τ_c and does not go to infinity at its center. For such a function, the required integral can be easily calculated; it is equal to 1/2. Hence, we find $I = \sigma \Delta t$ and, consequently[1],

$$K_1(x) = q(x) + \sigma g'(x)g(x) . \tag{9.1.22}$$

In a similar way we find the diffusion coefficient

$$K_2(x) = 2\sigma[g(x)]^2 . \tag{9.1.23}$$

One can easily verify that all other coefficients K_s vanish, so that the probability density $p(x, t)$ for this random process obeys the Fokker-Planck equation (8.3.15).

---

[1] The limit $\Delta t \to 0$ in (9.1.20) should now be interpreted in the sense of an intermediate asymptotic, as the limit of very small values of Δt still exceeding τ_c.

In some applications it is desirable to know the correlator $\langle x(t)f(t)\rangle$. This correlator can be calculated directly from the stochastic differential equation (9.1.10). Consider some earlier moment $t_0 = t - \Delta t$, with Δt much smaller than the characteristic variation time of x, but still large in comparison to the actual correlation time of the random force. Then, using (9.1.16), we can obtain $x(t)$ as a first iteration:

$$x(t) = q\big(x(t_0)\big)\big(t - t_0\big) + g\big(x(t_0)\big) \int_{t_0}^{t} f(\tau)d\tau \ . \tag{9.1.24}$$

Therefore, taking into account that the mean value of the random force is zero, we find

$$\langle x(t)f(t)\rangle = g(x(t)) \int_{t-\Delta t}^{t} \langle f(t)f(\tau)\rangle d\tau \ . \tag{9.1.25}$$

The pair correlation function $\langle f(t)f(\tau)\rangle$ is proportional to $\delta(t - \tau)$. Substituting this into (9.1.25), we see that we need to calculate the integral

$$\int_{t-\Delta t}^{t} \delta(t - \tau)d\tau \ , \tag{9.1.26}$$

where the upper integration limit coincides with the point of singularity.

Recall that actually the correlation function $\langle f(t)f(\tau)\rangle$ is smoothed and has a finite width about τ_c. This implies that $\delta(t - \tau)$ in (9.1.26) should be considered simply as a strong narrow peak with a width much smaller than Δt. With this convention, the integral (9.1.26) contains *half* of a delta function and is therefore equal to 1/2. From this we obtain

$$\langle x(t)f(t)\rangle = \sigma g(x(t)) \ . \tag{9.1.27}$$

Above we derived the expression for the drift coefficient $K_1(x)$ under the assumption that the multiplicative noise $f(t)$ has some vanishingly small, but still finite, correlation time τ_c. This additional convention is known as the *Stratonovich interpretation* [9.4, 10] of the stochastic differential equation (9.1.10). Since any realistic noise that enters into the dynamical equations usually has trajectories that are ultimately smooth and continuous, this interpretation is quite natural and should be applied in the majority of practical problems.

There is, however, a special class of problems where a different approach known as the *Ito interpretation* [9.11] should be used. In this interpretation, the drift coefficient lacks the correction due to the multiplicative noise, i.e.

$$K_1(x) = q(x) \ . \tag{9.1.28}$$

The Ito interpretation is applicable [9.12, 13] when the original description is constructed for discrete time and the noise is genuinely uncorrelated even for the closest time moments.

Suppose, for instance, that X_n is a population of insects in the nth generation and that their population in the next generation (e.g., next year) is determined by the random map

$$X_{n+1} = X_n + Q(X_n)\xi_n , \qquad (9.1.29)$$

where ξ_n are independent Gaussian random variables,

$$\langle \xi_n \rangle = 0 , \quad \langle \xi_n \xi_m \rangle = 0 , \quad n \neq m . \qquad (9.1.30)$$

If the population X_n changes only slightly from one generation to the next, the discrete time dynamics of such a system can approximately be modeled by a stochastic differential equation of the type (9.1.10). However, in this case the Ito interpretation of this equation must be chosen.

When a stochastic differential equation is understood by the Ito interpretation, one cannot so freely use the standard rules of differential calculus as was done above in the case of the Stratonovich interpretation. There are mathematical concepts which have been specially developed to deal with this kind of equation. Nonetheless, once the transition to the Fokker-Planck equation has been performed, both cases can be treated in the same manner. For the Ito interpretation, the coefficients of the related Fokker-Planck equation are determined by (9.1.28) and (9.1.23). If the noise is additive, i.e. $g(x) = 1$, stochastic drift is always absent, as it evident from (9.1.22) and (9.1.28), and the difference between the two interpretations is then negligible.

The above results are easily generalized for dynamical systems with many degrees of freedom. The set of stochastic Langevin equations

$$\dot{x}_i = q_i(x) + g_i(x)f_i(t) , \quad i = 1, 2, \ldots, N \qquad (9.1.31)$$

with delta-correlated random forces $f_i(t)$, such that

$$\langle f_i(t) \rangle = 0 , \quad \langle f_i(t)f_j(t') \rangle = 2\sigma_{ij}\delta(t - t') , \qquad (9.1.32)$$

generates a continuous Markov process whose probability distribution $p(x, t)$ satisfies the Fokker–Planck equation

$$\frac{\partial p}{\partial t} = -\sum_i \frac{\partial}{\partial x_i}\left[K_i^{(1)}(x)p \right] + \frac{1}{2}\sum_{i,j} \frac{\partial^2}{\partial x_i \partial x_j}\left[K_{i,j}^{(2)}(x)p \right] . \qquad (9.1.33)$$

When the Ito interpretation of the stochastic differential equations (9.1.31) is used, the coefficients of the Fokker–Planck equation are

$$K_i^{(1)}(x) = q_i(x) , \quad K_{ij}^{(2)}(x) = 2\sigma_{ij}g_i(x)g_j(x) . \qquad (9.1.34)$$

In the Stratonovich interpretation, these coefficients have the form

$$K_i^{(1)}(x) = q_i(x) + \sum_j \sigma_{ij}\frac{\partial g_i(x)}{\partial x_j}g_j(x) , \quad K_{ij}^{(2)}(x) = 2\sigma_{ij}g_i(x)g_j(x) . \quad (9.1.35)$$

Hence, the drift coefficient $K_j^{(1)}$ then includes a contribution due to multiplicative noises.

9.2 Internal Noise

It was tacitly assumed in Sect. 9.1 that noise has an external origin. In fact, we considered a dynamical system that was part of some larger system (i.e. of the "environment") with intrinsically chaotic behavior. Since the small subsystem could not exhibit any significant backward influence, the action of the fluctuating environment was described as some noise with the given, externally defined, statistical properties. Such noise induces fluctuations in the system variables.

In contrast to this, *internal noise*, discussed below, does not correspond to the action of any real chaotic environment. Its introduction is merely a convenient mathematical trick which allows us, in some cases, to describe probabilistic discrete automata as if they were dynamical systems with continuous variables subject to external chaotic influence.

The concept of internal noise can be illustrated by several characteristic examples. In Sect. 8.1 we described a stochastic oscillator representing a two-state probabilistic automaton with discrete time dynamics. Now we want to consider a large population of such stochastic oscillators. It will be assumed that the transition probabilities w^+ and w^- of upward and downward transitions depend on the average activity $u = M/N$ of the population at a given time (here M is the number of active automata and N is the total number of automata in the whole population). We show in a straightforward manner (following [9.14]) that the behavior of such a large interacting population can be approximately described by an iterative map with some weak "internal" noise.

Let $P_n(M)$ be the probability of finding M active elements at some moment n. This same probability at the next moment $n + 1$ is given by

$$P_{n+1}(M) = \sum_{M'} W(M|M')P_n(M') , \qquad (9.2.1)$$

where $W(M|M')$ is the transition probability. We can estimate $W(M|M')$ from simple combinatorial arguments. Suppose that $M > M'$, i.e. the number of active elements increases between time instants n and $n + 1$. The change from M' to M in the total activity can be achieved in many different ways. Indeed, we can choose at random any k active elements and require that they go into the passive state, while all other $M' - k$ active elements should remain in the active state. These downward transitions should be accompanied by exactly $M - M' + k$ upward transitions in the originally passive elements.

Therefore the global transition probability $W(M|M')$ for $M > M'$ is given by

$$W(M|M') = \sum_{k=0}^{N-M} (w^-)^k (1 - w^-)^{M'-k} \frac{M'!}{k!(M'-k)!}$$
$$\times (w^+)^{M-M'+k}(1 - w^+)^{N-M-k} \frac{(N-M')!}{(N-M-k)!(M-M'+k)!}$$
$$(9.2.2)$$

where we have taken into account that there are $M'!/[k!(M'-k)!]$ different ways to choose k of M' elements. The respective transition probability for $M < M'$ is

$$W(M|M') = \sum_{k=0}^{M}(w^-)^{M'-M+k}(1 - w^-)^{M-k}\frac{M'!}{(M-k)!(M'-M+k)!}$$

$$\times (w^+)^k(1 - w^+)^{N-M'+k}\frac{(N-M')!}{k!(M-M'+k)!} \ . \tag{9.2.3}$$

We can now notice that, if the total number of automata is very large ($N \gg 1$), the average activity $u = M/N$ changes almost continuously. Hence, the equation for the probability distribution $p_n(u)$ can be approximately written in the form

$$p_{n+1}(u) = \int_0^1 G(u|u')p_n(u')du' \ . \tag{9.2.4}$$

We can derive the transition probability $G(u|u')$ for $u' > u$ from (9.2.3) applying the approximation $n! \approx (2\pi n)^{1/2}(n/e)^n$. This yields

$$G(u|u') = \int_0^{1-u}[2\pi u'(1 - u')w^+(1 - w^+)w^-(1 - w^-)]^{-1/2}$$

$$\times \exp\left\{-\frac{N}{2}\left(\frac{[u - u' + y - (1 - u')w^-]^2}{(1 - u')w^+(1 - w^+)} + \frac{(y - u'w^-)^2}{u'w^-(1 - w^-)}\right)\right\}dy \ . \tag{9.2.5}$$

When $N \gg 1$, the exponent in (9.2.5) has a very sharp maximum at a point where the derivative of its argument vanishes. Hence, we can estimate this integral using the saddle-point approximation. It yields the following expression for $G(u|u')$:

$$G(u|u') = (2\pi N)^{1/2}[(1 - u')w^+(1 - w^+) + u'w^-(1 - w^-)]^{-1/2}$$

$$\times \exp\left\{-\frac{N}{2}\frac{[u - (u' + (1 - u')w^+ - u'w^-)]^2}{(1 - u')w^+(1 - w^+) + u'w^-(1 - w^-)}\right\} \ . \tag{9.2.6}$$

The same expression is obtained for $u' < u$. Note that the probabilities of upward and downward transitions depend on the average activity in the population, i.e. $w^+ = w^+(u')$ and $w^- = w^-(u')$.

Thus, time evolution of the probability distribution $p_n(u)$ for the average activity u in the population of stochastic two-state oscillators is described by (9.2.4) with the transition probability given approximately by (9.2.6). Moreover, when $N \gg 1$ this transition probability is vanishingly small everywhere except within some narrow interval about a point u (for a given u') where the exponent argument approaches zero. Therefore we expect that the time evolution of the activity u should be almost deterministic in the limit of large population numbers N.

Let us try to reproduce the behavior of the activity u by a map including some weak Gaussian noise,

$$u_{n+1} = F(u_n) + g(u_n)\xi_n \ , \tag{9.2.7}$$

where ξ_n are independent Gaussian variables with the probability distribution

$$p(\xi_n) = (2\pi)^{-1/2} \exp\left(-\frac{\xi_n^2}{2}\right) . \tag{9.2.8}$$

Since the variables ξ_n are not correlated for different moments of time, u_n represents a Markov process. Its probability distribution $p_n(u)$ should obey

$$p_{n+1}(u) = \int_0^1 \overline{W}(u|u')p_n(u')du' . \tag{9.2.9}$$

To determine the function $\overline{W}$, we note that $\overline{W}(u|u')du$ gives the probability of finding u_{n+1} in the interval $u < u_{n+1} < u + du$ provided that $u_n = u'$. Introducing $u_n = u'$ and $u_{n+1} = u$, we can express ξ_n from (9.2.7) as

$$\xi_n = \frac{u - F(u')}{g(u')} . \tag{9.2.10}$$

Since the distribution of ξ_n is known, we can find, using (9.2.8) and (9.2.10), the distribution of u for a given u', i.e. the transition probability for this Markov process:

$$\overline{W}(u|u') = \left[(2\pi)^{1/2}g(u')\right]^{-1} \exp\left\{-\frac{[u - F(u')]^2}{2[g(u')]^2}\right\} . \tag{9.2.11}$$

To put the random map (9.2.7) into approximate correspondence with the actual random process in the population of probabilistic automata, we can choose $F(u)$ and $g(u)$ in such a way that

$$\overline{W}(u|u') = G(u|u') . \tag{9.2.12}$$

This equality yields

$$F(u) = u + (1 - u)w^+ - uw^- , \tag{9.2.13}$$

$$g(u) = N^{-1/2}[(1 - u)w^+(1 - w^+) + uw^-(1 - w^-)] . \tag{9.2.14}$$

It follows from (9.2.14) that $g(u) \to 0$ in the limit $N \to \infty$. The map for the time evolution of the activity u then becomes purely deterministic,

$$u_{n+1} = F(u_n) . \tag{9.2.15}$$

At large but finite values of N, map (9.2.7) includes a small multiplicative internal noise $g(u_n)\xi_n$.

The above example illustrates how internal noise appears in the effective description of large populations of synchronous probabilistic automata. But the introduction of an effective internal noise can also be performed in situations where transitions are asynchronous, i.e. where they can occur at any randomly chosen time moments. A special (and very important) example of such situations is provided by chemically reacting systems.

Any single element of a reaction is a probabilistic event, leading to the spontaneous creation of new particles and/or the disappearance of existing ones. These events are the fundamental "transitions" of such systems. If we neglect the finite lifetime of a single reaction element and assume that its probability depends only on the relative positions of the particles, the reaction can be modeled as a discrete Markov process.

Hence, a comprehensive microscopic description of a reacting system consists in specifying the set of distribution functions $\{P_N/r_1, \ldots, r_N; t)\}$, each of which gives the probability of finding $N = 0, 1, 2, \ldots$ particles in the system at points with coordinates $r_1, \ldots, r_N$. The time evolution of these probability distributions due to reactions and the diffusion of particles should be described by a master equation. Usually, however, reacting systems are described by deterministic kinetic equations for continuous concentrations of reagents. This is possible because fluctuations often vanish after averaging over volume elements containing large numbers of particles.

Nonetheless, in some special circumstances, fluctuations can be selectively enhanced, resulting in the formation of macroscopic patterns in the spatial distribution of reagents. These processes may be studied directly using the framework of microscopic master equations [9.15–18]. However, a better approach would be to use an intermediate *mesoscopic* description, formulated in terms of smooth fluctuating concentrations of reagents. Then the fluctuations, caused by the atomistic nature of reactions, are modeled by introducing some internal noise into the kinetic equations.

We begin the analysis with the simplest case, namely, we assume that the reaction consists of the independent generation of single particles at an average rate of w_0 particles per unit time. In this case the probability $p(n, t)$ of finding n particles in the volume at time t obeys the following exact master equation (cf. Sect. 8.1):

$$\frac{\partial p(n, t)}{\partial t} = w_0 p(n - 1, t) - w_0 p(n, t) . \tag{9.2.16}$$

If the number of particles in the volume is sufficiently large, the quantity n can be approximated by a continuously varying variable, and we can write

$$p(n - 1) = p(n) - \frac{\partial p}{\partial n} + \frac{1}{2} \frac{\partial^2 p}{\partial n^2} + \ldots . \tag{9.2.17}$$

For smooth distributions $p(n)$ the terms with higher derivatives in expansion (9.2.17) can be ignored. Substituting (9.2.17) into (9.2.16) we find an approximate Fokker-Planck equation for this reaction:

$$\frac{\partial p}{\partial t} = -\frac{\partial}{\partial n} (w_0 p) + \frac{1}{2} \frac{\partial^2}{\partial n^2} (w_0 p) . \tag{9.2.18}$$

We can now try to find a stochastic differential equation that would produce this Fokker-Planck equation for the probability distribution. As follows from Sect. 9.1, this stochastic differential equation is

$$\dot{n} = w_0 + w_0^{1/2} f_0(t) ,\tag{9.2.19}$$

where $f_0(t)$ is white noise with unit intensity,

$$\langle f_0(t) f_0(t') \rangle = \delta(t - t') .\tag{9.2.20}$$

Next we consider the decay reaction. Let us denote by w_1 the decay rate, i.e. the probability that an individual particle will decay within unit time. The respective master equation is then

$$\frac{\partial p(n, t)}{\partial t} = w_1 (n + 1) p(n + 1, t) - w_1 n p(n, t) .\tag{9.2.21}$$

In the first term w_1 is multiplied by $n + 1$ because each of the available $n + 1$ particles can decay within a small time interval. The same argument explains the presence of the factor n in the last term. Carrying out an expansion of $p(n + 1)$ at large values of n by analogy with (9.2.17), we again find a Fokker-Planck equation

$$\frac{\partial p}{\partial t} = \frac{\partial (w_1 n p)}{\partial n} + \frac{1}{2} \frac{\partial^2 (w_1 n p)}{\partial n^2} .\tag{9.2.22}$$

It corresponds to the following stochastic differential equation (interpreted in the sense of Ito)

$$\dot{n} = -w_1 n + (w_1 n)^{1/2} f_1(t) ,\tag{9.2.23}$$

where $f_1(t)$ is again white noise with unit intensity.

Finally, for the reproduction reaction $X \rightarrow 2X$ the master equation is written as

$$\frac{\partial p(n, t)}{\partial t} = w_2 (n - 1) p(n - 1, t) - w_2 n p(n, t) ,\tag{9.2.24}$$

where w_2 is the probability of replication of an individual X particle per unit time. The respective approximate Fokker-Planck equation reads

$$\frac{\partial p}{\partial t} = -\frac{\partial (w_2 n p)}{\partial n} + \frac{1}{2} \frac{\partial^2 (w_2 n p)}{\partial n^2} .\tag{9.2.25}$$

It corresponds (in the Ito interpretation) to a stochastic differential equation

$$\dot{n} = w_2 n + (w_2 n)^{1/2} f_2(t)\tag{9.2.26}$$

with white noise $f_2(t)$ of unit intensity.

Note that the deterministic limits of (9.2.19), (9.2.23) and (9.2.26) with the noise terms omitted coincide with the standard "kinetic" equations for these three reactions. The introduction of internal noise, represented by the latter terms, allows us to approximate the effect of fluctuations due to the atomistic nature of these reactions.

In a similar way one can introduce internal noise for other kinds of reactions. When there are several different reactions, each of them gives an indepedent contribution to the noise term.

A distributed medium can be approximately described as a set of boxes inside which reactions occur. The particles can also move at random from one box to another, resulting in their diffusion. The internal noise corresponding to reactions in different boxes is not correlated because the elementary reaction events are local.

The diffusion process also gives rise to a certain internal noise that should be added to the usual diffusion equation for the local concentration of particles. Indeed, the usual equation

$$\dot{n} = D\Delta n \tag{9.2.27}$$

describes the damping with time of all concentration inhomogeneities. However, such inhomogeneities are permanently created by the diffusion process, even if the initial distribution was completely uniform. Since diffusion effectively results from the independent random wandering of individual particles, it can easily happen that at some time instant one of the boxes will be populated by a larger number of particles than another.

Below we derive, following *Mikhailov* [9.19], the stochastic differential equation that describes concentration fluctuations caused by the diffusion of particles.

We begin with a discrete version of the problem and first consider a one-dimensional medium, which can be represented as a linear chain of boxes with indices $j = 0, \pm 1, \pm 2, \ldots$ The state of such a system is specified by giving the set of numbers n_j of particles in each of the boxes. Suppose that, with a probability w per unit time, a particle can independently jump into one of the neighboring boxes in the chain. The master equation for this random process is

$$\frac{\partial p(\{n_j\})}{\partial t} = w \sum_j [(n_j + 1)p(/n_{j-1} - 1, n_j + 1/)$$
$$+ (n_j + 1)p(/n_j + 1, n_{j+1} - 1/) - 2n_j p(\{n_j\})] . \tag{9.2.28}$$

Here we are using the notation $p(/n_{j-1} - 1, n_j + 1/)$, which shows that in the set of occupation numbers $\{n_j\}$ there are changes of ± 1 only in the numbers of particles in the boxes $j - 1$ and j.

If the occupation numbers n_j are large, they can be treated as continuous variables and we have, approximately,

$$p(/n_{j-1} - 1, n_j + 1/)$$
$$\approx p - \frac{\partial p}{\partial n_{j-1}} + \frac{\partial p}{\partial n_j} + \frac{1}{2}\frac{\partial^2 p}{\partial n_j^2} + \frac{1}{2}\frac{\partial^2 p}{\partial n_{j-1}^2} - \frac{\partial^2 p}{\partial n_j \partial n_{j-1}} + \ldots . \tag{9.2.29}$$

A similar decomposition holds for $p(/n_j + 1, n_{j+1} - 1/)$. Substituting these expressions into (9.2.28) and keeping only terms up to the second derivatives, we obtain a Fokker–Planck equation for the distribution function:

$$\frac{\partial p}{\partial t} = - w \sum_j \frac{\partial}{\partial n_j} [(n_{j+1} + n_{j-1} - 2n_j)p]$$

$$+ \frac{1}{2} w \sum_j \frac{\partial^2}{\partial n_j^2} [(n_{j+1} + n_{j-1} + 2n_j)p]$$

$$- \frac{1}{2} w \sum_j \frac{\partial^2}{\partial n_j \partial n_{j-1}} (2n_j p) - \frac{1}{2} w \sum_j \frac{\partial^2}{\partial n_j \partial n_{j+1}} (2n_j p) . \qquad (9.2.30)$$

The next step is to switch from the discrete description, where the medium is partitioned into a sequence of boxes, to a continuous description in terms of a smooth concentration $n(x)$. The probability distribution function $p(\{n_j\})$ then converts into a functional $p[n(x)]$ that gives the probability density of various realizations of the concentration field $n(x)$. The multidimensional Fokker–Planck equation (9.2.30) transforms into a certain functional Fokker–Planck equation. As shown below, this equation is

$$\frac{\partial p}{\partial t} = - D \int dx \frac{\delta}{\delta n(x)} \left(\frac{\partial^2 n}{\partial x^2} p \right)$$

$$+ D \int\int dx dy \frac{\delta^2}{\delta n(x) \delta n(y)} \left(\frac{\partial^2}{\partial x \partial y} (n(x)\delta(x - y))p \right) , \qquad (9.2.31)$$

where $D = wl^2$ is the diffusion constant and l is the size of a box.

To validate (9.2.31), we construct a discrete version of this equation and demonstrate that it is equivalent to equation (9.2.30).

After introduction of a discrete description, the first term in (9.2.31) takes the form

$$D \int dx \frac{\delta}{\delta n(x)} \left(\frac{\partial^2 n}{\partial x^2} p \right) = w \sum_j \frac{\partial}{\partial n_j} [(n_{j+1} + n_{j-1} - 2n_j)p] \qquad (9.2.32)$$

which is identical to the first term on the right-hand side of (9.2.30).

The discretization of the last term in (9.2.31) is a more complicated procedure. It is convenient to introduce the operators

$$\hat{A}(x) \equiv \frac{\delta}{\delta n(x)} , \quad \hat{A}(y) \equiv \frac{\delta}{\delta n(y)} \qquad (9.2.33)$$

and to write this term as

$$J = D \int\int dx dy \hat{A}(x) \hat{A}(y) \frac{\partial^2}{\partial x \partial y} (n(x)\delta(x - y))p . \qquad (9.2.34)$$

Integrating by parts in (9.2.34), we find (assuming that the concentration $n(x)$ vanishes at $x = \pm\infty$)

$$J = D \int\int dx dy \frac{\partial \hat{A}(x)}{\partial x} \frac{\partial \hat{A}(y)}{\partial y} n(x)\delta(x - y)p = D \int dx \left(\frac{\partial \hat{A}}{\partial x} \right)^2 np . \qquad (9.2.35)$$

Note that we have

$$
\left(\frac{\partial \hat{A}}{\partial x}\right)^2 n = \frac{\partial \hat{A}}{\partial x}\frac{\partial}{\partial x}(\hat{A}n) - \frac{\partial \hat{A}}{\partial x}\hat{A}\frac{\partial n}{\partial x}
$$

$$
= \frac{\partial \hat{A}n}{\partial x}\frac{\partial}{\partial x}(\hat{A}) + \frac{1}{2}\hat{A}^2\frac{\partial^2 n}{\partial x^2} - \frac{1}{2}\frac{\partial}{\partial x}\left(\hat{A}^2\frac{\partial n}{\partial x}\right). \tag{9.2.36}
$$

The last term here represents a total derivative and it disappears after integration over x. Thus we see that, restoring the original notations, the last term in the functional Fokker–Planck equation (9.2.31) can also be written in the form

$$
J = D\int dx\frac{\partial}{\partial x}\frac{\delta}{\delta n(x)}\frac{\partial}{\partial x}\left(\frac{\delta}{\delta n(x)}n(x)\right)p
$$

$$
+ \frac{1}{2}D\int dx\frac{\delta^2}{\delta n^2(x)}\left(\frac{\partial^2 n}{\partial x^2}p\right), \tag{9.2.37}
$$

where the operators perform differentation with respect to x only on the immediately following expressions in the parentheses.

The discrete version of (9.2.37) is

$$
J = w\sum_j\left(\frac{\partial}{\partial n_{j+1}} - \frac{\partial}{\partial n_j}\right)\left(\frac{\partial}{\partial n_{j+1}}(n_{j+1}p) - \frac{\partial}{\partial n_j}(n_j p)\right)
$$

$$
+ \frac{1}{2}w\sum_j\frac{\partial^2}{\partial n_j^2}[(n_{j+1} + n_{j-1} - 2n_j)p]
$$

$$
= \frac{1}{2}\sum_j\frac{\partial^2}{\partial n_j^2}[(n_{j+1} + n_{j-1})p] + w\sum_j\frac{\partial^2}{\partial n_{j+1}^2}(n_{j+1}p)
$$

$$
- w\sum_j\frac{\partial^2}{\partial n_{j+1}\partial n_j}(n_j p) - w\sum_j\frac{\partial^2}{\partial n_j\partial n_{j+1}}(n_{j+1}p). \tag{9.2.38}
$$

It is easy to see that, after a change of the summation indices, this expression becomes the same as the last three terms in equation (9.2.30).

Hence, the functional Fokker–Planck equation (9.2.31) indeed describes fluctuations of the continuous population density field $n(x, t)$ caused by the atomistic nature of the diffusion process. Note that, although we have explicitly considered only the one-dimensional case, the same derivation also holds for systems of a higher dimensionality.

As we have seen in Sect. 9.1, there is a direct correspondence between stochastic Langevin equations and the Fokker–Planck equations describing the same continuous random Markov process. This relationship can be straightforwardly generalized to the case of fluctuating fields, where fluctuations are described by stochastic partial differential equations and the probability distribution for the fluctuations is given by a probability functional obeying the Fokker–Planck equation with the functional derivatives. Such a stochastic partial differential equation, corresponding to the Fokker–Planck equation (9.2.31), is

$$\dot{n} = D\nabla^2 n + \text{div}\,[(2Dn)^{1/2} \boldsymbol{f}(\boldsymbol{r}, t)]\,, \tag{9.2.39}$$

where the random Gaussian vector force $\boldsymbol{f}(\boldsymbol{r}, t)$ is characterized by the correlators

$$\langle f^{(\alpha)}(\boldsymbol{r}, t) f^{(\beta)}(\boldsymbol{r}', t')\rangle = \delta_{\alpha\beta}\delta(\boldsymbol{r} - \boldsymbol{r}')\delta(t - t') \tag{9.2.40}$$

and $\alpha, \beta = (x, y, z)$. This stochastic differential equation should be interpreted in the Ito sense.

The stochastic diffusion equation (9.2.39) can also be written as the continuity equation

$$\frac{\partial n}{\partial t} + \text{div}\,j = 0\,, \tag{9.2.41}$$

where the flux j is given by a sum of the deterministic and stochastic components,

$$j = -D\nabla n - (2Dn)^{1/2}\boldsymbol{f}(\boldsymbol{r}, t)\,. \tag{9.2.42}$$

The first term here represents the deterministic contribution due to a spatial gradient of the concentration. The second term is the internal noise of diffusion.

Above, we separately considered three types of processes, namely birth, death, and diffusion of particles, and determined the internal noises corresponding to each of them. When all of these processes are simultaneously present in the system, their contributions are additive. This means that the stochastic differential equation for the continuous concentration $n(\boldsymbol{r}, t)$ of particles X, taking part in the reactions $X \rightarrow 2X$, $X \rightarrow 0$ and performing random Brownian motion in the medium, has the form

$$\dot{n} = (w_2 - w_1)n + D\nabla^2 n + (w_1 n)^{1/2} f_1(\boldsymbol{r}, t) + (w_2 n)^{1/2} f_2(\boldsymbol{r}, t)$$
$$+ \text{div}\,[(2Dn)^{1/2}\boldsymbol{f}(\boldsymbol{r}, t)]\,, \tag{9.2.43}$$

where $f_1(\boldsymbol{r}, t)$, $f_2(\boldsymbol{r}, t)$ and $\boldsymbol{f}(\boldsymbol{r}, t)$ are independent Gaussian random forces whose correlation functions are given by

$$\langle f_1(\boldsymbol{r}, t) f_1(\boldsymbol{r}', t')\rangle = \delta(\boldsymbol{r} - \boldsymbol{r}')\delta(t - t')\,,$$
$$\langle f_2(\boldsymbol{r}, t) f_2(\boldsymbol{r}', t')\rangle = \delta(\boldsymbol{r} - \boldsymbol{r}'\delta(t - t') \tag{9.2.44}$$

and by equation (9.2.40).

Thus, the internal noise of reactions and diffusion is multiplicative; it is proportional to the square root of the local concentration of particles.

9.3 Optimal Fluctuations and Transition Probabilities

The action of noise gives rise to fluctuations in the behavior of a dynamical system, i.e., to deviation of its trajectories from the phase trajectory corresponding to the deterministic dynamics. When the influence of noise is investigated, it is usually assumed that this noise is sufficiently weak (otherwise it would completely overwhelm the original dynamics of the system). In this case, the typical fluctuating trajectories still lie close to the deterministic trajectory. The probability of other trajectories that represent large deviations from the deterministic path is extremely small.

Nonetheless, these nontypical and rare trajectories may play an important role in the evolution of a particular system. Such trajectories can visit regions of phase space that are never approached by the deterministic trajectory, even allowing a wide range of initial conditions. Visits to these regions might have drastic consequences, such as the extinction of one species and/or the explosive growth of others, or qualitative changes in the system's dynamics. Therefore, it is useful to have mathematical methods which allow us to estimate the probability of rare fluctuations caused by weak noises.

We consider in this section a dynamical system described by stochastic differential equations

$$\dot{x}_i = q_i(\boldsymbol{x}) + g_i(\boldsymbol{x})f_i(t) , \quad i = 1, 2, ..., N , \tag{9.3.1}$$

where the Gaussian noises $f_i(t)$ have finite correlation times (such noises are called *colored*). Since the correlation time is finite, (almost) all realizations of $f_i(t)$ are smooth functions. Hence, for any particular realization, equation (9.3.1) represents simply some set of ordinary differential equations which have phase trajectories specified by the initial conditions. The correlation functions of the noises are taken in the form

$$\langle f_i(t)f_i(t')\rangle = \varepsilon\sigma_i(t - t')\delta_{ij} , \tag{9.3.2}$$

where ε is a small parameter, $\varepsilon \ll 1$.

When $\varepsilon = 0$, equations (9.3.1) reduce to a system of deterministic differential equations

$$\dot{x}_i = q_i(\boldsymbol{x}) , \quad i = 1, 2, ..., N . \tag{9.3.3}$$

For fixed initial conditions, they specifiy a single phase-space trajectory $\boldsymbol{x} = \boldsymbol{x}_0(t)$ that constitutes the deterministic solution.

Suppose that an important outcome of the considered stochastic process consists in reaching, by time T, a certain point Q in the phase space that lies far from the deterministic trajectory $\boldsymbol{x}_0(t)$. There is a continuous family of trajectories that lead to Q from the given initial point. All these trajectories are (exponentially) rare in the limit $\varepsilon \to 0$ of the weak noise. However, among them one can find the most probable, or *optimal*, trajectory that would give the dominant contribution to the probability of reaching the distant phase-space point Q. To determine such a trajectory, we investigate below a general solution of the stochastic differential equation (9.3.1) constructed in the form of a path integral.

Path-integral solutions to stochastic differential equations have been introduced in various contexts by *Martin* et al. [9.20], *Graham* [9.21], *Janssen* [9.22], and *Phytian* [9.23] (see also *Hänggi* [9.24]). Such solutions for the problem with the colored noise have been constructed by *Pesquera* et al. [9.25] and for stochastic partial differential equations of reaction–diffusion systems by *Förster* and *Mikhailov* [9.26].

To derive the path-integral solution, we first replace the stochastic differential equations (9.3.1) by their finite-difference approximation,

$$x_{i,n} - x_{i,n-1} = q_i(\boldsymbol{x}_{n-1})\Delta t + g_i(\boldsymbol{x}_{n-1})\xi_{i,n} . \tag{9.3.4}$$

Here the time interval from 0 to T is divided into N subintervals $\Delta t = T/N$ by points $t_n = n\Delta t$, $n = 1, 2, ..., N$.

The noise is represented in (9.3.4) by the Gaussian random variables $\xi_{i,n}$ with the correlators

$$\langle \xi_{i,n}\xi_{j,n'} \rangle = \varepsilon\sigma_{nn'}^{(i)}\delta_{ij} , \tag{9.3.5}$$

where $\sigma_{nn'}^{(i)} = \sigma_i(t_n - t_{n'})\Delta t^2$. These variables have the joint probability distribution

$$P(\{\xi_{i,n}\}) = \prod_i (2\pi\varepsilon)^{N/2}(\det\hat{\sigma}_i)^{-1/2}\exp\left[-\frac{1}{2\varepsilon}\sum_{n.n'}(\hat{\sigma}_i^{-1})_{nn'}\xi_{i,n}\xi_{i,n'} \right] , \tag{9.3.6}$$

where $\hat{\sigma}_i$ is the matrix with the elements $\sigma_{nn'}^{(i)}$ and $\hat{\sigma}_i^{-1}$ is its inverse.

Now we can note that, in effect, equation (9.3.4) defines a transformation of variables $\{\xi_{i,n}\} \rightarrow \{x_{i,n}\}$ Hence, if we know the probability distribution for the variables $\xi_{i,n}$ we can find the probability distribution $P(\{x_{i,n}\})$ by using the equation

$$P(\{x_{i,n}\}) = P(\{\xi_{i,n}\})Y , \tag{9.3.7}$$

where $\xi_{i,n}$ should be expressed in terms of $x_{i,n}$ by means of (9.3.4) and Y is the Jacobian of this transformation,

$$Y = \prod_i \det\left\| \frac{\partial\xi_{i,n}}{\partial x_{i,n'}} \right\| . \tag{9.3.8}$$

Introducing delta functions, equation (9.3.7) can be written in an equivalent explicit form

$$P(\{x_{i,n}\}) = \int \prod_{i,n} d\xi_{i,n}$$
$$\times \delta\left(\xi_{i,n} - \frac{1}{g_i(\boldsymbol{x}_{n-1})}(x_{i,n} - x_{i,n-1} - \Delta t q_i(\boldsymbol{x}_{n-1})) \right) P(\{\xi_{i,n}\})Y . \tag{9.3.9}$$

Applying the property $\delta(y/a) = a\delta(y)$ of the delta function, we can also write this equation as

$$P(\{x_{i,n}\}) = \int \prod_{i,n} d\xi_{i,n} \delta(g_i(\boldsymbol{x}_{n-1})\xi_{i,n} - x_{i,n} + x_{i,n-1} + \Delta t q_i(\boldsymbol{x}_{n-1}))$$

$$\times P(\{\xi_{i,n}\}) \left[\prod_{i,n} g_i(\boldsymbol{x}_{n-1}) \right] Y \ . \tag{9.3.10}$$

The Jacobian Y is given by the determinant of the matrix with the elements $\frac{\partial \xi_{i,n}}{\partial x_{i',n'}}$. We can find them using equation (9.3.4), which expresses $\xi_{i,n}$ in terms of the variables x. Thus we obtain

$$\frac{\partial \xi_{i,n}}{\partial x_{i',n}} = \frac{1}{g_i(\boldsymbol{x}_{n-1})}\delta_{ii'} \ ,$$

$$\frac{\partial \xi_{i,n}}{\partial x_{i',n-1}} = \frac{\partial}{\partial x_{i',n-1}} \left\{ \frac{1}{g_i(\boldsymbol{x}_{n-1})}[x_{i,n} + x_{i,n-1} - \Delta t q_i(\boldsymbol{x}_{n-1})] \right\} \ , \tag{9.3.11}$$

$$\frac{\partial \xi_{i,n}}{\partial x_{i',n'}} = 0 \ , \quad \text{for } n' \neq n, n-1 \ .$$

The determinant of this matrix is equal to a product of the elements on the main diagonal, i.e.

$$Y = \prod_i \det \left\| \frac{\partial \xi_{i,n}}{\partial x_{i,n}} \right\| = \prod_{i,n} \frac{1}{g_i(\boldsymbol{x}_{n-1})} \ . \tag{9.3.12}$$

Substituting this into (9.3.10) yields

$$P(\{x_{i,n}\}) = \int \prod_{i,n} d\xi_{i,n}$$

$$\times \delta(g_i(\boldsymbol{x}_{n-1})\xi_{i,n} - x_{i,n} + x_{i,n-1} + \Delta t q_i(\boldsymbol{x}_{n-1}))P(\{\xi_{i,n}\}) \ . \tag{9.3.13}$$

Next we make use of the identity

$$\delta(\alpha) = (2\pi)^{-1} \int_{-\infty}^{\infty} e^{i\alpha p} dp \tag{9.3.14}$$

and, introducing additional integration variables $p_{i,n}$, write equation (9.3.13) as

$$P(\{x_{i,n}\}) = (2\pi)^{-N} \int_{-\infty}^{\infty} \prod_{i,n} d\xi_{i,n} \int_{-\infty}^{\infty} \prod_{i,n} dp_{i,n}$$

$$\times \exp\left\{ i \sum_{i,n} p_{i,n}[x_{i,n} - x_{i,n-1} - \Delta t q_i(\boldsymbol{x}_{n-1}) - g_i(\boldsymbol{x}_{n-1})\xi_{i,n}] \right\} P(\{\xi_{i,n}\}) \ . \tag{9.3.15}$$

Here the probability distribution $P(\{\xi_{i,n}\})$ is given by (9.3.6).

The multiple integral over the variables $\xi_{i,n}$ in (9.3.15) is Gaussian and therefore it can be calculated:

$$
\int_{-\infty}^{\infty} \prod_{i,n} d\xi_{i,n} \exp\left[-i \sum_{i,n} p_{i,n} g_i(\boldsymbol{x}_{n-1})\xi_{i,n} - \frac{1}{2\varepsilon} \sum_{i,n,n'} (\hat{\sigma}_i^{-1})_{nn'}\xi_{i,n}\xi_{i,n'} \right]
$$
$$
= \prod_i (2\pi\varepsilon)^{N/2}(\det\hat{\sigma}_i)^{1/2}
$$
$$
\times \exp\left[-\frac{1}{2} \sum_{n,n'} \varepsilon\sigma_{i,nn'} g_i(\boldsymbol{x}_{n-1})g_i(\boldsymbol{x}_{n'-1})p_{i,n}p_{i,n'} \right] . \tag{9.3.16}
$$

Thus, the probability distribution for the discrete version (9.3.4) of the considered system of stochastic differential equations (9.3.1) is

$$
P(\{x_{i,n}\})\prod_{i,n} dx_{i,n} = \int_{-\infty}^{\infty} \exp\left\{ i \sum_{i,n} p_{i,n}[x_{i,n} - x_{i,n-1} - \Delta t q_{i,n}(\boldsymbol{x}_{n-1})] \right\}
$$
$$
\times \exp\left[-\frac{1}{2} \sum_{i,n,n'} \varepsilon\sigma_{i,nn'} g_i(\boldsymbol{x}_{n-1})g_i(\boldsymbol{x}_{n'-1})p_{i,n}p_{i,n'} \right] \prod_{i,n} \frac{dp_{i,n}}{2\pi} \prod_{i,n} dx_{i,n} . \tag{9.3.17}
$$

Our aim is to construct the probability functional distribution $P[\boldsymbol{x}(t)]\,\mathcal{D}\boldsymbol{x}(t)$ that determines the probabilities of realization of various trajectories. We have shown in Sect. 8.2 that such functional probability distributions can be obtained as a formal limit of probability distributions for the discretized equations, when increasingly fine discretization is performed. Thus, in the formal limit of (9.3.17) for $\Delta t \to 0$ we obtain

$$
P(\{x_{i,n}\})\prod_{i,n} dx_{i,n} \to P[\boldsymbol{x}(t)]\mathcal{D}\boldsymbol{x}(t) \tag{9.3.18}
$$

where the probability functional $P[\boldsymbol{x}(t)]$ is given by a path integral over an additional set of variables $\boldsymbol{p}(t) = \{p_i(t)\}$, i.e.,

$$
P[\boldsymbol{x}(t)] = \int \mathcal{D}\boldsymbol{p}(t) \exp\left\{ i \sum_i \int_0^T p_i(t)[\dot{x}_i(t) - q_i(\boldsymbol{x}(t))]dt \right.
$$
$$
\left. -\frac{1}{2}\varepsilon \sum_i \int_0^T \int_0^T \sigma_i(t - t')g_i(\boldsymbol{x}(t))g_i(\boldsymbol{x}(t'))p_i(t)p_i(t')dt dt' \right\}. \tag{9.3.19}
$$

The path differentials are determined as the formal limits

$$
\prod_{i,n} dx_{i,n} \to \mathcal{D}\boldsymbol{x}(t) , \quad \prod_{i,n} \frac{dp_{i,n}}{2\pi} \to \mathcal{D}\boldsymbol{p}(t) . \tag{9.3.20}
$$

The integrations in (9.3.19) are performed over a set of real variables $p_{i,n}$. Equivalently, the integrand can be viewed as an analytical function of $p_i = \mathrm{Re}\, p_i + i\,\mathrm{Im}\, p_i$ with the integration running over a surface representing the plane $\mathrm{Im}\, p_i = 0$ in the space of complex variables $p_{i,n}$. As is well known from the theory of complex variables, we can deform the integration surface to which all paths $p(t)$ in (9.3.19) should belong without changing the value of this integral. Thus, instead of integration over all trajectories that lie in the real plane, we can also integrate over trajectories on a surface that coincides almost anywhere with the *imaginary* plane $\mathrm{Re}\, p_i = 0$ and approaches the real plane $\mathrm{Im}\, p_i = 0$ only in the distant region of large absolute values of p_i. Then it is convenient to replace variables $p_i(t)$ in (9.3.19) by $p_i(t) = i\varrho_i(t)$ and write this path integral in the form:

$$P[x(t)] = \int \mathcal{D}\varrho(t) \exp[-S] \,, \tag{9.3.21}$$

where the functional S is given by

$$S = \sum_i \left[\int_0^T dt\, \varrho_i(t)\dot{x}_i(t) - \frac{1}{2}\varepsilon \int_0^T \int_0^T dt_1 dt_2 \sigma_i(t_1 - t_2) \right.$$
$$\left. \times\, g_i(x(t_1))g_i(x(t_2))\varrho_i(t_1)\varrho_i(t_2) - \int_0^T dt\, \varrho_i(t)q_i(x(t)) \right] \,. \tag{9.3.22}$$

The most probable trajectory connecting two points in the phase space within time T corresponds to the minimum of the functional S with respect to the variables $x_i(t)$ and $\varrho_i(t)$. Hence, the functional derivatives of S over these variables should vanish when they are taken on such a trajectory, i.e., it is determined by equations

$$\frac{\delta S}{\delta x_i(t)} = 0 \,, \qquad \frac{\delta S}{\delta \varrho_i(t)} = 0 \,. \tag{9.3.23}$$

Substituting expression (9.3.22) for the functional S and taking functional derivatives, we obtain the equations[2]

$$\dot{x}_i = q_i(x) + \varepsilon g_i(x) \int_0^T \sigma_i(t - t')g_i(x(t'))\varrho_i(t')dt' \,. \tag{9.3.24}$$

$$\dot{\varrho}_i = \sum_j \frac{\partial q_j}{\partial x_i}\varrho_j - \varepsilon \sum_j \frac{\partial g_j}{\partial x_i}\varrho_j \int_0^T \sigma_j(t - t')g_j(x(t'))\varrho_j(t')dt' \,. \tag{9.3.25}$$

---

[2] They are valid if the random process is defined on the interval of time between $t = 0$ and $t = T$, i.e., if the initial and final time moments of the optimal trajectory coincide with the entire random process. The variational equations should be modified (see [9.27]) if the initial and final time moments of the transition caused by a colored noise lie inside the time interval where the random process is defined.

The optimal trajectory connecting points $x(t = 0) = x_0$ and $x(t = T) = x_Q$ is determined by equations (9.3.24) and (9.3.25) with the additional conditions $x = x_0$ at $t = 0$ and $x = x_Q$ at $t = T$.

Because the optimal trajectory gives the dominant contribution to the probability of transition, this probability can be approximately estimated as

$$\pi(x_0, 0 | x_Q, T) \propto \exp\{-S[x_{\text{opt}}(t), \varrho_{\text{opt}}(t)]\} \tag{9.3.26}$$

where $x_{\text{opt}}(t)$ and $\varrho_{\text{opt}}(t)$ represent the optimal trajectory. Substituting $\dot{x}_i$ from equation (9.3.24) into (9.3.22), we find that

$$S[x_{\text{opt}}(t), \varrho_{\text{opt}}(t)] = \frac{1}{2}\varepsilon \sum_i \int_0^T \int_0^T$$
$$\times \sigma_i(t - t')g_i(x(t))g_i(x(t'))\varrho_i(t)\varrho_i(t)\varrho_i(t')dtdt' , \tag{9.3.27}$$

where $x(t)$ and $\varrho(t)$ are taken along the optimal trajectory given by equations (9.3.24) and (9.3.25).

Note that our analysis is valid only for (exponentially) rare fluctuations and thus the condition $S[x_{\text{opt}}(t), \varrho_{\text{opt}}(t)] \gg 1$ must be satisfied in order to apply the estimate (9.3.26). As shown below, it indeed holds in the limit of weak noise when $\varepsilon \ll 1$.

Let us consider the behavior of (9.3.24) and (9.3.25) in the limit $\varepsilon \to 0$. If we simply drop the terms proportional to ε in equation (9.3.24), it would reduce to the deterministic equation for the variables x_i, which does not allow a transtion from x_0 to x_Q within time T. This means that, for an optimal trajectory, ϱ_i should increase when $\varepsilon \to 0$. If we put

$$\varrho_i(t) = \varepsilon^{-1}\mu_i(t) \tag{9.3.28}$$

and substitute this into (9.3.24) and (9.3.25), we find the equations

$$\dot{x}_i = q_i(x) + g_i(x)\int_0^T \sigma_i(t - t')g_i(x(t'))\mu_i(t')dt' , \tag{9.3.29}$$

$$\dot{\mu}_i = -\sum_j \frac{\partial q_j}{\partial x_i}\mu_j - \sum_j \frac{\partial g_j}{\partial x_i}\mu_j \int_0^T \sigma_j(t - t')g_j(x(t'))\mu_j(t')dt' , \tag{9.3.30}$$

which do not include the small parameter ε. Their solution yields an optimal trajectory in the space of variables $x(t)$ and $\mu(t)$ which does not depend on the noise intensity ε. If we take $\varrho_{\text{opt}}(t) = \varepsilon^{-1}\mu_{\text{opt}}(t)$, where $\mu_{\text{opt}}(t)$ is determined by this solution, and substitute this into (9.3.26) and (9.3.27), we obtain in the limit of weak noise the following estimate for the probability of rare transitions:

$$\pi(x_0, 0 | x_Q, T) \propto \exp\left(-\frac{\Phi}{\varepsilon}\right) , \tag{9.3.31}$$

where Φ does not depend on ε and is given by

$$\Phi = \frac{1}{2}\sum_i \int_0^T \int_0^T \sigma_i(t-t')g_i(\boldsymbol{x}(t))g_i(\boldsymbol{x}(t'))\mu_i(t)\mu_i(t')dtdt' \ . \tag{9.3.32}$$

Here $\boldsymbol{x}$ and $\boldsymbol{\mu}$ are taken along the optimal trajectory.

Thus, equations (9.3.31) and (9.3.32), together with the variational equations (9.3.29) and (9.3.30) with the conditions $\boldsymbol{x}(t=0) = \boldsymbol{x}_0$ and $\boldsymbol{x}(t=T) = \boldsymbol{x}_Q$, determine the probability of transitions for a general system of stochastic differential equations (9.3.1) in the limit of weak Gaussian colored noise. An example of the application of these equations in a particular problem is given by *Förster* and *Mikhailov* [9.28].

Now, when the solutions are constructed, we can also consider them when the correlation time of the noise goes to zero and hence

$$\langle f_i(t)f_j(t')\rangle = \varepsilon\sigma_i\delta(t-t')\delta_{ij} \ . \tag{9.3.33}$$

In this limit we have a set of stochastic Langevin equations in the Stratonovich interpretation. Now the variational equations for the optimal transition trajectory take the form

$$\dot{x}_i = q_i(\boldsymbol{x}) + \sigma_i[g_i(\boldsymbol{x})]^2\mu_i \tag{9.3.34}$$

$$\dot{\mu}_i = -\sum_j \frac{\partial q_j}{\partial x_i}\mu_j - \sum_j \sigma_j \frac{\partial g_j}{\partial x_i}g_j\mu_j^2 \ . \tag{9.3.35}$$

Note that these variational equations can also be written as the equations of motion

$$\dot{x}_i = \frac{\partial H}{\partial \mu_i} \ , \qquad \dot{\mu}_i = -\frac{\partial H}{\partial x_i} \tag{9.3.36}$$

of a mechanical system with coordinates x_i, momenta μ_i and the Hamilton function

$$H = \sum_i \left[q_i(\boldsymbol{x})\mu_i + \frac{1}{2}\sigma_i(g_i(\boldsymbol{x})\mu_i)^2 \right] \tag{9.3.37}$$

Therefore, these equations possess a conserved quantity $E = H(\boldsymbol{x}, \boldsymbol{\mu})$ which here plays the role of "energy".

When the optimal trajectory is known, the probability $\pi(\boldsymbol{x}_0, 0|\boldsymbol{x}_Q, T)$ of transition from point $\boldsymbol{x}_0$ to point $\boldsymbol{x}_Q$ within time T can be estimated as $\exp(-\Phi/\varepsilon)$, where

$$\Phi = \frac{1}{2}\sum_i \sigma_i \int_0^T [g_i(\boldsymbol{x}(t))\mu_i(t)]^2 dt \tag{9.3.38}$$

and the integral is taken along the optimal trajectory.

The above results refer to the limit of vanishingly small correlation times, i.e., they were derived for the Stratonovich interpretation. However, as we see by comparing equations (9.1.34) and (9.1.35), the difference between the predictions made using Ito and Stratonovich interpretations becomes negligible in the limit of weak noise. Therefore, the same estimate (9.3.38) for the transition probabilities and the same variational equations (9.3.34) and (9.3.35) are also expected to hold in the Ito interpretation.

Generally, we can also consider transitions between any two attractors of a system and determine the frequency of such transitions. Suppose that the original deterministic system (9.3.3) has two attractors Γ_1 and Γ_2, each with its own basin of attraction. When noise is applied, the system occasionally jumps from one attractor to another. The transition frequency $\omega(\Gamma_1, \Gamma_2)$ is the probability per unit time that a jump occurs from Γ_1 to Γ_2. Note that the actual duration T of a jump does not enter into this definition. To estimate $\omega(\Gamma_1, \Gamma_2)$, variational equations for the optimal trajectories can be used.

Looking at equations (9.3.34) and (9.3.35), we see that any trajectory that starts on the hyperplane $\mu = 0$ will lie completely on it. Moreover, the equations of motion on this plane are reduced to the deterministic equations of motion (9.3.1). Therefore any attractor Γ of the original deterministic dynamical system is at the same time an attractor for motion confined to the plane $\mu = 0$.

However, the equations (9.3.34) and (9.3.35) are Hamiltonian [see (9.3.36)]. They conserve the phase volume and no attractors are possible here. This means that any invariant set Γ which is attractive for motion on the plane $\mu = 0$, must be unstable with respect to motion in at least some of the directions that are orthogonal to this plane. Thus, any attractor of the original deterministic system corresponds to a *saddle* set of the enlarged system described by equations (9.3.34) and (9.3.35).

When we talk about a transition from one attractor to another, it is assumed that the system follows for a long time the deterministic motion on the attractor Γ_1 but then performs a transition and subsequently continues its motion on the attractor Γ_2. Hence, the optimal trajectory should satisfy the asymtotic conditions $x(t) \rightarrow \Gamma_1$ and $\mu(t) \rightarrow 0$ for $t \rightarrow -\infty$ and $x(t) \rightarrow \Gamma_2$ and $\mu(t) \rightarrow 0$ for $t \rightarrow +\infty$. Thus, it begins from the attractor Γ_1 on the plane $\mu = 0$, then deviates from this plane, so that finite values of μ are taken, but eventually returns to the initial plane $\mu = 0$ and converges there to the attractor Γ_2. In other words, it represents a *heteroclinic trajectory* that connects two saddle sets $(\Gamma_1, \mu = 0)$ and $(\Gamma_2, \mu = 0)$ of the enlarged dynamical system. The probability that such such a transition occurs is estimated as

$$\omega(\Gamma_1, \Gamma_2) \propto \exp\left[-\frac{1}{2\varepsilon}\sum_i \sigma_i \int_{-\infty}^{\infty} g_i(x)^2 \mu_i^2 dt\right]. \tag{9.3.39}$$

Although the integral in this equation has infinite limits, its value is still finite. Indeed, most of the time the optimal trajectory lies exponentially close to the plane $\mu = 0$, where the integrand vanishes; it deviates siginificantly from this plane only during the actual transition event. Note that our method is applicable so long as the frequency of transitions remains exponentially low; it does not allow one to determine the prefactor of the exponential term in the estimate (9.3.39).

To conclude our analysis we give, without a derivation (see [9.19]), the corresponding results for stochastic reaction–diffusion systems. Suppose that a system is described by *stochastic partial differential equations*.

$$\frac{\partial w_i}{\partial t} = q_i(\boldsymbol{w}) + D_i \nabla^2 w_i + g_i(\boldsymbol{w}) f_i(\boldsymbol{r}, t) , \quad i = 1, 2, \ldots, N , \tag{9.3.40}$$

where the random forces $f_i(\boldsymbol{r}, t)$ have correlators

$$\langle f_i(\boldsymbol{r}, t) f_j(\boldsymbol{r}', t') \rangle = \varepsilon \sigma_i \delta_{ij} \delta(t - t') \delta(\boldsymbol{r} - \boldsymbol{r}') \tag{9.3.41}$$

and the limit of weak noise $\varepsilon \to 0$ is considered. Then, the probability of various realizations of the field $\boldsymbol{w}(\boldsymbol{r}, t)$ is determined by a probability functional $P[\boldsymbol{w}(\boldsymbol{r}, t)]$ that is given by the path integral

$$P[\boldsymbol{w}(\boldsymbol{r}, t)] = \int \mathcal{D}\varrho(\boldsymbol{r}, t)$$

$$\times \exp \left\{ - \int d\boldsymbol{r} dt [-H + \sum_i \dot{w}_i(\boldsymbol{r}, t) \varrho_i(\boldsymbol{r}, t)] \right\} , \tag{9.3.42}$$

where

$$H = \sum_i \left\{ \varrho_i q_i(\boldsymbol{w}) + D_i \varrho_i \nabla^2 w_i + \frac{1}{2} \varepsilon \sigma_i \varrho_i^2 [g_i(\boldsymbol{w})]^2 \right\} . \tag{9.3.43}$$

Then, the probability of a transition from the initial spatial distribution $\boldsymbol{w} = \boldsymbol{w}_0(\boldsymbol{r})$ at $t = 0$ to the final spatial distribution $\boldsymbol{w} = \boldsymbol{w}_Q(\boldsymbol{r})$ at $t = T$ is determined by the optimal trajectory that obeys the following set of partial differential equations:

$$\dot{w}_i = q_i(\boldsymbol{w}) + D_i \nabla^2 w_i + \sigma_i [g_i(\boldsymbol{w})]^2 \mu_i , \tag{9.3.44}$$

$$\dot{\mu}_i = - \sum_j \frac{\partial q_j}{\partial w_i} \mu_j - D_i \nabla^2 \mu_i - \sum_j \sigma_j \frac{\partial g_j}{\partial w_i} g_j \mu_j^2 . \tag{9.3.45}$$

These equations include additional field variables $\mu(\boldsymbol{r}, t) = \varepsilon \varrho(\boldsymbol{r}, t)$. The optimal trajectory is given by the solution satisfying conditions $\boldsymbol{w} = \boldsymbol{w}_0(\boldsymbol{r})$ at $t = 0$ and $\boldsymbol{w} = \boldsymbol{w}_Q(\boldsymbol{r})$ at $t = T$. The transition probability can be estimated as

$$\pi(\boldsymbol{w}_0, 0 | \boldsymbol{w}_Q, T) \propto \exp(-\Phi/\varepsilon) , \tag{9.3.46}$$

where

$$\Phi = \frac{1}{2} \sum_i \sigma_i \int_0^T dt \int d\boldsymbol{r} [g_i(\boldsymbol{w}(\boldsymbol{r}, t)) \mu_i(\boldsymbol{r}, t)]^2 . \tag{9.3.47}$$

Here $\boldsymbol{w}$ and $\boldsymbol{\mu}$ are taken along the optimal trajectory.

Our analysis of rare fluctuations has focused on applications of path-integral methods. The advantage of such methods is that they not only yield the probability of transitions but also determine the respective optimal trajectory in the phase space. It should be noted, however, that calculation of transition probabilities is a classical problem in the theory of random processes and a number of other approaches, based on the Fokker–Planck equation, have been proposed and successfully used. Reviews of methods used to calculate the rates of escape from a metastable state have been given by *Landauer* [9.29] and *Hänggi* et al. [9.30].

10. Birth–Death Systems

The behavior of many systems in many different applications can be effectively described in terms of the reproduction, death, and diffusion of certain particles. In chemical kinetics this situation is realized for chemical chain reactions: Here the 'particles' represent free radicals. Nuclear fission reactions involve the reproduction of neutrons. In biology, the reproducing 'particles' could be micro-organisms, such as bacteria, insects, or other animals. As shown in Chap. 7 of the first volume, even problems in biological evolution and market economics can be analyzed in terms of populations of reproductive agents. On the other hand, reproduction of structural defects plays an important role in the transition to turbulence in distributed active systems (see Chap. 7 of the present volume). A general mathematical description for this class of processes is provided by birth–death models.

When explosive growth of a population occurs, the system is extremely sensitive to fluctuations in the local birth and death rates. They lead to the formation of intermittent population distributions, where small spatial regions with a very high population density are surrounded by large sparsely populated regions. Another important problem in the statistical theory of population explosions is to determine the explosion threshold in fluctuating media, e.g., in systems with a random distribution of breeding centers.

10.1 Stochastic Birth–Death Models

The reproduction of particles or biological species in real systems can be a very complicated process, depending not only on the properties of the medium but also on the state of the reproducing agent and its personal history. The simplest birth–death models neglect all this possible complexity. They treat each birth and death as an independent random event characterized by certain probabilities k_{birth} and k_{death} per unit time.

The time evolution of the total number n of particles in the system is thus described as a discrete Markov process and the probability $p(N, t)$ of finding N particles at time moment t satisfies the master equation

$$\frac{\partial p}{\partial t} = k_{\text{birth}}(N - 1)p(N - 1, t) + k_{\text{death}}(N + 1)p(N + 1, t)$$

$$- (k_{\text{birth}} + k_{\text{death}})Np(N, t) . \tag{10.1.1}$$

Equation (10.1.1) takes in account that the probability of finding N particles is increased due to birth events, where $N - 1 \rightarrow N$, and death events, where $N + 1 \rightarrow N$. Since each of the available particles can reproduce or die, the respective rates are given by $k_{\text{birth}}(N - 1)$ and $k_{\text{death}}(N + 1)$. On the other hand, the probability of finding N particles is decreased by birth events with $N \rightarrow N + 1$ and death events with $N \rightarrow N - 1$.

Multiplying both parts of (10.1.1) by N and summing over N from zero to infinity, the evolution equation for the mean number $\langle N \rangle$ of particles in the system is obtained as

$$\frac{d}{dt} \langle N \rangle = (k_{\text{birth}} - k_{\text{death}})\langle N \rangle \ . \tag{10.1.2}$$

Fluctuations around the mean value become relatively small in the limit of large numbers of particles N, and then the deterministic kinetic equation for the concentration $n = N/V$ of particles in the volume V is applicable:

$$\dot{n} = (k_{\text{birth}} - k_{\text{death}})n \ . \tag{10.1.3}$$

When an extended system is considered and the particles can perform Brownian motion in the medium, local concentrations can be introduced. We take an elementary volume ΔV, such that it still contains a large number of particles, count the number of particles $\Delta N(r, t)$ at time t within the elementary volume located at point r, and define the local concentration as $n(r, t) = \Delta N(r, t)/\Delta V$. In the deterministic limit, the time evolution of the concentration distribution obeys equation

$$\dot{n} = (k_{\text{birth}} - k_{\text{death}})n + D\Delta^2 n \ , \tag{10.1.4}$$

where D is the diffusion constant.

In a large system, the conditions for reproduction and/or death of particles may vary in space and over time. Therefore, the birth and death rates k_{birth} and k_{death} in equation (10.1.4) are generally some functions of r and t. These properties can also undergo random fluctuations, so that k_{birth} and k_{death} can represent certain random fields. Then (10.1.4) is a stochastic partial differential equation which can be conveniently written in the form

$$\dot{n} = \alpha n + f(r, t)n + D\nabla^2 n \ , \tag{10.1.5}$$

where $\alpha = \langle k_{\text{birth}} \rangle - \langle k_{\text{death}} \rangle$ is the difference between average birth and death rates while $f(r, t) = k_{\text{birth}}(r, t) - k_{\text{death}}(r, t) - \alpha$ is the difference in their fluctuating components.

Equation (10.1.5) describes the evolution of the birth–death system in the presence of an *external multiplicative noise* $f(r, t)$. The statistical properties of this noise are not universal: they depend on the particular physical processes which give rise to fluctuations in the birth and/or death rates.

When these fluctuations contain contributions from many independent weak local random processes, the noise is Gaussian and is characterized by its pair correlation function

$$\langle f(r,t)f(r',t')\rangle = S(|r - r'|, t - t') \tag{10.1.6}$$

where the function $S(x, t)$ vanishes for spatial separations exceeding a certain correlation radius r_c and for time intervals exceeding some correlation time τ_c.

If the correlation time τ_c of the external noise is much shorter than all other characteristic times of the problem, the delta-function approximation can be used, i.e., we can write

$$S(|r - r'|, t - t') = 2s(|r - r'|)\delta(t - t') . \tag{10.1.7}$$

Furthermore, if the noise correlation radius r_c is much shorter than all other characteristic lengths, we can model it as a white noise, i.e.,

$$S(|r - r'|, t - t') = 2s\delta(|r - r'|)\delta(t - t') . \tag{10.1.8}$$

The factor 2 is introduced in these definitions because it proves to be convenient in the subsequent analysis. Note that when the external noise is delta-correlated with respect to time, the Stratonovich interpretation of the stochastic differential equation (10.1.5) should be assumed.

Another interesting and practically important case is encountered when reproduction occurs only within some randomly distributed small regions in the medium (such spatial regions represent *breeding centers*), while the death rate is uniform over the system.

Though the stochastic differential equation (10.1.5) has a very simple form (it is even linear!), its statistical properties are far from trivial. The analysis, performed later in this chapter, reveals that it can generate strongly non-Gaussian intermittent population distributions. By a transformation of variables,

$$n(r, t) = \exp[\varphi(r, t) - \alpha t] , \tag{10.1.9}$$

it is reduced to the *Kardar–Parisi–Zhang equation* [10.1]

$$\dot{\varphi} = D(\nabla\varphi)^2 + D\nabla^2\varphi + f(r, t) , \tag{10.1.10}$$

which describes the growth of crystals. We see that, although equation (10.1.10) is nonlinear, it is equivalent to a linear stochastic differential equation (10.1.5).

Of course, the linear form of equation (10.1.5) describing the growth of a reproducing population is only an approximation. When the population density becomes large, nonlinear effects must come into play and they will saturate the population growth. The simplest nonlinear term able to produce the saturation is quadratic. Thus, we arrive at the stochastic differential equation of *logistic growth*

$$\dot{n} = \alpha n - \beta n^2 + f(r, t)n + D\nabla^2 n , \tag{10.1.11}$$

where β is a positive constant.

Another interesting nonlinear model is found when systems with *long-range inhibition* are considered. Suppose that the reproducing particles release some inhibitor substance which can also diffuse in the medium. The system is then described by a set of two coupled equations,

$$\dot{n} = \alpha n - \nu m n + f(\boldsymbol{r}, t)n + D\nabla^2 n \qquad (10.1.12)$$

$$\dot{m} = -\gamma m + \eta n + D_{\text{inh}}\nabla^2 m \ . \qquad (10.1.13)$$

Here m is the local inhibitor concentration and D_{inh} is its diffusion constant. We assume that the rate of inhibitor generation is proportional to the local population density n. The inhibitor decays at a rate γ which prevents its accumulation in the medium. In the presence of the inhibitor, the reproduction and the death rates change in such a way that their mean difference is $\alpha - \nu m$.

We consider the limiting case of long-range inhibition (see Chap. 5 of the first volume) in which the characteristic diffusion length $L_{\text{m}} = (D_{\text{inh}}/\gamma)^{1/2}$ is much larger than the linear size L of the medium where the reproduction process takes place (i.e., under the condition of ideal mixing for the inhibitor species). Then the inhibitor density m is uniform and equation (10.1.13) is equivalent to

$$\dot{m} = -\gamma m + \eta\langle n \rangle \ , \qquad (10.1.14)$$

where

$$\langle n \rangle = \frac{1}{V} \int_{(V)} n(\boldsymbol{r}, t)d\boldsymbol{r} \qquad (10.1.15)$$

is the spatial average of the population density n over the volume V of the medium.

If we further assume that the characteristic time scale for the inhibitor variation is short compared with that of the reproducing species (i.e., γ is large), then the inhibitor density m will adjust adiabatically to variations of the population density n, so that

$$m = (\eta/\gamma)\langle n \rangle \ . \qquad (10.1.16)$$

Substitution of (10.1.16) into equation (10.1.12) yields a nonlinear stochastic differential equation for the population density:

$$\dot{n} = \alpha n - \kappa\langle n \rangle n + f(\boldsymbol{r}, t)n + D\nabla^2 n \ , \qquad (10.1.17)$$

where $\kappa = \nu\eta/\gamma$.

At first glance, equation (10.1.17) appears to differ only slightly from the stochastic equation of logistic growth (10.1.11). The only difference is that the quadratic term in the latter equation is now replaced by the term with the spatial average $\langle n \rangle$ of the population density. However, it turns out that this single change is sufficient to dramatically alter the fluctuation behavior in the system.

The reason for this is that equation (10.1.17) can be reduced to a linear equation (10.1.5) for the population growth by a transformation of variables. Indeed, if we introduce a new variable $h(\boldsymbol{r}, t)$ by

$$n(\boldsymbol{r}, t) = h(\boldsymbol{r}, t) \exp\left(-\kappa \int_0^t \langle n(t')\rangle dt'\right) ,$$ (10.1.18)

we obtain for this new variable the evolution equation

$$\dot{h} = \alpha h + f(\boldsymbol{r}, t)h + D\nabla^2 h ,$$ (10.1.19)

which is identical to (10.1.5). When the solution $h(\boldsymbol{r}, t)$ of equation (10.1.19) is known, the population density can be found as

$$n(\boldsymbol{r}, t) = \frac{h(\boldsymbol{r}, t)}{1 + \kappa \int_0^t \langle h(t')\rangle dt'} ,$$ (10.1.20)

where $\langle h(t)\rangle$ is the spatial average of h at time t.

Besides fluctuations caused by external noise, birth–death systems also have intrinsic fluctuations. Indeed, even when the birth and death rates are constant, such systems have stochastic dynamics and obey the 'microscopic' master equation (10.1.1). The 'macroscopic' description in terms of continuous population densities is only an approximation. As shown in Sect. 9.2, intrinsic fluctuations can be incorporated into this description by introducing *internal noise*. Equation (9.2.43) takes into account internal noise sources due to reproduction, decay, and diffusion of particles. When the population density is low, such terms may also be added to equation (10.1.5) describing the effects of external noise.

Further stochastic models of biological and chemical birth–death processes are discussed in [10.2, 3].

10.2 The Ignition Problem

Even if a medium has all the necessary properties to support fast reproduction, it can still remain unpopulated. To initiate the reproduction chain, it must first be 'infected' by at least a small number of particles. In the theory of chemical explosions, the initial stage of a chain reaction leading to the appearance of macroscopic concentrations of reproducing particles is called *ignition*. Below we apply this term in the wider context of any birth–death system.

If the particles do not die, a single particle is sufficient to trigger the explosion. When, however, the particles can die, a reproduction chain initiated by the arrival of a given particle may be terminated. The explosion thus begins only when infinite reproduction chains become possible.

Every particle will eventually reproduce or die. If the probabilities of birth and death per unit time are k_{birth} and k_{death}, the life cycle of a given particle ends in reproduction with the probability $k_{birth}/(k_{birth} + k_{death})$ and in death with the probability $k_{death}/(k_{birth} + k_{death})$.

The probability ζ_1 that the reproduction chain terminates can be found from the following arguments. If the chain initiated by a given particle is later terminated, this might be an outcome of two independent events. Firstly, the initial particle can die without reproduction. Secondly, it can give rise to two new particles both of which themselves lead to terminated chains. Therefore, ζ_1 should satisfy the equation

$$\zeta_1 = \frac{k_{\text{death}}}{k_{\text{birth}} + k_{\text{death}}} + \frac{k_{\text{birth}}}{k_{\text{birth}} + k_{\text{death}}} \zeta_1^2 . \tag{10.2.1}$$

Its solution is

$$\zeta_1 = \begin{cases} k_{\text{death}}/k_{\text{birth}} , & \text{if } k_{\text{birth}} > k_{\text{death}} , \\ 1 , & \text{if } k_{\text{birth}} \le k_{\text{death}} . \end{cases} \tag{10.2.2}$$

Thus, for $k_{\text{birth}} < k_{\text{death}}$ all chains are eventually terminated ($\zeta_1 = 1$). Above the explosion threshold $k_{\text{birth}} = k_{\text{death}}$, however, there is a nonvanishing fraction of infinite reproduction chains. The probability that a given particle will give rise to an infinitely branching chain is $\zeta_0 = 1 - \zeta_1$ or, explicitly,

$$\zeta_0 = 1 - \frac{k_{\text{death}}}{k_{\text{birth}}} , \quad \text{for } k_{\text{birth}} > k_{\text{death}} . \tag{10.2.3}$$

Suppose that a medium occupying volume V is weakly infected by particles at a rate of ϱ particles per unit time per unit volume element. Then the initiation probability for an infinite reproduction chain in the entire volume is $\zeta_0 \varrho V$ and the *mean waiting time* for initiation of an infinite reproduction chain in the medium is $T = (\zeta_0 \varrho V)^{-1}$. Using (10.2.3), we obtain

$$T = \frac{1}{\varrho V} \frac{k_{\text{birth}}}{k_{\text{birth}} - k_{\text{death}}} . \tag{10.2.4}$$

We see that this time tends to infinity as the explosion threshold $k_{\text{birth}} = k_{\text{death}}$ is approached.

We say that ignition has taken place in the system if the population density of particles has everywhere exceeded a certain (macroscopic) level n_c. In a small volume, where diffusion maintains a statistically uniform spatial distribution of reproducing particles, the characteristic *ignition time* T_{ign} is the sum of two contributions, i.e., of the mean waiting time for the appearance of an infinite reproductive chain and of the time needed for this reproductive chain to yield $V n_c$ particles in the whole medium. The latter can be roughly estimated as $T' = \ln(V n_c)/(k_{\text{birth}} - k_{\text{death}})$. Thus we obtain the following estimate for the ignition time:

$$T_{\text{ign}} = \frac{1}{k_{\text{birth}} - k_{\text{death}}} \left[\frac{k_{\text{birth}}}{\varrho V} + \ln(V n_c) \right] . \tag{10.2.5}$$

This estimate is not valid, however, for large systems. When deriving equation (10.2.5), we assumed that the whole medium becomes populated from a single center, from which the first infinite reproductive chain appears. In reality, a large system would have many independent ignition centers.

The correct estimate of the ignition time for large volumes has been constructed by *Mikhailov* and *Zeldovich* [10.4]. It takes into account that each ignition center generates a spherical populated region that spreads over the medium. Propagation of the population front is described by the equation

$$\dot{n} = \alpha n + D\nabla^2 n \ , \tag{10.2.6}$$

where $\alpha = k_{\text{birth}} - k_{\text{death}}$. The velocity v_0 of the propagating front can be estimated (see Chap. 2 of the first volume) to have the order of magnitude

$$v_0 \sim (D\alpha)^{1/2} \ . \tag{10.2.7}$$

The ignition occurs when spreading populated regions (seats), produced by different ignition centers, merge and cover the whole medium.

Thus we have many centers created independently at random locations and at random times. Every center gives rise to a spherical spreading front. This front propagates until it comes into contact with another front produced by one of the neighboring ignition centers. If random generation of ignition centers begins at the initial time, the ignition time is determined as the time after which every point of the medium is populated, i.e., belongs to one of the ignition seats.

From a mathematical point of view, the ignition problem in a large volume is therefore equivalent to the problem of crystallization in a supercooled liquid, which was solved in 1937 by *Kolmogorov* [10.5]. In the latter case, crystallization centers are generated randomly in the liquid. Each of them produces a spherical solid region that grows at constant speed. The crystallization process is completed when the solid regions merge together and every point of the medium belongs to one of them.

As follows from the above analysis, the probability that, within a small time interval Δt, the volume is infected by a particle that gives rise to an infinite reproduction chain is $\zeta_0 \varrho V \Delta t$. This particle produces a spherical population seat that spreads at velocity v_0. At time t, an arbitrary point Q of the medium is found inside a population seat only if, at some previous time $t' < t$, an ignition center has been created at a point Q' located a distance less than $v_0(t - t')$ from the point Q. For a given time t' the total volume occupied by the points Q' satisfying this condition is

$$V'(t') = \frac{4\pi}{3} v_0^3 (t - t')^3 \ . \tag{10.2.8}$$

The probability of creation of an ignition center in such a volume within time Δt is $\zeta_0 \varrho V'(t')\Delta t$, while the probability that no centers appear in it within this time is $1 - \zeta_0 \varrho V'(t')\Delta t$ where $\zeta_0 = \alpha/k_{\text{birth}}$. Hence, the probability that no ignition centers were present within volumes $V'(t')$ at *any* moment $0 < t' < t$ is determined by the product

$$q(t) = \prod_{j=1}^{N} [1 - \zeta_0 \varrho V'(t_j)\Delta t] \ . \tag{10.2.9}$$

Here we have divided the time interval t into N small subintervals of length Δt, so that $t_j = j\Delta t$. The quantity $q(t)$ represents the joint probability that ignition centers were absent inside volumes V' for any of these small subintervals. It thus yields the probability that *no* population front has reached the point Q by time t.

Taking the logarithm of (10.2.9) and considering the limit $N \to \infty$ we obtain

$$\ln q(t) = - \sum_{j=1}^{0} \zeta_0 \varrho V'(t_j)\Delta t \to - \int_0^t \zeta_0 \varrho V'(t')dt'$$

$$= -\frac{4\pi}{4} v_0^3 \zeta_0 \varrho \int_0^t (t - t')^3 dt' = -\frac{\pi}{3} v_0^3 \zeta_0 \varrho t^4 . \tag{10.2.10}$$

The probability that an arbitrarily chosen point Q of the medium is populated by time t is $p(t) = 1 - q(t)$ or, using equation (10.2.10),

$$p(t) = 1 - \exp\left(-\frac{\pi}{3} v_0^3 \zeta_0 \varrho t^4\right) . \tag{10.2.11}$$

This probability determines the fraction of the medium occupied by populated regions at time t. If we introduce the characteristic time

$$T_c = (\zeta_0 \varrho v_0^3)^{-1/4} , \tag{10.2.12}$$

equation (10.2.11) can be written as

$$p(t) = 1 - \exp\left[-\frac{\pi}{3} \left(\frac{t}{T_c}\right)^4\right] . \tag{10.2.13}$$

Thus $p(t)$ exponentially close to unity and the entire medium is populated when $t > T_c$. Therefore, T_c represents the *ignition time* of explosion in large volumes. Taking into account (10.2.3) and (10.2.7), this characteristic time can be estimated as

$$T_c \sim \left(\frac{k_{birth}}{k_{birth} - k_{death}}\right)^{5/8} \frac{1}{\varrho^{1/4}(Dk_{birth})^{3/8}} . \tag{10.2.14}$$

We see that this estimate does not depend on the volume V of the medium.

Another important property of ignition in large systems is the total number of independent ignition centers in the medium. To determine this, we note that $Vq(t)$ gives us the part of the volume which remains unpopulated by time t. Because the probability of appearance of an ignition center, i.e., of a particle giving rise to an infinite reproduction chain, per unit time per unit volume is $\zeta_0 \varrho$, the number ΔM of ignition centers that would appear within a time interval Δt inside the previously unpopulated regions is therefore

$$\Delta M = \zeta_0 \varrho q V \Delta t . \tag{10.2.15}$$

The total number of ignition centers creating new population seats is given by the integral

$$M = \int_0^\infty \zeta_0 \varrho q(t) V dt , \qquad (10.2.16)$$

where $q(t)$ is determined by (10.2.10). Taking this integral, we find that

$$M \approx 0.9V \left(\frac{\zeta_0 \varrho}{v_0} \right)^{3/4} . \qquad (10.2.17)$$

The mean distance between independent ignition centers is $L_c \sim (V/M)^{1/3}$. Using equations (10.2.17), (10.2.7) and (10.2.3), it can be estimated as

$$L_c \sim \left[\frac{D k_{birth}^2}{\varrho^2 (k_{birth} - k_{death})} \right]^{1/8} . \qquad (10.2.18)$$

As follows from equations (10.2.7) and (10.2.12), this estimate can also be written in the form $L_c \sim v_0 T_c$. Hence, in its order of magnitude, the mean distance between the ignition centers is equal to the radius of a populated region that has been growing at speed v_0 over a time T_c.

As far as ignition is concerned, the volume V is small if the whole system would be ignited from just one center, i.e., if $V \ll L_c^3$. When this condition is satisfied, the earlier estimate (10.2.5) for the ignition time holds. In the opposite limiting case of large volumes, the ignition time is determined by equation (10.2.14).

The above analysis has been performed for three-dimensional media. When a system of arbitrary dimensionality d is considered, the ignition time in large systems is estimated as

$$T_c = \left(\frac{k_{birth}}{\alpha \varrho v_0^d} \right)^{1/(d+1)} , \qquad (10.2.19)$$

where $v_0 \sim (\alpha D)^{1/2}$ and $\alpha = k_{birth} - k_{death}$. The mean distance between ignition centers is about $L_c \sim v_0 T_c$.

10.3 Spatiotemporal Intermittency in Population Explosions

When ignition has occurred, the entire medium is occupied by reproducing particles. At the next stage, which can be characterized as *population explosion*, the population density rapidly increases with time. The explosion process is extremely sensitive to local fluctuations in birth and death rates. As noted by *Zeldovich* et al. [10.6], the spatial distribution of a population that undergoes explosion in a fluctuating medium is highly nonuniform. It includes rare strong population bursts alternating with large regions with relatively low population density. The whole pattern is not stationary and randomly evolves with time. This phenomenon is thus described as spatiotemporal *intermittency* of population explosions.

The intermittent behavior is intrinsically related to the fact that the noise enters into the stochastic differential equation (10.1.5) in a multiplicative manner. Its origin can be understood if we consider, following [10.6], a simple toy model with discrete dynamics.

Suppose that we have a spatial array of cell, inside which particles can reproduce or die. Diffusion is absent and the particle population in each of the cells is independent of that of its neighbors. Reproduction and death can occur only at discrete time moments. Furthermore, it is only possible that, at a given time moment, either all particles in a given cell reproduce (and hence its population is doubled) or all of them die (and therefore the cell becomes empty). We assume that both outcomes take place with the same probability 1/2. Initially each cell contains exactly one particle. The number Φ_N of particles in a cell at time moment N is then given by a product of N independent random factors ϕ_j,

$$\Phi_N = \prod_{j=1}^{N} \phi_j , \tag{10.3.1}$$

taking only two possible values of 0 and 2 with the same probability 1/2.

The population evolution in a cell is described by a realization of the discrete random process (10.1.3). It can immediately be seen that for almost all such realizations (namely, if at least one of the factors in the product is equal to zero) the product Φ_N vanishes and thus the population dies out. The population in a given cell survives after N time steps only if all N independent random variables ϕ_j have taken the value 2, i.e. if the population has doubled at each time step j. The probability of such an exceptional realization is very small, $p_N = 2^{-N}$. However, the surviving population is equal to 2^N and is therefore very large.

The statistical average of the population numbers over the ensemble of cells at time moment N is

$$\langle \Phi_N \rangle = \frac{\text{sum of all possible realizations}}{\text{number of all possible realizations}} = \frac{0 + 0 + \ldots + 0 + 2^N}{2^N} = 1 .$$

$$\tag{10.3.2}$$

Hence, the mean population density remains constant in time. However, in a great majority of cells, the population actually vanishes and this constant average of the population density is maintained only due to the rapid population growth in a few remaining populated cells. Remarkably, the higher statistical moments in this problem,

$$\langle (\Phi_N)^k \rangle = \frac{0 + 0 + \ldots + 0 + 2^{kN}}{2^N} = 2^{(k-1)N} , \tag{10.3.3}$$

grow indefinitely with time.

We see that, as time goes on, the total (statistically conserved) population tends to concentrate in increasingly rare still occupied cells. The mean distance between the occupied cells on the plane grows with time as $2^{N/2}$. Therefore, for sufficiently large times, any given finite area always eventually becomes void. Thus, in a spatially finite system the population dies out with probability one.

This simple example demonstrates the essential features of intermittency in systems with fluctuating birth and death rates. As shown below, this behavior is effectively retained when a continuous birth–death model described by the stochastic differential equation (10.1.5) is considered. Two important new properties will appear however. Since the fluctuating birth and death rates can now take a continuous range of values, the amplitudes of the population spikes are not all equal, but vary randomly from one spike to another. Furthermore, diffusion smears the spikes and makes their widths finite. It also leads to random wandering of the spikes through the medium.

As the next step, we analyze a model in which diffusion is still absent but the time evolution of the population in a cell is now described by a continuous variable obeying the stochastic differential equation

$$\dot{n} = \alpha n + f(t)n \ . \tag{10.3.4}$$

Here $f(t)$ is the white noise,

$$\langle f(t)f(t')\rangle = 2s\delta(t - t') \ , \tag{10.3.5}$$

and the Stratonovich interpretation is chosen.

The probability distribution $p(n, t)$ for a system described by equation (10.3.4) satisfies (see Sect. 9.1) the following Fokker–Planck equation:

$$\frac{\partial p}{\partial t} = -\frac{\partial}{\partial n}[(\alpha + s)np] + \frac{\partial^2}{\partial n^2}(sn^2p) \ . \tag{10.3.6}$$

To derive an evolution equation for the kth statistical moment $\langle n^k \rangle$, we multiply both sides of (10.3.6) by n^k and integrate over n from zero to infinity:

$$\frac{d}{dt}\int_0^\infty n^k p\,dn = -\int_0^\infty n^k \frac{\partial}{\partial n}[(\alpha + s)np]dn + \int_0^\infty n^k \frac{\partial^2}{\partial n^2}(sn^2p)dn \ . \tag{10.3.7}$$

Integrating by parts on the right-hand side of this equation and taking into account that the probability density p vanishes at infinity, we find

$$\frac{d}{dt}\int_0^\infty n^k p\,dn = \int_0^\infty k(\alpha + s)n^k p\,dn + \int_0^\infty k(k - 1)sn^k p\,dn \ , \tag{10.3.8}$$

which yields, after a simple transformation, the evolution equation

$$\frac{d}{dt}\langle n^k \rangle = k(\alpha + ks)\langle n^k \rangle \ . \tag{10.3.9}$$

Thus, the time dependence of the statistical moments of the population density n is given by

$$\langle n^k(t)\rangle \sim \exp[k(\alpha + ks)t] \ . \tag{10.3.10}$$

The mean population density $\langle n \rangle$ corresponds to the first statistical moment $(k = 1)$; it evolves with time as

$$\langle n(t) \rangle \sim \exp[(\alpha + s)t] . \tag{10.3.11}$$

If we define the explosion threshold by the condition that the mean population density in the ensemble remains constant, it follows from (10.3.11) that the explosion threshold is reached at $\alpha = -s$, i.e., when $\langle k_{\text{birth}} \rangle = \langle k_{\text{death}} \rangle - s$. Note that fluctuations lead to lowering of the explosion threshold – this effect will be discussed in detail in Sect. 10.4.

Comparing (10.3.11) and (10.3.10), we see that

$$\frac{\langle n^k(t) \rangle}{\langle n(t) \rangle^k} \sim \exp[k(k - 1)st] . \tag{10.3.12}$$

Hence, the higher statistical moments increase faster with time than the first moment, yielding the mean population density. Even when the condition $\alpha = -s$ is satisfied and the mean population remains constant, the growth of higher statistical moments continues.

This behavior of statistical moments closely resembles what has been found above in a simple discrete model. It indicates that the probability distribution $p(n, t)$ in this birth–death model is strongly non-Gaussian. In the course of time, the exploding population tends to concentrate in an increasingly small number of cells. The population level in other cells of the ensemble is much lower (but still nonvanishing, in contrast to the previously considered discrete model).

When, additionally, diffusion of reproducing particles is taken into account, it tends to diminish the difference between population densities in neighboring cells. The question is whether this effect is strong enough to suppress spatiotemporal intermittency in the system. Below we consider a model described by the partial stochastic differential equation (10.1.5) with external white noise that has the correlation function (10.1.7).

Any spatially distributed system can be viewed as the limit of a system with a large number of variables. Indeed, we can always introduce a sufficiently fine lattice, replace all spatial differential operators by the respective difference expressions, and thus arrive at a system of dynamical equations where each variable corresponds to a particular lattice site. Therefore, we can apply here the results obtained in Chap. 9 for systems of stochastic Langevin equations.

The state of a distributed system is described by its population field $n(r)$ at a given time moment. When a stochastic system is considered, each possible state is realized with a certain probability. Hence, a probability functional $P[n(r)]$ can be constructed that puts into correspondence with each field $n(r)$ a certain number, i.e., the probability measure of such a realization.

The probability functional for the system (10.1.5) obeys the functional Fokker–Planck equation:

$$\frac{\partial}{\partial t} P[n(r), t] = -\int dr \frac{\delta}{\delta n(r)} \{[\alpha n + D\nabla^2 n + s(0)n]P\}$$

$$+ \iint dr dr' \frac{\delta^2}{\delta n(r)\delta n(r')} [s(|r - r'|)n(r)n(r')P] . \quad (10.3.13)$$

It can be viewed as a variant of the Fokker–Planck equation (9.1.33), provided we treat each coordinate vector r as an 'index' i specifying a certain variable in this equation. Summation over discrete indices is replaced then by integration over r. The coefficients of the functional Fokker–Planck equation (10.3.13) are obtained by applying equations (9.1.35).

Single-time correlation functions of the population density $n(r, t)$ are defined as the averages

$$M_k(t; r_1, r_2, \ldots, r_k) = \langle n(r_1, t)n(r_2, t) \ldots n(r_k, t) \rangle . \quad (10.3.14)$$

The evolution equation for these functions can be derived using the functional Fokker–Planck equation, similar to the above derivation of the evolution equation for statistical moments in the model without diffusion. We multiply both sides of (10.3.13) by $n(r_1)n(r_2) \ldots n(r_k)$ and integrate over all possible fields $n(r)$,

$$\frac{\partial}{\partial t} \int \mathcal{D}n(r)n(r_1)n(r_2) \ldots n(r_k)P[n(r), t]$$

$$= -\int \mathcal{D}n(r) \int dr n(r_1)n(r_2) \ldots n(r_k) \frac{\delta}{\delta n(r)} \{[\alpha n + D\nabla^2 n + s(0)n]P\}$$

$$+ \int \mathcal{D}n(r) \iint dr dr' n(r_1)n(r_2) \ldots$$

$$\ldots n(r_k) \frac{\delta^2}{\delta n(r)\delta n(r')} [s(|r - r'|)n(r)n(r')P] . \quad (10.3.15)$$

The integral on the left-and side yields the time derivative of the correlation function. The integrals on the right-hand side can be transformed if we use integration by parts. The first of them can be written in the equivalent form

$$-\int \mathcal{D}n(r) \int dr n(r_1)n(r_2) \ldots n(r_k) \frac{\delta}{\delta n(r)} \{[\alpha n + D\nabla^2 n + s(0)n]P\}$$

$$= \int \mathcal{D}n(r) \int dr [\alpha n + D\nabla^2 n + s(0)n]P \frac{\delta}{\delta n(r)} [n(r_1)n(r_2) \ldots n(r_k)]$$

$$= \int \mathcal{D}n(r) \left\{ \left[k\alpha + ks(0) + D\sum_{i=1}^{k} \frac{\partial^2}{\partial r_i^2} \right] n(r_1)n(r_2) \ldots n(r_k) \right\} P$$

$$= k(\alpha + s(0))M_k + D\sum_{i=1}^{k} \frac{\partial^2 M_k}{\partial r_i^2} . \quad (10.3.16)$$

The second integral on the right-hand side of equation (10.3.15) yields

$$\int \mathcal{D}n(r) \int\int dr dr' n(r_1)n(r_2)\ldots n(r_k)\frac{\delta^2}{\delta n(r)\delta n(r')}[s(|r-r'|)n(r)n(r')P]$$

$$= \int \mathcal{D}n(r) \int\int dr dr' s(|r-r'|)n(r)n(r')$$

$$\times P\frac{\delta^2}{\delta n(r)\delta n(r')}[n(r_1)n(r_2)\ldots n(r_k)]$$

$$= \sum_{\substack{i,j=1 \\ i\neq j}}^{k} s(r_i - r_j)M_k \; . \qquad (10.3.17)$$

Thus, the correlation functions $M_k = \langle n(r_1)\ldots n(r_i)\ldots n(r_j)\ldots n(r_k)\rangle$ obey the following evolution equation:

$$\dot{M}_k = k\alpha M_k + \sum_{i,j=1}^{k} s(r_i - r_j)M_k + D\sum_{i=1}^{k}\frac{\partial^2 M_k}{\partial r_i^2} \; . \qquad (10.3.18)$$

For the first statistical moment $M_1 = \langle n(r,t)\rangle$ this equation reduces to

$$\dot{M}_1 = \alpha M_1 + s(0)M_1 + D\nabla^2 M_1 \; . \qquad (10.3.19)$$

Therefore, the explosion threshold in this model is given by the condition $\alpha = -s(0)$ and does not depend on the diffusion constant D.

To perform the analysis of higher statistical moments, it is convenient to write (10.3.18) in a slightly different form, i.e., as

$$\dot{M}_k = (\alpha + s(0))kM_k - \hat{L}M_k \; , \qquad (10.3.20)$$

where

$$\hat{L} = -\sum_{\substack{i,j=1 \\ i\neq j}}^{k} s(r_i - r_j) - D\sum_{i=1}^{k}\frac{\partial^2}{\partial r_i^2} \; . \qquad (10.3.21)$$

The linear operator $\hat{L}$ is formally identical to the Hamilton operator for a system of k interacting quantum particles. When this analogy is explored, the first term in (10.3.21) describes pair interactions between the particles with the potential $u(r) = -s(r)$ where r is the distance between two particles. Since we assume that the spatial correlation function $s(r)$ of noise in equation (10.1.5) is positive, the effective interaction potential $u(r)$ between particles is attractive. It represents a potential well of depth $s(0)$ with a characteristic radius r_c [see (10.1.7)]. The diffusive term in equation (10.3.21) corresponds to the kinetic energy term in the quantum Hamilton operator.

The general solution of equation (10.3.20) can be written as

$$M_k(t; r_1, \ldots, r_k) = \sum_{l} C_l \exp[k(\alpha + s(0))t - \lambda_l t]\psi_l(r_1, \ldots, r_k) \; , \qquad (10.3.22)$$

where λ_l and ψ_l are eigenvalues and eigenfunctions of the operator $\hat{L}$. For the continuous part of the spectrum, the sum in (10.3.22) should be replaced by an integral.

Note that eigenfunctions with negative eigenvalues λ_l, corresponding to bound states of particles in the respective quantum mechanical problem, give rise to exponentially increasing contributions to the correlation functions M_k. In the long time limit, the behavior of M_k is dominated by the most rapidly growing contribution from the largest negative eigenvalue λ of the linear operator $\hat{L}$ (which corresponds to the deepest energy level in the quantum problem).

Below we estimate the largest negative eigenvalue of $\hat{L}$ in the two limiting cases of large and small diffusion constants D, and thus determine the asymptotic behavior of the correlation functions.

If the diffusion constant D in (10.3.21) is very small, the particles in the quantum analogy are heavy and therefore the localization radius of the deepest bound state of k particles is much smaller than the spatial size of the potential well. Near to its minimum, the potential well $u(r) = -s(r)$ can always be approximated by a harmonic oscillator potential, i.e., using the expansion

$$s(r) = s(0)\left[1 - \frac{1}{2}\left(\frac{r}{r_0}\right)^2\right] , \qquad (10.3.23)$$

where the characteristic radius r_0 has the same order of magnitude as the correlation radius r_c of the noise.

When a system of k quantum particles, interacting via the harmonic pair potential given by (10.3.23), is considered, the energy of the ground state of such a system can be determined and hence the largest negative eigenvalue of the operator $\hat{L}$ can be estimated (see [10.7]) as

$$\lambda = -k(k-1)s(0) + d(k-1)\left(\frac{ks(0)D}{r_0^2}\right)^{1/2} , \qquad (10.3.24)$$

where d is the dimensionality of the medium. This expression is valid provided the condition

$$ks(0) \gg \frac{D}{r_0^2} \qquad (10.3.25)$$

is satisfied.

Hence, in the case of weak diffusion the time dependence of the correlation functions in the long-time limit is

$$M_k \propto \exp\left[k(\alpha + ks(0))t - d(k-1)\left(\frac{ks(0)D}{r_0^2}\right)^{1/2}\right] . \qquad (10.3.26)$$

When diffusion is absent, i.e, $D = 0$, this result coincides with the estimate (10.3.10) for the statistical moments of the population density. We see that diffusion decreases the rates of growth of higher statistical moments. However, when it is weak, so that the condition (10.3.25) is satisfied, diffusion does not prevent the development of intermittency. At the threshold $\alpha = -s(0)$, the average population density $\langle n \rangle$ remains constant, whereas the higher moments still grow indefinitely with time.

Next we discuss the opposite case of strong diffusion. Perhaps this is able to suppress intermittency? If we look again at equation (10.3.22), we notice that the kth statistical moment increases exponentially with time at the explosion threshold $\alpha = -s(0)$, provided that the linear operator $\hat{L}$ has at least one negative eigenvalue. In the framework of our quantum-mechanical analogy, this occurs when the system of interacting particles has at least one bound state with negative energy.

In one-dimensional systems, even very shallow potential wells are able to bind a particle [10.8] and hence a negative eigenvalue λ is always present in the spectrum. When $D \gg ks(0)r_0^2$, the localization radius of the bound state is much larger than the characteristic radius r_c of the potential well. When this condition is satisfied, we can replace $s(x - x')$ in the operator $\hat{L}$ by its delta-function approximation

$$s(x - x') \approx s\delta(x - x') \,, \tag{10.3.27}$$

where

$$s = \int_{-\infty}^{\infty} s(x)dx \,. \tag{10.3.28}$$

The spectrum of the linear operator (10.3.21) with the function $s(x)$ given by (10.3.27) has been determined exactly by *Berezin* et al. [10.9]. Its largest negative eigenvalue (which corresponds to a bound state of all k interacting particles) is given by

$$\lambda = -\frac{1}{12}\frac{s^2}{D}(k^3 - k) \,. \tag{10.3.29}$$

Therefore, the asymptotic time dependence of the density correlation functions in the long-time limit under the condition $D \gg ks(0)r_0^2$ in a one-dimensional system is

$$M_k \propto \exp\left[(\alpha + s(0))kt + \frac{1}{12}\frac{s^2}{D}(k^3 - k)t\right] \,. \tag{10.3.30}$$

We see that even at the explosion threshold, i.e., when $\alpha = -s(0)$, higher statistical moments with $k > 1$ grow indefinitely with time and thus intermittency remains present.

The two-dimensional case is marginal. Although, as known in quantum mechanics [10.8], a shallow potential well then contains a discrete energy level corresponding to the bound state of a quantum particle, the respective negative eigenvalue λ of the operator $\hat{L}$ is exponentially small. Therefore, the development of intermittency is an extremely slow process.

In three dimensions, only sufficiently strong potential wells can support bound states. If $D \gg s(0)r_0^2$, the linear operator $\hat{L}$ has no negative eigenvalues for $k = 2$. This means that the pair density correlation function, M_2, now has the same asymptotic law of growth as the square of the first statistical moment, M_1, i.e., the square of the mean population density $\langle n \rangle$. Therefore, spatiotemporal intermittency is not observed in the lower statistical moments.

Note, however, that even when there are no bound states for two particles, bound states can still be formed by a larger number of particles. Indeed, if we have k particles within the same spatial region of size about r_c, all respective potential wells overlap. Then, an individual particle finds itself in an effective well that is about k times deeper than the single well produced by the attractive interaction of two particles. When $ks(0)$ is about D/r_c^2, a bound state of k particles becomes possible, and obviously persists for higher values of $ks(0)$.

If we look at equations (10.3.20) and (10.3.21), which describe the time evolution of the density correlation functions, we can see that the bound states of k' particles are relevant only for the correlation functions of order k' and higher. Therefore, if a bound state is first formed by an agglomerate of k' particles, the negative eigenvalues λ_l will be present only in the spectra of the linear operators $\hat{L}$ for $k \geq k'$. This means that all lower correlation functions M_k with $k < k'$ will behave as $\langle n \rangle^k$ in the long-time limit. However, the higher-order correlation functions with $k \geq k'$ will have a stronger growth law. They will continue to grow at the explosion threshold, where the mean population density $\langle n \rangle$ remains constant. In the limit $k \to \infty$ the condition (10.3.25) is always satisfied and therefore the dependence (10.3.26) holds.

Thus we have derived estimates for single-time density correlation functions in a reproducing population. We have shown that high-order correlation functions and high statistical moments $\langle n^k \rangle$ grow faster with time than the respective power $\langle n \rangle^k$ of the mean population density. Because the population density is a non-negative random variable, this behavior is only possible if the spatial population distribution becomes strongly nonuniform and is concentrated, to a large extent, inside small spatial regions. Such intermittent population distributions can therefore be viewed as consisting of a number of strong *spikes* surrounded by a sparsely populated background.

Since the spikes are produced by random fluctuations in birth and death rates, their properties, such as their speed of growth and characteristic spatial size, are subject to statistical variations. Moreover, the spikes can also move through the medium. They have finite lifetimes. While some spikes disappear, new ones are being created.

Since the spikes are very rare, we can consider them individually. An important question then is: what is the *probability* of finding a spike with certain properties? Furthermore, we can also investigate the *typical* shape of a spike characterized by a given set of parameters.

To examine these questions, the path-integral solution of the stochastic differential equation (10.1.5) can be used. When the noise is Gaussian and delta-correlated in time and space, the general solution is given by equation (9.3.42). For the one-dimensional stochastic birth–death system described by equations (10.1.5), (10.1.6) and (10.1.8), this solution takes the form [10.10, 11]

$$P[n(x,t)] = \int \mathcal{D}\varrho(x,t) \, \exp\left[-\int dt \left(-H + \int dx \dot{n}\varrho\right)\right] , \qquad (10.3.31)$$

where H is given by

$$H = \int dx \left(\alpha\varrho n + D\varrho\frac{\partial^2 n}{\partial x^2} + s\varrho^2 n^2\right) . \qquad (10.3.32)$$

The optimal trajectories corresponding to the minima of the 'action'

$$S = \int dt \left(-H + \int dx \dot{n}\varrho\right) \qquad (10.3.33)$$

obey the variational equations

$$\frac{\partial n}{\partial t} = \frac{\delta H}{\delta \varrho(x)}, \quad \frac{\partial \varrho}{\partial t} = -\frac{\delta H}{\delta n(x)} . \qquad (10.3.34)$$

Explicitly, these equations read

$$\dot{n} = \alpha n + 2s\varrho n^2 + D\frac{\partial^2 n}{\partial x^2} , \qquad (10.3.35)$$

$$\dot{\varrho} = -\alpha\varrho - 2s\varrho^2 n - D\frac{\partial^2 \varrho}{\partial x^2} . \qquad (10.3.36)$$

To specify an optimal trajectory, we should fix its initial and final points. For a distributed system this means fixing initial and final spatial distributions, i.e., $n(x, t = 0) = n_0(x)$ and $n(x, t = T) = n_T(X)$. Note that the equations determining the optimal trajectory are *nonlinear* although we started from a linear stochastic birth–death model (10.1.5). This is a consequence of the fact that noise enters multiplicatively into equation (10.1.5).

The probability p of a rare statistical event can be approximately estimated as $p \propto \exp(-S)$ where S is the 'action' (10.3.33) for the respective optimal trajectory. Using the variational equations (10.3.35) and (10.3.36), we obtain [cf. (9.3.46) and (9.3.47)]

$$p \propto \exp\left[-s \int dx dt (\varrho n)^2\right] . \qquad (10.3.37)$$

As noted by *Mikhailov* [10.10, 11], the variational equations (10.3.35) and (10.3.36) are closely related to the nonlinear Schrödinger equation

$$i\frac{\partial\psi}{\partial t} = -\frac{\partial^2\psi}{\partial\xi^2} - 2\psi^2\psi^* \qquad (10.3.38)$$

for a complex variable $\psi(\xi, t)$. To reveal this analogy, we write the respective equation for the complex conjugate variable

$$-i\frac{\partial \psi^*}{\partial t} = -\frac{\partial^2 \psi^*}{\partial \xi^2} - 2\psi^{*2}\psi \tag{10.3.39}$$

and introduce new variables ξ, u and v by

$$\xi = D^{-1/2}x, \quad u = e^{-\alpha t}n, \quad v = se^{\alpha t}\varrho, \tag{10.3.40}$$

so that equations (10.3.35) and (10.3.36) become

$$\dot{u} = 2vu^2 + \frac{\partial^2 u}{\partial \xi^2}, \tag{10.3.41}$$

$$-\dot{v} = 2v^2u + \frac{\partial^2 v}{\partial \xi^2}. \tag{10.3.42}$$

We see then that if, in equations (10.3.38) and (10.3.39), we transform to imaginary time, $t \to -it$, and treat ψ and ψ^* as two independent real variables, i.e., put $\psi \to u$ and $\psi^* \to v$, these equations are transformed into equations (10.3.41) and (10.3.42).

As shown by *Zakharov* and *Shabat* [10.12], the nonlinear Schrödinger equation is completely integrable. It describes a system of solitons that behave in many respects as independent particles. The soliton solutions are given by

$$\psi(\xi, t) = \frac{2H \exp[-4i(G^2 - H^2)t + 2iG\xi + i\Delta]}{\cosh[2H(\xi - \xi_0) - 8HGt]}, \tag{10.3.43}$$

where H, G, Δ and ξ_0 are free parameters. The soliton has velocity $4G$, width $2(H)^{-1}$ and amplitude $2H$.

Since the transformation that generates equations (10.3.41) and (10.3.42) from the nonlinear Schrödinger equation is known, it can be employed to construct solutions of these variational equations. By putting $t \to -it$, $G = ig$, and $\Delta = i\delta$ in (10.3.43) we obtain

$$u(\xi, t) = \frac{2H \exp[4(g^2 + H^2)t - 2g\xi - \delta]}{\cosh[2H(\xi - \xi_0) - 8Hgt]}. \tag{10.3.44}$$

Furthermore, taking the complex conjugate of (10.3.43) and applying the transformation $t \to -it$, $G = ig$, and $\Delta = i\delta$ yields

$$v(\xi, t) = \frac{2H \exp[-4(g^2 + H^2)t - 2g\xi + \delta]}{\cosh[2H(\xi - \xi_0) - 8Hgt]}. \tag{10.3.45}$$

Equations (10.3.44) and (10.3.45) represent special solutions of the variational equations (10.3.41) and (10.3.42), as can be verified by their direct substitution.

Returning to the original variables, the solutions (10.3.44) and (10.3.45) can be written in the form [10.10, 11]

$$n(x,t) = \frac{C \exp\left[-\frac{V}{2D}(x - Vt - x_0)\right] \exp\left[\left(\alpha + q - \frac{V^2}{4D}\right)t\right]}{\cosh\left[\left(\frac{q}{D}\right)^{1/2}(x - Vt - x_0)\right]} , \tag{10.3.46}$$

$$\varrho(x,t) = \frac{\left(\frac{q}{sC}\right) \exp\left[\frac{V}{2D}(x - Vt - x_0)\right] \exp\left[-\left(\alpha + q - \frac{V^2}{4D}\right)t\right]}{\cosh\left[\left(\frac{q}{D}\right)^{1/2}(x - Vt - x_0)\right]} . \tag{10.3.47}$$

Here we have $V = 4g\sqrt{D}$, $q = 4H^2$ and $C = 2He^{-\delta}$. Note that, since g, H and δ are arbitrary parameters, we could equally well choose V, q and C as three independent parameters specifying a particular solution (another free parameter x_0 determines the pattern's location at the initial time moment).

Equation (10.3.46) describes a spike that moves at constant velocity V and grows exponentially with time at a rate $Q = \alpha + q - V^2/4D$. When $V = 0$, the spike is standing. As time increases, it grows exponentially with time while retaining its bell-shaped profile (Fig. 10.1a). The width of a standing spike is $\delta x_q = 2(D/q)^{1/2}$. Hence, more rapidly growing spikes are narrower.

The moving spikes with $V \neq 0$ are not symmetric (Fig. 10.1b). They are steeper at the front than at the back. When the propagation velocity V of the spike approaches the critical value $V_c = 2(qD)^{1/2}$, the slope at the back decreases until the profile transforms into a moving front for $V = V_c$. No spatially localized solutions with $V > V_c$ are possible. The width of a spike moving at velocity V is estimated as

$$\delta x_{V,q} = \frac{\delta x_q}{1 - (V/V_c)^2} . \tag{10.3.48}$$

The associated pattern of the field $\varrho(x,t)$, described by equation (10.3.47), represents a pulse which travels in the same direction as the spike of the population density n. However, the amplitude of this pulse is exponentially *decreasing* with time and the pulse is steeper at the back than at the front.

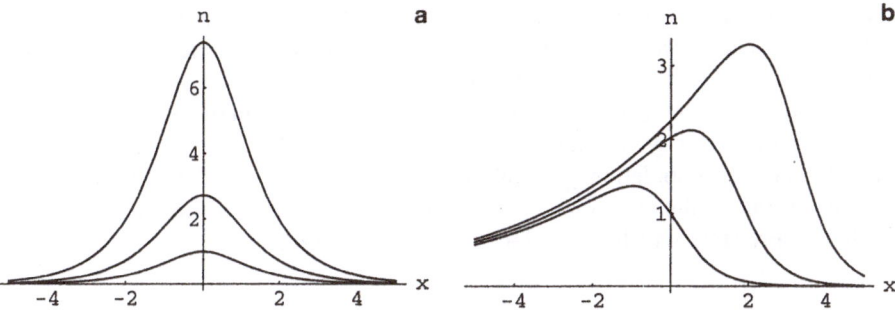

Fig. 10.1. Profiles of (a) standing ($V = 0$) and (b) moving ($V = 1.5$) population spikes at successive time moments $t_0 = 0$, $t_1 = 1$ and $t_2 = 2$; other parameters are $\alpha = 0$, $q = 1$, $D = 1$, $C = 1$, and $x_0 = 0$

Substituting (10.3.46) and (10.3.47) into (10.3.37), we can estimate the probability of finding a spike which exist at time T, moves at a velocity V, and has a growth rate Q:

$$p \propto \exp\left[-\frac{4}{3s} \left(Q - \alpha + \frac{V^2}{4D} \right)^{3/2} D^{1/2}T \right] . \qquad (10.3.49)$$

This estimate holds so long as spike formation is a rare statistical event, i.e., provided the probability p is exponentially small.

Of course, the solutions of the variational equations (10.3.35) and (10.3.36) describe only optimal spikes. The actual population distribution is subject to random fluctuations. However, it can still be conveniently described in terms of moving spikes. The main effect of the fluctuations would be that the parameters q and V of a spike are not constant in time, but rather undergo a slow diffusion-like evolution, so that the velocity and the width of the spike vary randomly in time. Furthermore, fluctuations provide the possibility of spontaneous creation and death of spikes.

Above, we have considered only the case of a one-dimensional medium. However, a similar procedure can also be developed for systems of higher dimensions. Then the variational equations would again be given by (10.3.35) and (10.3.36), with x replaced by the coordinate vector r. They would again be closely related to the nonlinear Schrödinger equation, but for a d-dimensional medium. It is known that for $d > 1$ this equation has no soliton solutions. Instead it describes the phenomena of wave collapse. The collapse solutions of the nonlinear Schrödinger equation could be used to construct the burst solutions of the variational equations (10.3.35) and (10.3.36). The bursts have a different growth law than the spikes in the one-dimensional case: their amplitudes diverge within finite time intervals.

The stochastic differential equation (10.1.5), which describes the development of intermittency in population explosions, is linear. Therefore, formation of population spikes in this birth–death model is a process which is principally different to pattern formation in a nonlinear distributed system, as discussed in Vol. I of this book. Here, traveling or standing spikes represent rare statistical events that result from a random local increase in the birth rate (or a decrease in the death rate) persisting for some period of time.

A population explosion cannot continue for an indefinitely long time: When the population density becomes very high, the nonlinear effects come into play and saturate the population growth. As a consequence, a statistically stationary populated state of the medium is established. Fluctuations of the population density in this steady state will be considered later in Sect. 12.4.

10.4 Explosions in Media with Random Breeding Centers

Until now we have assumed that fluctuations in birth and death rates are Gaussian and that their correlation time is vanishingly small, so that the approximation of white noise can be made. To analyze the effects lying beyond this assumption, we consider in this section (following *Mikhailov* and *Uporov* [10.13] and *Mikhailov* [10.14]) a stochastic model with random breeding centers.

Suppose that reproduction of particles can take place only within certain small spatial regions (we call them *breeding centers*), because energy, light, nutrients, etc. are present only inside such domains. Each breeding center exists for some time and then disappears. The centers are independently generated at random time moments at random locations in the medium. As the generation rate of breeding centers is increased while the death rate of particles is kept constant, a population explosion begins when a certain threshold is reached. The problem we address in this section is how to determine the explosion threshold for systems with random breeding centers.

We assume that, at least initially, the population is present with a small but nonvanishing density everywhere in the medium (i.e., ignition has already occurred) and consider a stochastic birth–death model described by the following equation for the local population density:

$$\dot{n} = -k_{\text{death}}n + k_{\text{birth}}(\boldsymbol{r}, t)n + D\nabla^2 n \ . \tag{10.4.1}$$

Since reproduction is possible only within breeding centers, the birth rate $k_{\text{birth}}(\boldsymbol{r}, t)$ is given by a sum

$$k_{\text{birth}}(\boldsymbol{r}, t) = \sum_j \chi(\boldsymbol{r} - \boldsymbol{r}_j, t - t_j) \ . \tag{10.4.2}$$

Here $\boldsymbol{r}_j$ and t_j denote random spatial locations and moments of appearance of breeding centers. All such centers are identical; a single center is described by

$$\chi(r, t) = J\phi(r)\theta(t) \ , \tag{10.4.3}$$

where the coefficient J characterizes the reproduction intensity on the center, while the functions $\phi(r)$ and $\theta(t)$ specify the spatial shape of a center and its time evolution. We assume that the function $\phi(r)$ falls rapidly to zero when $r \gg r_0$, where r_0 is the radius of a breeding region, and reaches $\phi(0) = 1$ in the center of this region. The lifetime of a center is τ_0 and the function $\theta(t)$ is chosen as $\theta(t) = 1$ for $\tau_0 > t > 0$ and $\theta(t) = 0$ outside this interval. The centers are independently generated at a rate of m centers per unit volume per unit time and hence their distribution in the medium is Poissonian.

Note that the centers can overlap and thus, as follows from (10.4.2), the local birth rate is given by the sum of contributions from all overlapping centers. If the dimensionless generation rate of breeding centers is introduced as

$$c = mr_0^d\tau_0 \ , \tag{10.4.4}$$

where d is the dimensionality of the medium, then c yields the mean number of such centers lying within the spatiotemporal element $r_0^d \tau_0$ occupied by a given center. When $c \gg 1$ the centers overlap strongly and the fluctuating field $k_{\mathrm{birth}}(r, t)$ is approximately Gaussian.

If the condition $r_0 \gg (D\tau_0)^{1/2}$ is satisfied, reproducing particles do not diffuse away from the breeding center within its lifetime. Such breeding centers have *short* lifetimes. On the other hand, if $r_0 \ll (D\tau_0)^{1/2}$ the particles are able to leave the reproductive spatial domain and form a cloud around it within the lifetime of a single center. These centers have *long* lifetimes.

A further important property of a breeding center is the relative increase of the population produced by this center within its lifetime. The increase in population density at a single center is described by the equation

$$-\dot{n} = -D\nabla^2 n - J\phi(r)n , \quad 0 < t < \tau_0 . \tag{10.4.5}$$

We have multiplied both sides by -1 so that it becomes formally identical with the Schrödinger equation for a quantum particle in the potential $U = -J\phi(r)$ if imaginary times are considered. Its general solution is

$$n(r, t) = \sum_l C_l \psi_l(r) e^{\lambda_l t} + \int d\lambda C(\lambda)\psi_\lambda(r) e^{\lambda t} , \tag{10.4.6}$$

where the sum is taken over the discrete spectrum and the integral is taken over the continuous spectrum of the linear operator

$$\hat{A} = D\nabla^2 + J\phi(r) . \tag{10.4.7}$$

For breeding centers we have $J > 0$ and therefore the eigenvalues λ in the discrete spectrum are positive. They correspond to negative energy levels of the quantum particle in the potential well U. The coefficients C are determined by the initial conditions.

Suppose that the linear operator $\hat{A}$ has positive eigenvalues and λ_0 is the largest of them. According to equation (10.4.6), the respective contribution to the population density grows with time as $\exp(\lambda_0 t)$. Hence, the properties of a breeding center essentially depend on the product $\lambda_0 \tau_0$.

If the condition $\lambda_0 \tau_0 \gg 1$ holds, each individual breeding center gives rise to a 'microexplosion', i.e., to an exponential increase of the population in the surrounding medium within the lifetime of this center. Such centers can be called *strong*. In the opposite case, when $\lambda_0 \tau_0 \ll 1$, or when the linear operator $\hat{A}$ has no positive eigenvalues at all, a single breeding center does not produce a significant increase of the local population density. These centers are termed *weak*.

We can estimate the largest positive eigenvalue of the operator (10.4.7) using the fact that it formally coincides (up to its sign) with the Hamiltonian operator of a quantum particle in a potential well. Referring to the analysis of the respective quantum problem [10.8], two limiting cases can be distinguished. If $J \gg Dr_0^2$, the potential well is deep and hence the estimate $\lambda_0 \approx J$ holds.

When $J \ll Dr_0^2$, the potential well is shallow. Then the results depend on the dimensionality of the problem. In the one-dimensional system, even very shallow wells are able to bind a quantum particle, i.e., they have a discrete energy level. Its value is determined using perturbation theory (see [10.8]). Employing the quantum analogy, we find

$$\lambda_0 \approx J \left(\frac{Jr_0^2}{D} \right)^{1/2} , \quad d = 1 . \tag{10.4.8}$$

The two-dimensional case is marginal: even very shallow potential wells have bound states but the binding energies are then exponentially small. Perturbation theory yields an estimate

$$\lambda_0 \approx \frac{D}{r_0^2} \exp\left(-\frac{\nu_0 D}{Jr_0^2} \right) , \quad d = 2 , \tag{10.4.9}$$

where ν_0 is a numerical constant of order unity. In three dimensions, shallow potential wells cannot bind a particle. Hence, at $d = 3$ the linear operator (10.4.7) has no positive eigenvalues if the condition $J \ll Dr_0^2$ is realized.

Applying the above estimates, an explicit criterion for strong breeding centers can be constructed. A center is strong if its intensity J is much larger than a certain critical intensity J^*. For the short-lived ($\tau_0 \ll r_0^2/D$) centers we have

$$J^* = \tau_0^{-1} , \quad d = 1, 2, 3 . \tag{10.4.10}$$

If the breeding centers are long-lived ($\tau_0 \gg r_0^2/D$), the critical intensity J^* depends on the dimensionality d of the medium,

$$J^* = \left(\frac{D}{r_0^2 \tau_0} \right)^{1/2} , \quad d = 1 , \tag{10.4.11}$$

$$J^* = \frac{D}{r_0^2} \left[\ln \left(\frac{D\tau_0}{r_0^2} \right) \right]^{-1} , \quad d = 2 , \tag{10.4.12}$$

$$J^* = \frac{D}{r_0^2} , \quad d = 3 . \tag{10.4.13}$$

When individual breeding centers are *weak* (i.e., $J \ll J^*$), we can take, as a first approximation for the explosion threshold, the condition that the mean birth rate $\langle k_{\text{birth}}(r, t) \rangle$ is equal to the death rate k_{death}. The mean birth rate is obtained by taking the average of equation (10.4.2), which yields

$$\langle k_{\text{birth}} \rangle = mJ \int \chi(r,t) dr\, dt = cJa_1 , \tag{10.4.14}$$

where the coefficient a_1 is

$$a_1 = r_0^{-d} \int \chi(r) dr . \tag{10.4.15}$$

This *mean-field approximation* determines the critical generation rate of breeding centers as

$$c_{\text{crit}} = \frac{k_{\text{death}}}{a_1 J} \; . \tag{10.4.16}$$

To evaluate the fluctuational corrections to this simple result, we take the average of both parts in equation (10.4.1) over the entire volume of the system and obtain

$$\frac{d}{dt} \langle n \rangle = (\langle k_{\text{birth}} \rangle - k_{\text{death}}) \langle n \rangle + \langle \delta k_{\text{birth}} \delta n \rangle \; , \tag{10.4.17}$$

where $\delta k_{\text{birth}}(\boldsymbol{r}, t) = k_{\text{birth}}(\boldsymbol{r}, t) - \langle k_{\text{birth}}(\boldsymbol{r}, t) \rangle$. Thus, the estimate (10.4.16) for the explosion threshold is derived by neglecting fluctuations in the population density.

The fluctuation corrections can be systematically evaluated using the diagrammatic perturbation method for stochastic differential equations proposed by *Wild* [10.15] (see also [10.16]). This analysis has been performed in Refs. [10.13, 17]. Its main result is that the weakness of breeding centers (i.e., the inequality $J \ll J^*$) does not guarantee that the mean-field approximation is applicable. Fluctuations near the explosion threshold are small and hence the estimate (10.4.16) holds if the additional condition

$$c(J\tau_0)(J/J^*) \ll 1 \tag{10.4.18}$$

is satisfied.

When the condition (10.4.18) is violated, fluctuations may lead to significant lowering of the explosion threshold. *Mikhailov* and *Uporov* [10.13, 14] have shown that in this case an important role is played by *clusters* of breeding centers. Although each individual center is weak, a group of such objects may effectively represent a strong breeding center, i.e., lead to an exponentially large local increase of the population density. Hence, strong local 'microexplosions' would take place on the background characterized by much slower growth of the population density. Such strong clusters appear very rarely but they can still give an essential contribution to the average rate of growth of the total population.

Since formation of a strong cluster is a rare statistical event, its probability can be estimated using variational methods, similar to those described in Sect. 9.3. When the probability distribution for clusters is known, their mean contribution to the rate of growth of the population in the considered medium can be evaluated and compared with the contribution from the more uniform background (see [10.13,14]). This analysis yields the same condition (10.4.18) that has been independently derived by estimating diagrams of the infinite perturbation series.

The principal difference in the case of *strong* individual breeding centers (with $J \gg J^*$) is that each such center already produces a 'microexplosion' in the medium. When these microexplosions are well separated in space and in time (see the discussion below), diffusion is able to homogenize the population distribution after each single microexplosion. Then every subsequent microexplosion begins from a uniform level of the population density.

Under these conditions, the evolution of the average population density is determined in the mean-field approximation by the equation

$$\frac{d}{dt}\langle n \rangle = -k_{\text{death}}\langle n \rangle + m\Delta N_1 \ , \qquad (10.4.19)$$

where ΔN_1 represents a population increase on a single strong breeding center and m is the rate of generation of such centers.

In this approximation, when a new breeding center appears, the population density is constant and equal to the spatial average $\langle n \rangle$. Therefore the population increase on the center is determined by a general solution (10.4.6) where the decomposition coefficients C_l are given by

$$C_l = \langle n \rangle \int \psi_l(\boldsymbol{r})d\boldsymbol{r} \qquad (10.4.20)$$

if the orthogonal eigenfunctions $\psi_l(\boldsymbol{r})$ of the linear operator (10.4.7) are normalized as

$$\int \psi_l(\boldsymbol{r})\psi_{l'}(\boldsymbol{r})d\boldsymbol{r} = \delta_{ll'} \ . \qquad (10.4.21)$$

If, in the decomposition (10.4.6), we keep only the fastest growing exponential term with $l = 0$ which corresponds to the largest positive eigenvalue λ_0, the total increase of the population at a strong breeding center within its lifetime τ_0 is approximatly given by

$$\Delta N_1 \approx \langle n \rangle C_0 \exp(\lambda_0\tau_0) \int \psi_0(\boldsymbol{r})d\boldsymbol{r} = \langle n \rangle r_1^d \exp(\lambda_0\tau_0) \ , \qquad (10.4.22)$$

where r_1 is the localization radius of the eigenfunction $\psi_0(\boldsymbol{r})$ defined as

$$r_1 = \left[\int \psi_0(\boldsymbol{r})d\boldsymbol{r} \right]^{2/d} \ . \qquad (10.4.23)$$

This radius is estimated as $r_1 \approx r_0$ for $J \gg D/r_0^2$ and $r_1 \approx (D/\lambda_0)^{1/2}$ for $J \ll D/r_0^2$ if $d = 1, 2$.

Hence, as follows from (10.4.19), the explosion threshold is reached when the dimensionless generation rate $c = m r_0^d \tau_0$ of breeding centers is

$$c_{\text{crit}} = k_{\text{death}}\tau_0 \left(\frac{r_0}{r_1} \right)^d \exp(-\lambda_0\tau_0) \ . \qquad (10.4.24)$$

This estimate for the explosion threshold for strong breeding centers is derived in the mean-field approximation, i.e., by neglecting interference of different breeding centers. Such interference can be indeed neglected if the volume Ω of the space–time time region occupied by an enhanced-density spot created by a breeding center is much smaller than average space–time volume per single center, i.e., if the condition $m\Omega \ll 1$ is satisfied.

To estimate the volume Ω, we consider first the density at the central point of a spot as a function of time. As the breeding center disappears, it leaves in the medium an enhanced-density spot of linear size about r_1 with a maximum population density (in the central point) of about $\delta n \sim \langle n \rangle \exp(\lambda_0 \tau_0)$. After that, the spot begins to spread due to diffusion and the density at the central point decreases with time as

$$\delta n(t) \approx \frac{r_1^d \delta n(0)}{(4\pi Dt)^{d/2}} \sim \langle n \rangle \frac{r_1^d}{(4\pi Dt)^{d/2}} \exp(\lambda_0 \tau_0) . \tag{10.4.25}$$

The enhanced-density spot exists for as long as it remains distinguishable from its background, i.e. while the condition $\delta n(t) \gg \langle n \rangle$ is satisfied. Therefore, we can estimate the lifetime T^* of a single enhanced-density spot using equation (10.4.25) as

$$T^* = \frac{D}{r_1^2} \exp\left(\frac{2\lambda_0 \tau_0}{d}\right) . \tag{10.4.26}$$

Since the spot spreads diffusively, its linear size by the time moment $t = T^*$ is about $(DT^*)^{1/2}$. Therefore, the characteristic space–time volume Ω occupied by a single enhanced-density spot is $\Omega = T^*(DT^*)^{d/2}$. Substituting this estimate into the condition $m\Omega \ll 1$ and using equation (10.4.26), we find that the overlaps between the enhanced-density spots produced by different breeding centers can be neglected if their dimensionless generation rate c is sufficiently small, i.e., if

$$c \ll \frac{D\tau_0}{r_1^2} \exp\left[-(1 + \frac{2}{d})\lambda_0 \tau_0\right] . \tag{10.4.27}$$

Hence, the mean-field estimate (10.4.24) for the explosion threshold is valid only if the critical generation rate c_{crit} of the breeding centers satisfies condition (10.4.27).

What happens if condition (10.4.27) is violated near the explosion threshold?

Note, that besides individual microexplosions caused by single breeding centers, there are also *chains* of such local population explosions. In a chain, subsequent microexplosions start within the enhanced-density spots left by previous local explosions. Therefore, the total production of the new population on a chain is much larger than in the same number of separated breeding centers.

Generally, chains of different lengths, consisting of two, three, etc., subsequent microexplosions, are present in the medium and the probability distribution over chain lengths can be defined. If $m\Omega \ll 1$, formation of even the simplest (and the most probable) chain of length two is an extremely rare event and the contribution from such chains can be neglected (this contribution has been analytically determined in [10.13]).

When, however, the generation rate m is increased and the condition $m\Omega \sim 1$ is approached, so that the overlaps become much more frequent, the chains cease to represent rare events. Now the chains of local microexplosions begin to give an essential contribution to the total increase of the population in the medium, leading to a significant lowering of the explosion threshold, as compared with the mean-field estimate (10.4.24). Further increase of the generation rate would result in the formation, with a nonvanishing probability, of *infinite* chains. Along such chains, a rapid growth of the population density would take place.

The process leading to the formation of infinitely long chains represents a variant of the *percolation* phenomenon. A typical percolation problem is formulated as follows:

Suppose that we have a set of randomly distributed points (sites). Each site is surrounded by a sphere of the same radius. Any two sites are called linked if their corresponding spheres overlap. Linked sites form clusters. If we view the spheres as pores in a solid material, these clusters would represent channels through which liquid can penetrate. Let us assume that we permanently supply liquid to a certain pore. Then we will find the liquid in all the pores that are connected with the first one, i.e., in all pores belonging to the cluster that includes the initial site. If the sites are rare and distances between them are large, only small clusters are likely. However, when the number of sites in the medium increases (or the radius of spheres becomes larger), the typical size of a cluster grows and, at a certain threshold frequency of sites, infinite clusters appear with a nonvanishing probability. The presence of an infinite cluster in the medium means that, if liquid is supplied to one of its sites, it can spread through this cluster to reach even infinitely distant parts of the medium. An introduction to percolation theory is given by *Stauffer* [10.18].

For explosions caused by strong breeding centers, the role of 'spheres' in a standard percolation problem is played by enhanced-density spots while the 'sites' represent breeding centers randomly distributed in space and time. The analogy is clearer still if we treat time as an additional coordinate in the problem. Then the chains of microexplosions would look like clusters in the enlarged space of $d+1$ dimensions. However, there is a difference: only forward percolation along the time axis is allowed. Such *directed* percolation is known to have special properties (see, e.g. [10.19]).

As already noted, the nature of the 'particles' is irrelevant for the general stochastic birth–death models studied in this chapter. They can represent actual physical particles, such as molecules, or entire biological organisms, such as bacteria. Our analysis of spatiotemporal chaos in Chap. 7 has revealed that the breakdown of spatial coherence in distributed-active systems is intimately related to the appearance of multiple structural defects. These defects behave in many respects like individual particles. They can die and reproduce in the medium. The transition to developed turbulence is accompanied by a sharp increase in the density of defects.

We have seen that turbulence in reaction–diffusion systems is often characterized by spatiotemporal intermittency. An intriguing question is whether this phenomenon has the same general origin as the intermittency in population explosions discussed in the present chapter. Although this has not yet been convincingly demonstrated, it seems to be very probable. In the case of the complex Ginzburg–Landau equation, local fluctuations in the reproduction rates of defects could be provided by phase turbulence, thus playing the role of external noise in a stochastic birth–death model. Even the intermittency in hydrodynamic turbulence, described by the Navier–Stokes equations, possesses common features with the intermittency in birth–death models as is revealed, for instance, by investigations of *cluster statistics* in both systems that were performed by *Sanada* [10.20] and by *Zanette* and *Mikhailov* [10.21].

The processes involving topological defects, such as (cores of) spiral waves in two-dimensional oscillatory media, have, however a special property. Such defects can disappear only in pairs, when two defects with opposite topological charges (i.e., rotation directions) collide. This means that, instead of death events for individual particles, processes of binary annihilation take place in the system. The effects related to such processes are investigated in the next chapter.

11. Extinction and Complex Relaxation

Relaxation is a process by which a system approaches one of its attractors. When this attractor is a trivial unpopulated state, the term 'extinction' is applied. In nonlinear and nonuniform distributed systems, both relaxation and extinction may strongly deviate from a simple exponential relaxation law.

Population extinction in a medium with randomly distributed traps is an example of a situation where the hierarchy of characteristic relaxation times for different spatial modes reveals itself in complex relaxation laws. Similar effects are found when binary annihilation of diffusing particles is considered.

If conserved quantities are present, the relaxation process in a reaction–diffusion system becomes diffusion controlled in its last stage and this leads to a slow power-law relaxation of the population densities.

11.1 Diffusion with Random Traps

The simplest situation in which complicated spatial patterns are produced in the process of extinction, is found if the particles, wandering at random through the medium, are captured (or destroyed) by a set of randomly distributed static *traps* (sinks). This problem arises in many applications [11.1–7]. In the continuum limit, the population density of particles obeys

$$\dot{n} = -\alpha(r)n + D\Delta n . \tag{11.1.1}$$

We assume that traps are located at randomly distributed independent points r_j in the medium, so that

$$\alpha(r) = \sum_j \chi(r - r_j) , \tag{11.1.2}$$

where the function $\chi(r)$ describes an individual trap. This function reaches very large positive values within a small region of radius r_0 (the size of a trap) and vanishes outside it. Traps can overlap, and the mean number of traps per unit volume is m. Note that (11.1.1) can also be interpreted as a model with randomly distributed *decay centers*.

Since there is no reproduction, the mean particle density $\langle n \rangle$, averaged over the volume, goes to zero with time. At first glance, it seems that the extinction process should be exponential, i.e. that

$$\langle n(t) \rangle = n_0 \exp(-\langle \alpha \rangle t) , \qquad (11.1.3)$$

where $\langle \alpha \rangle$ is the mean decay rate and n_0 is the initial population density of particles.

Actually, this prediction is wrong in the long-time limit. *Balagurov* and *Vaks* [11.8] showed in 1973 that in this limit the law (11.1.3) is replaced by a much slower time dependence, due to the contribution from large rare voids. Later, this result was independently discovered by *Grassberger* and *Procaccia* [11.9].

Note that, when the traps are distributed randomly, it is always possible to find large enough regions that are free of traps. Particles in the interior of such large voids disappear only when they arrive, by means of diffusion, at the walls. The larger the void, the longer the life-time of particles inside it. In the long-time limit almost all surviving particles are found in large voids. The minimum size of the trap-free regions populated by particles increases with time. On the other hand, very large voids are extremely rare. The effective law of extinction is thus determined by the competition between the very long life-times in large voids and the very small probabilities of finding such voids in the medium.

To derive the correct long-time extinction law, *Balagurov* and *Vaks* [11.8] used the analogy between (11.1.1) and the quantum-mechanical Schrödinger equation with a random stationary potential. Indeed, if we write (11.1.1) in the form

$$-\dot{n} = \hat{L} n \qquad (11.1.4)$$

with the linear operator

$$\hat{L} = -D\Delta + \alpha(r) , \qquad (11.1.5)$$

the problem is reduced to the study of the Schrödinger equation with imaginary time. However, in contrast to the situation with breeding centers investigated in Sect. 10.3, the "potential" $\alpha(r)$ is non-negative. This implies that all eigenvalues λ_k of the operator $\hat{L}$, defined by

$$\hat{L}\psi_k = \lambda_k \psi_k , \qquad (11.1.6)$$

are also non-negative, i.e. the spectrum of $\hat{L}$ is bounded from below by $\lambda = 0$. This property of the spectrum holds for any random realization of $\alpha(r)$. According to the terminology used in the theory of the Schrödinger equation with a random potential [11.10], $\lambda = 0$ represents the *finite fluctuational boundary* of the spectrum.

The spectrum of the operator $\hat{L}$ is discrete near the $\lambda = 0$ boundary. Eigenvalues λ_k lying near this boundary are determined by the fluctuations of the potential for which α is close to zero in large regions of space, i.e. by large voids. The probability of finding a fluctuation with very small λ is exponentially small. This probability was estimated by *Lifshitz* [11.1, 12] as

$$p(\lambda) \sim \exp\left[-\mu_d m \left(\frac{\lambda}{D}\right)^{-d/2}\right] . \tag{11.1.7}$$

The coefficient μ_d is given by

$$\mu_d = \frac{\pi^{d/2}}{\Gamma(d/2+1)} \varepsilon_d^{d/2} , \tag{11.1.8}$$

where ε_d is the ground-state energy in the d-dimensional potential well of unit radius with infinitely high walls ($\varepsilon_1 = \pi^2/8$, $\varepsilon_2 = 2.88 \ldots$, $\varepsilon_3 = \pi^2/2$) and $\Gamma(x)$ is the gamma-function. An optimal fluctuation that gives rise to the eigenvalue λ represents a spherical void of radius $R \sim (D/\lambda)^{1/2}$.

The general solution of (11.1.1) can be written as a superposition of eigenfunctions ψ_k:

$$n(\boldsymbol{r},t) = \sum_k C_k \psi_k(\boldsymbol{r}) \exp(-\lambda_k t) . \tag{11.1.9}$$

The coefficients C_k are determined by the initial condition $n(\boldsymbol{r},0) = n_0$. If the eigenfunctions ψ_k are orthonormal we have

$$C_k = n_0 \int \psi_k(\boldsymbol{r}) \, d\boldsymbol{r} . \tag{11.1.10}$$

At large times the dominant contribution to (11.1.9) is given by terms with very small values of λ_k. The eigenfunctions corresponding to such λ_k''s are localized in rare separate fluctuations of the field $\alpha(\boldsymbol{r})$, which represent large voids. The mean population density $\langle n \rangle$ by time t can then be estimated as

$$\langle n(t) \rangle \sim n_0 \int_0^\infty p(\lambda) \exp(-\lambda t) \, d\lambda , \tag{11.1.11}$$

where $p(\lambda)$ is given by (11.1.7).

The integrand of (11.1.11) has a sharp maximum at the point $\lambda = \lambda_c$ which can easily be found from (11.1.7) and (11.1.11):

$$\lambda_c = D^{d/(d+2)} \left(\frac{d\mu_d m}{2t}\right)^{2/(d+2)} . \tag{11.1.12}$$

Approximation of (11.1.11) by its value at λ_c yields

$$\langle n(t) \rangle \sim n_0 \exp\left[-\nu_d \left(D t m^{2/d}\right)^{d/(d+2)}\right] , \tag{11.1.13}$$

where ν_d is a numerical factor ($\nu_1 = 1.61 \ldots$, $\nu_2 = 6.01 \ldots$, $\nu_3 = 13.54 \ldots$).

The asymptotic extinction law (11.1.13), derived in [11.8,9], is much slower than the exponential dependence (11.1.3). However, it should be remembered that this law becomes valid only at sufficiently large times. Numerical simulations show that the crossover from the standard exponential dependence (11.1.3) to the law (11.1.13) in three-dimensional media is found only at exceptionally small concentrations $\langle n \rangle$. In one- and two-dimensional media this asymptotic law is more easily observed.

It is quite remarkable that (11.1.13) does not depend on the properties of an individual trap, such as function $\chi(r)$ in (11.1.2). Only the statistics of the formation of large trap-free regions is relevant for the derivation of this result. *Kayser* and *Hubbard* [11.14, 15] showed that the extinction law (11.1.13) remains valid when the traps are not allowed to overlap. They also provided a proof that the long-time exponent $d/(d+2)$ is exact. *Blumen* et al. [11.16] and *Klafter* and *Blumen* [11.17] generalized the above analysis to diffusion on fractal structures (such as the Serpinski gasket, see Sect. 4.1). They found that the law (11.1.13) remains valid in this case if we interpret d as a so-called "spectral dimension" of a fractal pattern.

Note that in the late stages of extinction the spatial distribution of particles becomes strongly nonuniform. *Kayser* and *Hubbard* [11.15] obtained a lower bound on the root mean square (rms) relative fluctuation $\delta n(t)/\langle n(t) \rangle = \langle (n - \langle n \rangle)^2 \rangle^{1/2}/\langle n \rangle$ of the population density at long times:

$$\frac{\delta n(t)}{\langle n(t) \rangle} > \exp\left[+\left(1 - 2^{-2/(d+2)}\right)\nu_d\left(Dtm^{2/d}\right)^{d/(d+2)}\right] . \tag{11.1.14}$$

We see that the relative fluctuations are greatly enhanced in the late stages, which reflects the fact that the long-time transients are dominated by a highly improbable class of configurations.

Actually, particles that have survived up until time t would be concentrated in randomly scattered patches, each representing a large trap-free region. The minimum spatial extent of a patch can be estimated as $R_c(t) \sim (D/\lambda_c)^{1/2}$, where $\lambda_c(t)$ is given by (11.1.12); $R_c(t)$ is the radius of the voids that make the dominant contribution to $\langle n(t) \rangle$. The average distance between the patches increases with time as $[p(\lambda_c(t))]^{-1/d}$ where $p(\lambda)$ is the probability (11.1.7) of finding a void with the lowest eigenvalue λ.

11.2 Irreversible Annihilation

In this section we consider the reaction of irreversible annihilation of particles A and B, which perform a diffusive random walk in the medium,

$$A + B \to C . \tag{11.2.1}$$

The standard kinetic equations [11.18, 19] for this reaction are simply

$$\dot{n}_A = -\kappa n_A n_B + D\Delta n_A , \qquad \dot{n}_B = -\kappa n_A n_B + D\Delta n_B . \tag{11.2.2}$$

For simplicity we assume that the diffusion constants for both types of particles are the same, $D_A = D_B = D$. The solution which satisfies the initial conditions $n_A(0) = n_B(0) = n_0$ is

$$n_A(t) = n_B(t) = n_0\left(1 + \kappa n_0 t\right)^{-1} .$$

$$(11.2.3)$$

Hence, according to standard kinetics, in the long-time limit, the concentrations of both reagents should follow the asymptotic law

$$n_A(t) = n_B(t) = (\kappa t)^{-1} .$$

$$(11.2.4)$$

However, this prediction is in fact wrong. The correct asymptotic behavior of the concentrations $n_A(t)$ and $n_B(t)$ in the limit $t \to \infty$ can only be found by using *fluctuational kinetics*, which takes into account stochastic nonuniformity of the distributions of A and B.

Let us examine more closely the initial state of our system before the annihilation reaction starts. Note that, even if the initial distribution is *statistically* homogeneous, there are still some nonhomogeneous fluctuations present. The properties of these fluctuations depend on the procedure that was used to prepare the initial state of the system. For example, one can generate A and B particles in pairs and at a fixed distance from one another. This would lead to a strong correlation between the positions of A and B particles.

Another possibility is to assume that initially all A and B particles were randomly distributed in the medium, independently of one another. This initial state is described by the Poisson distribution of particles within any given element of the medium. Note that this type of distribution develops if the A and B particles are allowed to diffuse in the absence of reaction.

In the presence of the annihilation reaction, the population densities of A and B particles evolve according to (11.2.2). Therefore, the evolution of the local difference $\eta = n_A - n_B$ of the two population densities obeys a pure diffusion equation

$$\dot{\eta} = D\Delta\eta .$$

$$(11.2.5)$$

Because of random inhomogeneities, in the initial state of the system there are some regions with an excess of A particles (where $\eta > 0$) and other regions (where $\eta < 0$) with a dominant population of B particles. We can expect that, in the long-time limit, the B particles will disappear completely from regions with an initial excess of A particles, so that in these regions $n_A = \eta$, $n_B = 0$. In other regions, only B particles will remain, i.e. $n_A = 0$, $n_B = \eta$. Consequently, the entire medium in which the annihilation reaction proceeds will be partitioned into regions of two types: those filled exclusively with A particles and those filled exclusively with B.

How can we find the average dimensions of these regions and the average A and B concentrations in each type of region? Note that, according to (11.2.5), the annihilation reaction does not influence the temporal evolution of the difference in population densities η. It evolves as

$$\eta(\boldsymbol{r}, t) = \int G_{\mathrm{d}}(\boldsymbol{r} - \boldsymbol{r}', t)\eta(\boldsymbol{r}', 0)\, d\boldsymbol{r}' ,$$

$$(11.2.6)$$

where

$$G_d(\varrho, t) = (2\pi D t)^{-d/2} \exp\left(-\frac{\varrho^2}{4Dt}\right) \tag{11.2.7}$$

is the Green's function of the diffusion equation (11.2.5). Hence, the mean square of η at time t is given by

$$\langle \eta^2(t) \rangle = \iint d\mathbf{r}_1 \, d\mathbf{r}_2 \, G_d(\mathbf{r} - \mathbf{r}_1, t) G_d(\mathbf{r} - \mathbf{r}_2, t) \, \langle \eta(\mathbf{r}_1, 0)\eta(\mathbf{r}_2, 0) \rangle . \tag{11.2.8}$$

Since $\eta = n_A - n_B$, we have

$$
\begin{aligned}
&\langle \eta(\mathbf{r}_1, 0)\eta(\mathbf{r}_2, 0) \rangle \\
&= \langle n_A(\mathbf{r}_1, 0)n_A(\mathbf{r}_2, 0) \rangle + \langle n_B(\mathbf{r}_1, 0)n_B(\mathbf{r}_2, 0) \rangle \\
&\quad - \langle n_A(\mathbf{r}_1, 0)n_B(\mathbf{r}_2, 0) \rangle - \langle n_B(\mathbf{r}_1, 0)n_A(\mathbf{r}_2, 0) \rangle .
\end{aligned}
\tag{11.2.9}
$$

In the initial state of the system, all particles are randomly and independently distributed in the medium. Since the distribution of particles in each volume is statistically independent, the correlation function of density fluctuations $\delta n_A = n_A - \langle n_A \rangle$ should vanish for noncoinciding arguments, i.e.

$$\langle \delta n_A(\mathbf{r}_1, 0)\delta n_A(\mathbf{r}_2, 0) \rangle = Q\delta(\mathbf{r}_1 - \mathbf{r}_2) . \tag{11.2.10}$$

Furthermore, the statistical uniformity of the distribution implies that this correlation function is invariant under spatial translations. Hence, Q in (11.2.10) cannot depend on $\mathbf{r}_1$ or $\mathbf{r}_2$. The explicit value of Q can be found from the following arguments: In the absence of reactions, the total number N of A particles inside any volume element Ω obeys a Poisson distribution, so that for fluctuations $\delta N = N - \langle N \rangle$ from the average number $\langle N \rangle$ we have

$$\frac{\langle \delta N^2 \rangle}{\langle N \rangle} = 1 . \tag{11.2.11}$$

On the other hand, $\langle N \rangle = \Omega \langle n_A \rangle$ and

$$\langle \delta N^2 \rangle = \int_{(\Omega)} \int \langle \delta n_A(\mathbf{r}_1)\delta n_A(\mathbf{r}_2) \rangle \, d\mathbf{r}_1 \, d\mathbf{r}_2 = Q\Omega . \tag{11.2.12}$$

Comparing (11.2.11) and (11.2.12), we find $Q = \langle n_A \rangle$. Hence, the density-density correlation function for independently distributed particles is

$$\langle n_A(\mathbf{r}_1, 0)n_A(\mathbf{r}_2, 0) \rangle = \langle n_A \rangle^2 + \langle n_A \rangle \delta(\mathbf{r}_1 - \mathbf{r}_2) . \tag{11.2.13}$$

Evidently, the same result holds for B particles. Moreover, since the distributions of A and B are independent,

$$\langle n_A(\mathbf{r}_1, 0)n_B(\mathbf{r}_2, 0) \rangle = \langle n_A \rangle \langle n_B \rangle . \tag{11.2.14}$$

If we take into account that the initial mean concentrations of A and B particles are the same ($\langle n_A \rangle = \langle n_B \rangle = n_0$) and use (11.2.13) and (11.2.14), we find from (11.2.9) that

$$\langle \eta(\boldsymbol{r}_1, 0) \eta(\boldsymbol{r}_2, 0) \rangle = 2n_0 \delta(\boldsymbol{r}_1 - \boldsymbol{r}_2) . \tag{11.2.15}$$

This expression can be further substituted into (11.2.8), yielding

$$\langle \eta^2(t) \rangle = 2n_0 \int G_d^2(\varrho, t) \, d\varrho = 2n_0 (2\pi Dt)^{-d/2} . \tag{11.2.16}$$

On the other hand, the definition of η implies $\langle \eta^2 \rangle = \langle n_A^2 \rangle + \langle n_B^2 \rangle - 2 \langle n_A n_B \rangle$. Moreover, in the long-time limit A and B particles are separated in different spatial regions, and we have $\langle n_A n_B \rangle \approx 0$. Hence, using (11.2.16) and the initial conditions, we find in this limit

$$\langle n_A^2 \rangle = \langle n_B^2 \rangle \approx n_0 (2\pi Dt)^{-d/2} . \tag{11.2.17}$$

Because $\langle n_A \rangle^2 \leq \langle n_A^2 \rangle$, this result provides an upper bound for the mean population density of reacting particles at time t, i.e.

$$\langle n_A \rangle = \langle n_B \rangle \sim n_0^{1/2} (Dt)^{-d/4} . \tag{11.2.18}$$

This estimate was first suggested by *Zeldovich* [11.20] and later derived in a more rigorous way by *Ovchinnikov* and *Zeldovich* [11.21] and by *Toussaint* and *Wilczec* [11.22]. It was shown in [11.23] that, in the long-time limit, the lower bound on $\langle n_A \rangle$ and $\langle n_B \rangle$ coincides with (11.2.18) and therefore (11.2.18) gives an actual law of extinction of particles due to the annihilation reaction. We see that it is fundamentally different from the prediction (11.2.4) of classical kinetics, which does not take into account fluctuations of the population densities.

Regions that are occupied predominantly by either A or B particles (below we shall call these A and B regions) form a complicated geometric pattern which is determined by the random function $\eta(r, t)$. When crossing a boundary that separates any two A and B regions, this function changes its sign. Hence, the boundaries are those surfaces (lines or points, depending on the dimensionality of the medium) where $\eta(r, t) = 0$.

To find the velocity at which the boundary moves, we can note [11.24] that in the one-dimensional case the position x_0 of the boundary at time t is determined by the equality $\eta(x_0(t), t) = 0$. Differentiating this with respect to time, we obtain

$$\frac{dx_0}{dt} \frac{\partial \eta}{\partial t} \bigg|_{x_0} + \frac{\partial \eta}{\partial t} \bigg|_{x_0} = 0 . \tag{11.2.19}$$

On the other hand, η obeys a diffusion equation, so we can write

$$\frac{\partial \eta}{\partial t} \bigg|_{x_0} = D \frac{\partial^2 \eta}{\partial x^2} \bigg|_{x_0} . \tag{11.2.20}$$

Substituting (11.2.20) into (11.2.19), we find the displacement velocity of the boundary:

$$\frac{dx_0}{dt} = -D \frac{\partial^2 \eta / \partial x^2}{\partial \eta / \partial x} \Big|_{x=x_0(t)} . \qquad (11.2.21)$$

To obtain an estimate of the displacement velocity, we consider the boundary separating an A from a B region. The corresponding maximum population densities are denoted n_A and n_B and the characteristic linear dimensions L_A and L_B. This situation can be modeled, for example, by the dependence

$$\eta = \frac{n_A \left[\exp(x/L_A) - 1 \right] - n_B \left[\exp(-x/L_B) - 1 \right]}{\exp(x/L_A) + \exp(-x/L_B)} . \qquad (11.2.22)$$

Using (11.2.22) and (11.2.21), we obtain

$$\frac{dx_0}{dt} = -\frac{D}{2} \frac{n_A - n_B}{n_A L_B + n_B L_A} . \qquad (11.2.23)$$

As an order-of-magnitude estimate, this expression obviously holds even when the detailed dependencies of the population densities in adjoining A and B regions are different. According to (11.2.23), if the maximum population density n_A in the A region is higher than the maximum population density n_B, the boundary will shift to the right and the A region will grow at the expense of the B region. If the inequality $n_A < n_B$ holds, the boundary will move in the opposite direction. Since the population density of particles in a region is usually higher when the linear dimension of the region is larger, we conclude that, as time elapses, the large regions will eat up the small regions, and the total number of regions will decrease.

In the two- and three-dimensional cases, the motion of a boundary between A and B regions will also be affected by the curvature of this boundary. To estimate the correction in the two-dimensional case, we can introduce a local coordinate system whose z axis lies perpendicular to the boundary and whose x axis is tangential to the boundary line at a given point. Simple considerations then show that the displacement velocity of the boundary should be

$$\frac{dz_0}{dt} = -D \frac{\partial^2 \eta / \partial z^2}{\partial \eta / \partial z} \Big|_{z=z_0(t)} - \frac{D}{R} , \qquad (11.2.24)$$

where R is the radius of curvature of the interface at this point (R is taken to be positive if the interface is convex in the direction towards the z axis).

In the three-dimensional case, the displacement velocity is

$$\frac{dz_0}{dt} = -D \frac{\partial^2 \eta / \partial z^2}{\partial \eta / \partial z} \Big|_{z=z_0(t)} - D \left(\frac{1}{R_1} + \frac{1}{R_2} \right) , \qquad (11.2.25)$$

where R_1 and R_2 are the two principal radii of curvature of the interface surface.

The effect of the terms which depend on the curvature in (11.2.24) and (11.2.25) is the suppression of sharp A or B tentacles that penetrate into regions occupied by the opposite type of particles, and also rapid wiping out of small A islands inside B regions or vice versa. This results in a smoothing of the interface boundaries, a general increase in the sizes of the regions and a decrease in the total number of A and B regions as time elapses.

Our strict partitioning into A and B regions is of course an idealization. At any moment, A and B particles continue to annihilate and this reaction is obvioiusly only possible if there is some overlap at the boundary. This overlap between two annihilating components which are transported by diffusion from the exterior into the reaction zone was investigated by *Zeldovich* [11.25] in 1949 as part of a study of the combustion of unmixed gases (see also [11.26, 27]).

At the boundary we have $\eta = 0$, so we can set $\eta = \mu x$ in the immediate vicinity. Since the reaction terms do not appear in (11.2.5), the motion of the boundary and the change in the coefficient μ are determined only by the slow diffusion process. If the annihilation rate κ is large enough, we can assume that the distribution of reacting particles in the overlap zone adjusts adiabatically to the position of the boundary and the value of μ at a given moment in time.

For the sum $q = n_A + n_B$ we find from (11.2.2)

$$\frac{\partial q}{\partial t} = -\frac{1}{4}\kappa(q^2 - \eta^2) + D\frac{\partial^2 q}{\partial x^2} . \tag{11.2.26}$$

In a stationary state, the behavior of q in the overlap zone is described by the equation

$$D\frac{d^2 q}{dx^2} = \frac{1}{4}\kappa(q^2 - \mu^2 x^2) \tag{11.2.27}$$

with the boundary condition $q \to \mu|x|$ as $x \to \pm\infty$.

Introducing the dimensionless variables θ and ξ,

$$q = \left(\frac{D\mu^2}{\kappa}\right)^{1/3} \theta , \qquad x = \left(\frac{D}{\kappa\mu}\right)^{1/3} \xi , \tag{11.2.28}$$

we can put this equation into the form

$$\theta'' = \tfrac{1}{4}(\theta^2 - \xi^2) , \tag{11.2.29}$$

where $\theta \to |\xi|$ as $\xi \to \pm\infty$.

The width of the overlap zone is thus given in order of magnitude by

$$l_0 = \left(\frac{D}{\kappa\mu}\right)^{1/3} , \tag{11.2.30}$$

and the population densities of A and B particles in this zone (at $x = 0$) are given, again in order of magnitude, by

$$n_A(0) = n_B(0) = \frac{q(0)}{2} = \left(\frac{D\mu^2}{\kappa}\right)^{1/3} .$$

(11.2.31)

When the annihilation rate κ increases, the overlap zone becomes narrower and the population densities of A and B particles in it become smaller.

Equations (11.2.30) and (11.2.31) contain the quantity μ, which can be determined using the following considerations. The characteristic linear dimension of the A and B regions at a time t after the annihilation reaction is switched on will be $L(t) \approx (Dt)^{1/2}$, while the characteristic population densities of A and B in these regions will be $n_A = n_B \approx n_0^{1/2}(Dt)^{-3/4}$ [cf. (11.2.18)]. Adopting n_A/L as an estimate of μ, we find

$$\mu(t) \approx n_0^{1/2}(Dt)^{-5/4} .$$

(11.2.32)

By time t the width of the overlap zone thus becomes

$$l_0 \approx \left(\frac{D^2}{\kappa^2 n_0}\right)^{1/6} (Dt)^{5/12} .$$

(11.2.33)

As time goes by, the overlap zone broadens, but at a slightly slower pace than the rate of growth of the linear dimensions of the A and B regions:

$$\frac{l_0}{L} \sim \left(\frac{D^2}{\kappa^2 n_0}\right)^{1/6} (Dt)^{-1/12} .$$

(11.2.34)

It should be emphasized that (11.2.34) applies only to a three-dimensional medium. For a two-dimensional medium ($d = 2$) we would have

$$l_0 \sim \left(\frac{D^2}{\kappa^2 n_0}\right)^{1/6} (Dt)^{1/3} ,$$

(11.2.35)

$$\frac{l_0}{L} \sim \left(\frac{D^2}{\kappa^2 n_0}\right)^{1/6} (Dt)^{-1/6} .$$

(11.2.36)

In the one-dimensional case ($d = 1$) we find

$$l_0 \sim \left(\frac{D^2}{\kappa^2 n_0}\right)^{1/6} (Dt)^{1/4} ,$$

(11.2.37)

$$\frac{l_0}{L} \sim \left(\frac{D^2}{\kappa^2 n_0}\right)^{1/6} (Dt)^{-1/4} .$$

(11.2.38)

Hence, we can conclude (following [11.24]) that the approximate description in terms of distinct A and B regions becomes progressively more accurate in the long-time limit, since the relative width of the overlap zone decreases with time.

The spatial structure formed by such regions has a hierarchical self-similar nature. Inside a closed B region one may find closed A regions; inside the latter one may find even smaller closed B regions, etc., down to a length scale $L = (Dt)^{1/2}$

which represents the minimum size of a closed region after time t. Therefore, in the sense of intermediate asymptotic behavior, the structure of the nested A and B regions is fractal (this aspect is further discussed in [11.24, 28]).

11.3 Conserved Quantities and Long-Time Relaxation

We see that the long-time stage of the irreversible annihilation reaction is *controlled by diffusion*, in the sense that the bare reaction rate κ does not enter into the asymptotic law (11.2.18) governing the disappearance of reaching particles. Such behavior is expected for both fast and slow annihilation reactions (with large and small values of κ, respectively). However, the slower the reaction, the longer the time it would take to arrive at the final diffusion-controlled asymptotic.

The asymptotic law (11.2.18) is closely related to the fact that this reaction possesses a conserved quantity,

$$M = \int \eta(r, t)\, dr \, , \tag{11.3.1}$$

which is simply the difference in the total numbers of A and B particles in the medium. Note that conservation of the quantity M holds only when applied to the entire reacting medium. If we consider a section of it, the quantity M given by the integral over this section can vary with time. However, these variations are caused only by the diffusion flow of reacting particles through the boundaries of the section and *not* by the reaction itself (since the difference in the numbers of reacting particles is conserved in any single annihilation event). Hence, the local density $\eta(r, t)$ of this conserved quantity obeys a pure diffusion equation (11.2.5). For a diffusion process, the relaxation times of different spatial modes depend strongly on the characteristic lengths associated with the modes. The larger the characteristic length l, the larger the corresponding relaxation time $t_{\text{rel}} = l^2/D$. If the initial distribution of A and B particles included homogeneities of arbitrarily large length scales, its diffusional relaxation will follow a power law. A particular example of this phenomenon is provided by (11.2.18), which gives the law of disappearance of particles for an initial condition that includes, with equal probabilities, the spatial modes of any length scales.

It should be emphasized that conserved quantities play an important role not only in the considered case of the irreversible annihilation reaction. To demonstrate this we first take an example [11.29, 30] of a reversible recombination reaction

$$A + B \leftrightharpoons C \, . \tag{11.3.2}$$

The classical kinetic equations for this reaction scheme read

$$\begin{aligned}
\dot{n}_A &= -\kappa_1 n_A n_B + \kappa_2 n_C \, , \\
\dot{n}_B &= -\kappa_1 n_A n_B + \kappa_2 n_C \, , \\
\dot{n}_C &= \kappa_1 n_A n_B - \kappa_2 n_C \, ,
\end{aligned} \tag{11.3.3}$$

where κ_1 and κ_2 are the rates of the forward and reverse reactions.

Examination of equations (11.3.3) reveals that they have a conserved quantity,

$$w = n_A + n_B + 2n_C . \tag{11.3.4}$$

This also follows directly from the reaction scheme (11.3.2). According to this, in an elementary event of the forward reaction, two particles A and B transform into one particle C, while in an elementary event of the reverse reaction the opposite transformation occurs.

Hence, if we take into account diffusion of the reacting particles and assume for simplicity that their diffusion constants are the same, we find that the evolution of this globally conserved quantity is again described by a pure diffusion equation

$$\dot{w} = D\Delta w . \tag{11.3.5}$$

According to this equation, any initial inhomogeneity in w will spread and diminish with time, until a perfectly homogeneous state of the system is reached. We know, however, that new inhomogeneities are permanently being created in the system, because of the stochastic nature of reactions and diffusion. This process can be taken into account by introducing additional random terms into the kinetic equations.

Since w is conserved in any elementary event of the reaction, the stochasticity of the reaction cannot influence the evolution of this quantity. Therefore, only fluctuations due to diffusion will contribute to random terms that should be included in (11.3.5).

Suppose that we have a system of identical particles that perform random Brownian motion. Then the fluctuating population density $n(r, t)$ of these particles should satisfy the following differential equation (Sect. 9.2):

$$\dot{n} = \text{div}[D \, \text{grad} \, n + j(r, t)] , \tag{11.3.6}$$

where $j(r, t)$ is the stochastic component of the diffusion flux, such that

$$j(r, t) = (2Dn)^{1/2} f(r, t) , \tag{11.3.7}$$

$$\langle f_\alpha(r, t) f_\beta(r', t') \rangle = \delta_{\alpha\beta}\delta(r - r')\delta(t - t') , \qquad \alpha, \beta = 1, 2, 3 . \tag{11.3.8}$$

Terms of this type should be added to the right-hand sides of all the kinetic equations (11.3.3). Consequently, (11.3.5) will acquire the form

$$\dot{w} = \text{div}\left[D \, \text{grad} \, w + j_w(r, t)\right] , \tag{11.3.9}$$

where

$$j_w(r, t) = (2D)^{1/2}\left[n_A^{1/2} f_A(r, t) + n_B^{1/2} f_B(r, t) + 2n_C^{1/2} f_C(r, t)\right] . \tag{11.3.10}$$

Here f_A, f_B and f_C are the independent random forces satisfying (11.3.8).

In the steady state of the reversible recombination reaction, fluctuations of population densities around their stationary values are small and, to a good approximation, we can replace n_A, n_B and n_C in (11.3.10) by the corresponding stationary values[1]. Then $j_w(r, t)$ is a delta-correlated random force with the correlation functions

$$\left\langle j_w^{(\alpha)}(r, t) j_w^{(\beta)}(r', t') \right\rangle = 2D(n_A + n_B + 4n_C)$$
$$\times \delta_{\alpha\beta}\delta(r - r')\delta(t - t') ; \quad \alpha, \beta = 1, 2, 3 . \quad (11.3.11)$$

If we decompose $w(r, t)$ into a sum of plane waves,

$$w(r, t) = V^{-1/2} \sum_k w_k(t) \exp(-ik \cdot r) , \quad (11.3.12)$$

and use (11.3.9) and (11.3.11), we find

$$\frac{d}{dt} \left\langle |w_k|^2 \right\rangle = -2Dk^2 \left(\left\langle |w_k|^2 \right\rangle - \left\langle |w_k|^2 \right\rangle_{eq} \right) , \quad (11.3.13)$$

where

$$\left\langle |w_k|^2 \right\rangle_{eq} = n_A + n_B + 4n_C \quad (11.3.14)$$

is the equilibrium spectrum of fluctuations of w. Therefore,

$$\left\langle |w_k|^2 \right\rangle (t) = \left\langle |w_k|^2 \right\rangle_{eq} + \left[\left\langle |w_k|^2 \right\rangle (0) - \left\langle |w_k|^2 \right\rangle_{eq} \right]$$
$$\times \exp(-2Dk^2 t) . \quad (11.3.15)$$

Equation (11.3.15) implies that, if the fluctuation spectrum $\left\langle |w_k|^2 \right\rangle$ at the initial time $t = 0$ had no singularity at $k = 0$, the relaxation of the mean-square fluctuation[2]

$$\left\langle w^2 \right\rangle = \frac{1}{V} \int w^2 \, dr = \frac{1}{V} \sum_k \left\langle |w_k|^2 \right\rangle \quad (11.3.16)$$

will follow an asymptotic law in the long-time limit:

$$\left\langle w^2 \right\rangle (t) = \left\langle w^2 \right\rangle_{eq} + \text{const} \, (Dt)^{-3/2} . \quad (11.3.17)$$

---

[1] Note that the fluctuating diffusion flux can also be added into (11.2.5) for irreversible annihilation. However, such a correction would not change the final long-time asymptotic. Since the population densities of A and B particles rapidly decrease, the intensity of this fluctuating flux also diminishes with time. As a result, the most important role is played by random inhomogeneities in the initial state of the system.

[2] Formally, this sum diverges at large k. However, it is actually terminated at some finite k, since the continuous description is valid only at distances much larger than the average distance between the particles.

Since w is a certain linear combination of the population densities n_A, n_B and n_C, the same long-time asymptotic should hold for the fluctuations of these population densities. We conclude that the final long-time state of relaxation to the state of equilibrium for the reversible recombination reaction $A + B \leftrightarrows C$ is *diffusion controlled*. At large time, the exponential relaxation is replaced by the power law (11.3.17).

This is not a special property of the reversible recombination reaction. In effect, the derivation of (11.3.17) used only the assumption that there is a conserved quantity w which obeys a pure diffusion equation. Simple considerations show that a certain conserved quantity can be constructed for *any* reversible reaction. Therefore, as was first noted by *Burlatskii* [11.31] (see also [11.32, 33]) the long-time asymptotic (11.3.17) is universal for such reactions.

11.4 Stochastic Segregation

A special situation is found when the reaction includes both a stage that conserves some quantity and another stage which gives rise to its stochastic wandering. As an example of such a situation, we consider below the irreversible annihilation accompanied by continual generation of the reacting particles,

$$A + B \rightarrow C , \qquad Q \rightarrow A , \qquad Q \rightarrow B . \qquad (11.4.1)$$

Let us assume that the rates of generation of particles A and B are equal; the average number of A (or B) particles generated per unit time per unit volume is w_0.

Introducing again the difference $\eta = n_A - n_B$ of the population densities of A and B particles, we find that it obeys the stochastic differential equation[3]

$$\dot{\eta} = D\Delta\eta + f(r, t) . \qquad (11.4.2)$$

Here $f(r, t)$ is the difference between the two random processes f_A and f_B that describe the internal noise due to the generation of A and B particles [see (9.2.19)]. Therefore, it can be modeled by white noise with

$$\langle f(r, t) f(r', t') \rangle = 2w_0 \delta(t - t') \delta(r - r') . \qquad (11.4.3)$$

Since the average generation rates of A and B are the same, $\langle f(r, t) \rangle = 0$. Consequently, there is no deterministic drift of the quantity η. However, this quantity performs stochastic wandering.

Note that the random force $f(r, t)$ has a flat spectrum, while the rates of diffusive damping of the spatial Fourier modes η_k decrease as Dk^2 for modes with

---

[3] We omit here the term corresponding to the random diffusional fluxes of A and B particles. This term becomes vanishingly small for spatial modes with very large length scales.

smaller wave numbers k (i.e. for modes with larger characteristic length scales). We can thus expect that this random force will induce stronger fluctuations for modes with larger length scales. At very small length scales, random diffusion fluxes become important, a fact that could be taken into account by adding the respective random terms to (11.4.2). We do not introduce these terms here, however, because we are mainly interested in the behavior at large scales.

A simple consideration shows that in the steady state we have[4]

$$\left\langle |\eta_k|^2 \right\rangle = 2w_0 \left(Dk^2\right)^{-1} . \tag{11.4.4}$$

This expression holds for $k \ll (w_0/Dn_A)^{1/2}$, where n_A is the stationary population density of A particles in the steady state (we assume that $n_A = n_B$).

Note that, if we suddenly switch off the generation of particles, the subsequent evolution of the system will be different from that found earlier for the initial state with a flat spectrum. Instead of the law (11.2.18) we would find [11.34] that in the long-time limit

$$\langle n_A \rangle (t) \approx \left(\frac{w_0}{D}\right)^{1/2} (Dt)^{-1/4} . \tag{11.4.5}$$

Let us further investigate (following [11.35], see also [11.36]) the statistical properties of the steady distribution. To be more general, we consider below the reaction scheme that includes additionally the decay processes for A and B particles; the decay rates for both types of particles are equal to α. In the absence of reactions, the total number N of A particles inside any volume element Ω obeys a Poisson distribution, so that for the fluctuations $\delta N = N - \langle N \rangle$ from the average number $\langle N \rangle$ we have

$$\frac{\langle \delta N^2 \rangle}{\langle N \rangle} = 1 . \tag{11.4.6}$$

Reactions between particles give rise to deviations from the law (11.4.6), i.e. to the suppression or enhancement of particle-number fluctuations. These deviations are determined by the normalized pair correlation function

$$g(r - r') = \frac{\langle n_A(r) n_A(r') \rangle}{\langle n_A \rangle^2} - 1 . \tag{11.4.7}$$

It can be shown that

$$\frac{\langle \delta N^2 \rangle}{\langle N \rangle} = 1 + \langle n_A \rangle \, \Omega^{-1} \int_{(\Omega)} g(r - r') \, dr \, dr' . \tag{11.4.8}$$

If we assume that the decay rate α is small compared to the annihilation rate (i.e. $\alpha \ll \kappa n_A$), the normalized correlation function in the steady state [11.35] is

---

[4] The larger the medium, the longer the time required for it to reach the steady state. This time diverges as $T \approx L^2/D$ for large dimensions L of the medium.

$$g(r) = (32\pi \langle n_A \rangle rr_c^2)^{-1} \left[\exp\left(-\frac{r}{r_\alpha}\right) - \exp\left(-\frac{r}{r_c}\right) \right] , \tag{11.4.9}$$

where $r_\alpha = (D/\alpha)^{1/2}$ and $r_c = (D/2\langle n_A \rangle \kappa)^{1/2}$. Then, after a simple but tedious calculation using (11.4.8) and (11.4.9), one obtains the following expression for the mean fluctuation of the number of particles inside a spherical region of radius R:

$$\frac{\langle \delta N^2 \rangle}{\langle N \rangle} = 1 + \frac{1}{8} \left(\frac{r_\alpha}{r_c}\right)^2 F\left(\frac{R}{r_\alpha}\right) - \frac{1}{8} F\left(\frac{R}{r_c}\right) , \tag{11.4.10}$$

where the function $F(x)$ is defined as (because of a mistake in [11.35], a slightly different expression for $F(x)$ was given there)

$$F(x) = 1 + \frac{3}{2x^3} \left[1 - x^2 - (1+x)^2 e^{-2x} \right] . \tag{11.4.11}$$

Examination of (11.4.10) shows that the spatial fluctuations are *enhanced* in comparison with the Poisson distribution. Such enhancement begins at $R \ll r_c$, but is still fairly small:

$$\frac{\langle \delta N^2 \rangle}{\langle N \rangle} \approx 1 + \frac{1}{24} \left(\frac{R}{r_c}\right)^3 . \tag{11.4.12}$$

Enhancement of the fluctuations becomes large for regions with $r_c \ll R \ll r_\alpha$;

$$\frac{\langle \delta N^2 \rangle}{\langle N \rangle} \approx \frac{1}{20} \left(\frac{R}{r_c}\right)^2 . \tag{11.4.13}$$

In even bigger regions with $R \gg r_\alpha$, the growth is saturated and we have:

$$\frac{\langle \delta N^2 \rangle}{\langle N \rangle} \approx \frac{1}{8} \left(\frac{r_\alpha}{r_c}\right)^2 . \tag{11.4.14}$$

The enhancement of spatial fluctuations in the particle number can be interpreted as an effect of *stochastic segregation* of particles for the recombination scheme (11.4.1). This effect has also been observed in computer simulations [11.37–39].

When the particles are produced independently and at random in the medium, it may happen that within some region there will be an excess of one kind of particle (say, A) over the other. Then the B particles will quickly find "partners" and undergo annihilation. As a result, only A particles will be left in the region. In this way, new clusters of A and B particles are permanently created in the medium. The clusters have finite lifetimes because of diffusion and slow one-particle decay. Diffusion leads to the spreading and disappearance of clusters. However, the greater the spatial dimension of a cluster, the slower is its diffusive spreading. Consequently, diffusion cannot stabilize the distribution of clusters at very large length scales. The essential role here is played by the process of slow one-particle decay which provides a constant lifetime independent of the cluster size.

Fluctuation effects in stochastic segregation for two kinds of particles that have different diffusion coefficients D_A and D_B, decay rates α_A and α_B and different average steady-state population densities n_A and n_B were studied in [11.35]. Stochastic segregation on fractals was investigated in [11.40].

To conclude this section, we consider the fluctuation behavior for irreversible annihilation of *identical* particles, i.e. for the reaction

$$A + A \rightarrow 0 . \tag{11.4.15}$$

The analysis by *Gutin* et al. [11.35] showed that in this case the normalized pair correlation function is

$$g(r) = - \left(16 \pi r r_c^2 n \right)^{-1} \exp \left(-\frac{r}{r_c} \right) \tag{11.4.16}$$

where $r_c = (D/2\kappa n)^{1/2}$ and n is the population density of A particles in the steady state. Deviations from the Poisson distribution are given by

$$\frac{\langle \delta N^2 \rangle}{\langle N \rangle} = 1 - \frac{1}{4} F \left(\frac{R}{r_c} \right) \tag{11.4.17}$$

where the function $F(x)$ is again defined by (11.4.11).

Fluctuations of the number of particles in sufficiently small regions with $R \ll r_c$ are

$$\frac{\langle \delta N^2 \rangle}{\langle N \rangle} \approx 1 - \frac{1}{10} \frac{R^2}{r_c^2} , \tag{11.4.18}$$

whereas for regions of large radius ($R \gg r_c$) we have

$$\frac{\langle \delta N^2 \rangle}{\langle N \rangle} \approx \frac{3}{4} . \tag{11.4.19}$$

Hence, annihilation of identical particles leads to the suppression of large-scale fluctuations of the population density, and to formation of a *sub-Poissonian* distribution of reacting particles. At the same time, the particle number fluctuations in a region with dimensions smaller than the correlation length r_c approximately obey the Poisson distribution, i.e. at these scales particles behave as if no reaction is present.

These profound differences in the fluctuation behavior for annihilation of identical particles are related to the fact that, for such a reaction scheme, no conserved quantity exists. This is also reflected in the long-time relaxation phenomena [11.35]. For the reaction scheme $A + A \rightarrow 0$, $Q \rightarrow A$, the exponential relaxation law

$$\langle \delta n^2 \rangle (t) \sim (Dt)^{-3/2} \exp(-2n\kappa t) \tag{11.4.20}$$

holds, whereas the long-time relaxation for the reaction scheme $A + B \rightarrow 0$, $Q \rightarrow A$, $Q \rightarrow B$ is diffusion controlled and obeys (11.3.17).

12. Nonequilibrium Phase Transitions

Usually, when a control parameter is gradually varied, attractors of a distributed dynamical system undergo a continuous variation. However, at certain critical values of the control parameter qualitative changes in the system's attractors can take place. Then the old attractors disappear and new ones are established. From the viewpoint of the theory of nonlinear dynamical systems, these effects represent various bifurcations and crises.

Because the behavior of a system near a critical point is very labile, fluctuations caused by external or internal noises play an important role in these regions. Their magnitude may become greatly enhanced and they can even produce transitions between different attractors. In analogy with statistical physics, the effects of fluctuations near bifurcations in distributed dynamical systems are called *nonequilibrium phase transitions*.

There is an essential difference between soft and hard transitions. If a transition is hard, the existing (stationary or time-dependent) pattern is abruptly replaced by a completely new one when the control parameter exceeds a certain threshold. In this case, even below the threshold the old pattern is already unstable with respect to sufficiently strong fluctuations. An example of a hard transition is provided by first-order phase transitions in physical systems. Below we do not discuss such hard transitions (also called *catastrophes*). First-order phase transitions in reaction–diffusion systems have been considered in Chap. 1 of the first volume. The probability of rare strong fluctuations triggering a transition to a new attractor can be estimated as shown in Sect. 9.3.

Soft transitions are not accompanied by abrupt changes of the existing pattern. They correspond to bifurcations where the old attractor becomes unstable and gives birth to new attractors that initially lie in close proximity to the old attractive set and gradually separate from it as the control parameter is further increased. Because of the branching observed in soft transitions, the fluctuations become greatly enhanced in the critical region. They determine which of the new attractors is finally adopted the system. The simplest soft transition represents branching of a uniform steady state, i.e., a pitchfork bifurcation of an attractive fixed point. This phenomenon is known as a *second-order phase transition*.

12.1 Second-Order Phase Transitions

We begin this chapter with discussion of fluctuations at second-order phase transitions. Such transitions are possible and have been thoroughly investigated for physical systems at thermal equilibrium. But the respective analysis also applies to other systems which are far from equilibrium or where the 'thermal equilibrium' is not even defined (e.g., for ecosystems or societies).

The generic mathematical model for such phase transitions is provided by the time-dependent Ginzburg–Landau equation [see eq. (12.1.10)]. In statistical physics, this equation is used to describe equilibrium systems that undergo a transition to a new phase when temperature is decreased. To illustrate the scope of validity of the Ginzburg–Landau equation, we derive it below for an ecological model describing a transition from symmetrical coexistence to domination in a system of two mutually inhibiting (i.e., 'hostile') species (see also [12.1]).

Suppose that we have two species A and B with local population densities n_A and n_B. In the absence of the antagonist, each species reproduces itself and its population grows until saturation is reached due to the exhaustion of resources. If both species are present, they decrease the reproduction rate (or increase the death rate) of each species' antagonist. We consider the case in which all reproduction properties of the two species are identical. Then the system is described by

$$\dot{n}_A = gn_A - \beta n_A^2 - \kappa n_A n_B + D\Delta n_A \,,$$
$$\dot{n}_B = gn_B - \beta n_B^2 - \kappa n_A n_B + D\Delta n_B \,. \qquad (12.1.1)$$

It is assumed that individuals of both species wander diffusively in the medium with diffusion constant D.

The asymptotic state of such a system depends on the level of "hostility" between the populations, i.e. on the parameter κ that specifies the strength of mutual inhibition. This is most easily seen in the situation of ideal mixing (i.e. of very large diffusion constant D) when the population densities n_A and n_B are spatially homogeneous. In this case (12.1.1) reduces to a set of ordinary differential equations which has three fixed points:

$$n_A = n_B = \frac{g}{(\beta + \kappa)} \,, \qquad (12.1.2)$$

$$n_A = \frac{g}{\beta} \,, \qquad n_B = 0 \,, \qquad (12.1.3)$$

$$n_A = 0 \,, \qquad n_B = \frac{g}{\beta} \,. \qquad (12.1.4)$$

The stability analysis reveals that the symmetric coexistence point (12.1.2) is stable for $\kappa \leq \beta$; the other two fixed points are stable for $\kappa > \beta$.

Figure 12.1 shows the phase plane of (12.1.1) in the case of ideal mixing. At sufficiently low levels of reciprocal hostility (when $\kappa \leq \beta$), the fixed point (12.1.2) corresponding to coexistence of the two populations is attractive. When

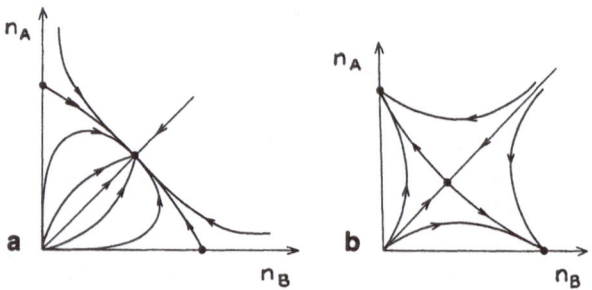

Fig. 12.1. Phase plane of (12.1.1) for (**a**) $\kappa < \beta$ and (**b**) $\kappa > \beta$

the level of hostility between the populations is increased, coexistence abruptly ceases at $\kappa = \beta$. At higher values of κ, the character of the asymptotic solution is changed: only one of the species survives. Note that both fixed points (12.1.3) and (12.1.4) are then attractive and, hence, either one of the two species can survive.

To obtain a soft transition, we additionally assume that the individuals can occasionally change their affiliation from one species to another, i.e., *mutate*. When this process is taken into account, evolution of the population densities of the two species is described by the equations

$$\dot{n}_A = g n_A - \beta n_A^2 - \kappa n_A n_B + D \Delta n_A + \mu(n_B - n_A)$$
$$\dot{n}_B = g n_B - \beta n_B^2 - \kappa n_A n_B + D \Delta n_B + \mu(n_A - n_B) \;,$$

$$(12.1.5)$$

where μ is the mutation rate. It is convenient to introduce new variables

$$\eta = n_A - n_B \;, \qquad q = n_A + n_B \;,$$

$$(12.1.6)$$

which allow (12.1.5) to be written as

$$\dot{\eta} = g\eta - \beta \eta q + D \Delta \eta - 2\mu\eta \;,$$
$$\dot{q} = g q - \tfrac{1}{2}\beta(q^2 + \eta^2) - \tfrac{1}{2}\kappa(q^2 - \eta^2) + D \Delta q \;.$$

$$(12.1.7)$$

The stationary uniform solutions of these equations are (assuming $\mu \ll g$)

$$\eta = 0 \;, \qquad q = \frac{2g}{\beta + \kappa} \;,$$

$$(12.1.8)$$

$$\eta = \pm \frac{g}{\beta}\left(\frac{\kappa - \kappa_{\mathrm{cr}}}{\kappa - \beta}\right)^{1/2} \;, \qquad q = \frac{g}{\beta} \;,$$

$$(12.1.9)$$

where $\kappa_{\mathrm{cr}} = \beta(1 + 4\mu/g)$.

The stability analysis shows that the symmetric state (12.1.8) is stable only for $\kappa < \kappa_{\mathrm{cr}}$. It is replaced at $\kappa \geq \kappa_{\mathrm{cr}}$ by two asymmetric solutions (12.1.9). However, in contrast to the previous model, these new solutions branch continuously from the symmetric solution as the control parameter κ is increased (Fig. 12.2). Hence, the transition becomes soft when mutations are taken into account. In mathematical

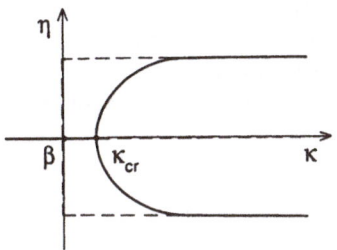

Fig. 12.2. Behavior of η as a function of κ for the fork bifurcation. The dashed line shows the corresponding dependence for the model without mutations

terms, such a system undergoes a *pitchfork bifurcation* when κ exceeds κ_{cr}. In this bifurcation, an attractive fixed point loses its stability and gives birth to two new attractive fixed points.

Examination of (12.1.7) shows that near the critical point κ_{cr}, fluctuations of η have very large relaxation times while the variations of q from its stationary value are damped with a much smaller characteristic time of about g^{-1}. Therefore, such variations would adiabatically follow the slow time dependence of η. This allows us to derive a closed equation for the evolution of the difference $\eta = n_A - n_B$ between the population densities of the two species:

$$\dot{\eta} = a\eta - b\eta^3 + D\Delta\eta , \qquad \text{where} \qquad (12.1.10)$$

$$a = \frac{g}{2\beta}(\kappa - \kappa_{cr}) , \qquad b = 2\mu\left(\frac{\beta}{g}\right)^2 . \qquad (12.1.11)$$

Although we have derived equation (12.1.10) for a particular model, it has much wider applicability. Actually, it provides an approximate description of any reaction–diffusion system near the point of a pitchfork bifurcation. It was originally used by L.D. Landau and V.L. Ginzburg in their studies of equilibrium physical phase transitions and therefore bears the name *time-dependent Ginzburg–Landau equation*. The variable η in this equation is called the *order parameter*. Note that if the coefficients in (12.1.10) are made complex, this yields the complex Ginzburg–Landau equation investigated in Sect. 7.4. The latter represents a normal form for dynamical systems near an Andronov–Hopf bifurcation, leading to the emergence of a stable limit cycle.

Consider a system described by (12.1.10) below the critical point, where $a < 0$. We can then neglect the cubic term and obtain simply

$$\dot{\eta} = a\eta + D\Delta\eta . \qquad (12.1.12)$$

Suppose that we have created some initial spatially uniform perturbation of η. According to (12.1.12), it will fade away within the characteristic lifetime $\tau_c = 1/|a|$, which is called the *correlation time* for fluctuations of η. Since a vanishes at the critical point, τ_c diverges at $a = 0$. This effect is known as *critical slowing-down*.

Suppose further that there is a permanent source placed at some point r_0 of the medium. According to (12.1.12), the perturbation of η introduced by such a local source fades away at a distance of about $r_c \sim (D/|a|)^{1/2}$ from the source. The characteristic distance r_c is called the *correlation radius* (see also the discussion below). It should be noted that r_c diverges at the critical point $a = 0$.

Above the critical point (i.e. at $a > 0$) there are two stable steady solutions of (12.1.10)

$$\eta_0 = \pm \left(\frac{a}{b}\right)^{1/2} .$$

(12.1.13)

Small deviations $\delta\eta = \eta - \eta_0$ from (12.1.13) are again described by a linear equation

$$\delta\dot{\eta} = -2a\delta\eta + D\Delta\delta\eta .$$

(12.1.14)

It follows from (12.1.14) that critical slowing-down and divergence of the correlation radius are also observed when the critical point is approached from above.

Because of critical slowing-down, the distributed system with a pitchfork bifurcation is extremely sensitive to external perturbations near its critical point. Small fluctuations of the system parameters can be greatly enhanced there, producing complex large-scale patterns.

As an example, let us assume that some of the parameters in the original model (12.1.5) with two hostile species fluctuate in space and time, having vanishingly short memory and a very small correlation radius. In the vicinity of the critical point, such variations can be modeled by introducing weak additive white noise to the right-hand side of equation (12.1.10).

Thus we arrive at the general time-dependent Ginzburg–Landau equation with fluctuating forces

$$\dot{\eta} = a\eta - b\eta^3 + D\Delta\eta + f(r, t) ,$$

(12.1.15)

where

$$\langle f(r, t) \rangle = 0 , \qquad \langle f(r, t) f(r', t') \rangle = 2S\delta(r' - r)\delta(t' - t) .$$

(12.1.16)

The same equation also holds for physical systems near equilibrium second-order phase transitions; then it describes the effects of thermal noise. Its application in studies of systems far from thermal equilibrium has been advocated by *Haken* [12.2] and by *Nitzan* et al. [12.3] (see also the review [12.4]). Our further analysis in this and the next section will be based on this general equation.

We expect that the fluctuations will become very strong in the immediate vicinity of the critical point, i.e. when $|a| \ll a_f$ (where a_f is as yet unknown). Outside such a *fluctuational region*, the linearized version of (12.1.15) can be used to describe the fluctuation behavior (which means that there we can neglect the interactions between fluctuations).

Consider first the fluctuation behavior in the symmetric state (when $a \ll -a_f$). The linearized version of (12.1.15) is then simply

$$\dot{\eta} = -|a|\eta + D\Delta\eta + f(r, t) .$$

(12.1.17)

We can define the pair correlation function $G(r, t)$ as

$$G(r, t) = \langle (r, t)\eta(0, t) \rangle \, . \tag{12.1.18}$$

Using (12.1.17), one can show that this function satisfies the equation

$$\frac{1}{2} \frac{\partial G}{\partial t} = -|a|G + D\Delta G + S\delta(r) \, . \tag{12.1.19}$$

Comparing (12.1.19) and (12.1.12), we see that G obeys the equation for the order parameter with a permanent point-like source placed at the origin of the coordinates. In the steady regime, the pair correlation function $G(r)$ falls exponentially with a characteristic decay length r_c, which is the correlation radius. If we define the correlation radius by

$$r_c^2 = \frac{\int r^2 G(r)\, dr}{\int G(r)\, dr} \, , \tag{12.1.20}$$

simple calculations yield

$$r_c = \left(\frac{2dD}{|a|} \right)^{1/2} \, , \tag{12.1.21}$$

where d is the dimensionality of the medium.

It is also possible to estimate the intensity of fluctuations of the order parameter. Let us introduce the order parameter averaged over some volume element V,

$$(\eta)_V = \frac{1}{V} \int_{(V)} \eta(r + \varrho)\, d\varrho \, . \tag{12.1.22}$$

Using expression (12.1.17), one can show that its mean square for a volume element V, larger than the correlation volume $V_c \sim r_c^d$, is given approximately by

$$\langle (\eta)_V^2 \rangle \approx \frac{1}{V} \int G(r)\, dr \, , \tag{12.1.23}$$

where integration is performed over the entire medium. An approximate evaluation of (12.1.23) yields

$$\langle (\eta)_V^2 \rangle \approx \frac{S}{|a|V} \, . \tag{12.1.24}$$

Hence, this mean value depends significantly on the averaging volume V and, to specify the intrinsic intensity of fluctuations of the order parameter, we should advocate a particular choice of the averaging element V. It turns out[1] that an appropriate choice is to take V as the correlation volume $V_c \sim r_c^d$, since the correlation radius r_c gives a natural length scale associated with the transition.

---

[1] A rigorous justification of this choice is provided by diagramatic perturbation theory [12.5].

Fluctuations of the order parameter η at any two points that are separated by a distance much larger than r_c are statistically independent.

After substitution of $V_c \sim r_c^d$ into (12.1.24) we obtain[2]

$$\langle (\eta)_{V_c}^2 \rangle \sim S|a|^{(d-2)/2} D^{-d/2} . \tag{12.1.25}$$

We now have an estimate of the fluctuation intensity which allows us to determine the limits of validity of the linear approximation. The cubic term in the time-dependent Ginzburg-Landau equation (12.1.15) is negligibly small in comparison with the first (linear) term in (12.1.15) provided the condition

$$b \langle (\eta)_{V_c}^2 \rangle \ll |a| \tag{12.1.26}$$

is satisfied. Using (12.1.15), this can be cast into the form $a \ll -a_f$, where the characteristic value

$$a_f = \left(\frac{bS}{D^{d/2}} \right)^{2/(4-d)} \tag{12.1.27}$$

determines the boundary of the region with strong fluctuations.

In a similar way, we can investigate the behavior of fluctuations above the critical point for positive control parameters a. The only difference is that in this case we should replace η in (12.1.17–26) by the deviation $\delta\eta = \eta - \eta_0$ from the equilibrium value given by (12.1.13) and replace $|a|$ by $2a$. Note that (12.1.26) is the equivalent to the inequality

$$\langle (\delta\eta)_{V_c}^2 \rangle \ll \eta_0^2 . \tag{12.1.28}$$

This means that the fluctuations are small in comparison to the deterministic value of the order parameter.

Close to the critical point, where $|a| \ll a_f$, fluctuations are strong and their behavior can be accurately described only by a complete nonlinear Ginzburg-Landau equation (12.1.15). The *renormalization group theory*, based on scaling considerations (e.g. [12.5]), predicts that inside this region the correlation radius r_c continues to grow as some inverse power of $|a|$ when the critical point $a = 0$ is approached. The mean-square intensity of fluctuations $\langle (\delta\eta)_V^2 \rangle$ within any fixed volume V also diverges as some inverse power of $|a|$ at this point.

Hence, even very weak noise can be enhanced in the vicinity of the critical point of a second-order phase transition to produce large-scale fluctuations of the order parameter with long characteristic lengths.

Although the above analysis was carried out only in the case of a real scalar order parameter η, it can easily be generalized to describe situations where η is a complex quantity or a vector.

Examination of (12.1.25) and (12.1.27) shows that the fluctuation behavior in the vicinity of a second-order phase transition is sensitive to the dimensionality

---

[2] Although (12.1.24) holds only for volumes larger than V_c, we can still use it to estimate the order of magnitude of fluctuations within the correlation volume V_c.

d of the medium. Obviously, the dimensionality of any realistic system never exceeds $d = 3$. Nonetheless, in abstract terms we can also consider fluctuation behavior in spaces with $d > 3$.

Note that (12.1.27) is not defined at $d = 4$. Returning to the condition (12.2.26), we find that for $d > 4$ it yields

$$|a|^{(d-4)/2} \ll \frac{D^{d/2}}{bS} . \tag{12.1.29}$$

Hence, in spaces with dimensionality higher than four, the fluctuations are always weak when the system is sufficiently close to the critical point. At $d = 4$ the fluctuations are weak if $D^{d/2}/bS \gg 1$.

Coming back to our initial model of two hostile populations, we can conclude that near the bifurcation point $\kappa = \kappa_{cr}$ (which corresponds to $a = 0$), this system should exhibit strong fluctuations in the local difference of the population densities of the two species. The characteristic spatial size of such fluctuations and the characteristic relaxation time of an individual fluctuation diverge at the critical point.

12.2 Sweeping Through the Critical Region

Above the critical point of a second-order phase transition described by the Ginzburg-Landau equation (12.1.16) there are two stationary uniform states with $\eta = \pm(a/b)^{1/2}$. Both are stable, and the choice of a particular state is determined by the evolution history of the system.

Suppose that initially the system is far below the critical point $a = 0$ in the symmetric state $\eta = 0$, and we then gradually increase the control parameter a and pass through the critical region. If we go sufficiently slowly[3], the system has enough time to adjust adiabatically to the changing control parameter. The system is then constantly in the equilibrium state with the instantaneous value of the control parameter $a(t)$. As long as $a(t)$ is less than $-a_f$, where a_f is given by (12.1.27), the fluctuations are small. When a is further increased, so that the fluctuational regime $|a(t)| < a_f$ is entered, the intensity of the fluctuations becomes much larger. Beyond this regime, the medium splits into a mosaic of well-defined domains, each occupied by one of the two distinct and opposite "phases" (with either $\eta \approx +(a/b)^{1/2}$ or $\eta \approx -(a/b)^{1/2}$ throughout the entire domain). If $a \gg a_f$, fluctuations of η around these two possible values in each domain are small and can be neglected.

The interfaces separating different domains are not static. They move slowly with time, changing the shapes and sizes of the domains. This process can be described in complete analogy with the propagation of trigger waves in bistable media, as discussed in Chap. 2 of the first volume.

---

[3] The nearer we are to the critical point $a = 0$, the slower we must approach it.

A flat interface is stationary. The velocity v of the motion of a curved interface is determined by its local curvature K as $v = -DK$. This effect leads to the shrinking of small domains immersed in larger regions of the opposite phase, and results in a smoothing of the interface boundary. If we wait for a sufficiently long time, only the largest macroscopic domains of each phase will be left in the medium.

Suppose now that we quickly sweep at a constant rate c through the critical region, so that $a = ct$. We start sweeping at large negative times $t \to -\infty$ and pass the point $a = 0$ at $t = 0$. What fluctuation phenomena would we expect under these conditions? We discuss this problem below, following *Grossmann* and *Mikhailov* [12.6].

Since the relaxation time τ_c diverges at the critical point, one must wait for increasingly longer times to achieve the equilibrium probability distribution for fluctuations as one approaches the critical point. When one sweeps through the critical region at a constant rate, the system does not have enough time to relax and nonadiabatic effects are expected. As we shall see, these effects suppress the growth of fluctuations and prevent divergence of the correlation radius.

Suppose that the sweeping rate is so large that the fluctuations remain weak as the critical point is approached (the precise condition is formulated below). This allows us to neglect the cubic term in the Ginzburg-Landau equation (12.1.15) and to write

$$\dot\eta = ct\eta + D\Delta\eta + f(r, t) .\tag{12.2.1}$$

Using (12.2.1), we find the following evolution equation for the pair correlation function

$$\frac{1}{2}\frac{\partial G}{\partial t} = ctG + D\Delta G + S\delta(r) .\tag{12.2.2}$$

The pair correlation function G contains important information about the statistical properties of the process. If we introduce its zeroth and second moments by

$$\chi_0(t) = \int G(r, t)\, dr , \qquad \chi_2(t) = \int r^2 G(r, t)\, dr ,\tag{12.2.3}$$

the mean-square $\langle(\eta)_V^2\rangle$ of the order parameter averaged over a sufficiently large volume element V at time t can be estimated simply as

$$\langle(\eta)_V^2\rangle = \frac{1}{V}\chi_0(t) .\tag{12.2.4}$$

Moreover, the time-dependent correlation radius $r_c(t)$ will be given by

$$r_c^2(t) = \frac{\chi_2(t)}{\chi_0(t)} .\tag{12.2.5}$$

The dynamical equations for the moments $\chi_0(t)$ and $\chi_2(t)$ can be derived directly from (12.2.2):

$$\dot{\chi}_0 = 2ct\chi_0 + 2S , \tag{12.2.6}$$

$$\dot{\chi}_2 = 2ct\chi_2 + 4dD\chi_0 . \tag{12.2.7}$$

First we consider (12.2.6). Its exact solution is

$$\chi_0(t) = 2S \exp(ct^2) \int_{-\infty}^{t} \exp(-ct_1^2) \, dt_1 . \tag{12.2.8}$$

Specifically, we obtain

$$\chi_0(0) = \frac{2S}{c^{1/2}} \int_{0}^{\infty} \exp(-u^2) \, du = \frac{\pi^{1/2}S}{c^{1/2}} . \tag{12.2.9}$$

It follows from (12.2.6) that two regions with different behavior of the zeroth moment $\chi_0(t)$ can be distinguished.

In the *adiabatic* region we can neglect the term with the time derivative and find

$$\chi_0(t) \approx \frac{S}{c|t|} . \tag{12.2.10}$$

Then $d\chi_0/dt = -S/ct^2$ and we see that it can indeed be neglected in comparison with S or $ct\chi_0$ in (12.2.6), provided that $t \ll -t^*$, where $t^* = c^{-1/2}$. Hence $-t^*$ sets the upper boundary of the region. In the *nonadiabatic region* $|t| \ll t^*$ we have approximately

$$\chi_0(t) \approx 2SDt + \chi_0(0) . \tag{12.2.11}$$

Since $\langle(\eta)_V^2\rangle = (1/V)\chi_0(t)$, we see from (12.2.10) and (12.1.24) that the mean-square of the order parameter in the adiabatic region is the same as that in the equilibrium state with the instantaneous value $a = ct$. In the nonadiabatic region which surrounds the point $a = 0$, the behavior of $\langle(\eta)_V^2\rangle$ is different. Instead of diverging at $a = 0$, this quantity remains finite and grows linearly with time. The maximum value of $\langle(\eta)_V^2\rangle$ is reached at the right-hand boundary $t \approx t^*$ of the nonadiabatic region,

$$\langle(\eta)_V^2\rangle (t^*) \approx \frac{2S}{V} \left(\frac{\pi}{c}\right)^{1/2} . \tag{12.2.12}$$

In a similar way we can find the solution of (12.2.7) for the second moment:

$$\chi_2(t) = 8dDS \exp(ct^2) \int_{-\infty}^{t} dt_1 \int_{-\infty}^{t_1} dt_2 \exp(-ct_2^2) . \tag{12.2.13}$$

Explicitly, we obtain

$$\chi_2(0) = \frac{8dDS}{c} \int_{-\infty}^{0} du_1 \int_{-\infty}^{u_1} du_2 \exp(-u_2^2) = \frac{4dDS}{c} . \tag{12.2.14}$$

In the adiabatic region ($t \ll -t^*$) the second moment is the same as at equilibrium, i.e.,

$$\chi_2(t) \approx \frac{2dD}{c|t|}\chi_0(t) , \qquad (12.2.15)$$

while in the nonadiabatic region ($|t| \ll t^*$) we have

$$\chi_2(t) \approx 4dD\chi_0(0)t + \chi_2(0) . \qquad (12.2.16)$$

When these two moments are known, the instantaneous correlation radius r_c can be found from (12.2.5). In the adiabatic region it is the same as at equilibrium,

$$r_c^2(t) = \frac{2dD}{c|t|} , \qquad (12.2.17)$$

while in the nonadiabatic region ($|t| \ll t^*$) we obtain approximately

$$r_c^2(t) \approx r_c^2(0) + \xi_d Dt , \qquad (12.2.18)$$

where the numerical factor is $\xi_d = 4d[1 - (2/\pi)]$. Thus, instead of divergence of the correlation radius at $a = 0$, we find that it grows linearly with time and remains bounded in the critical region.

In the above we have kept only the linear terms in the time-dependent Ginzburg-Landau equation (12.1.15). This is justified in the left-hand adiabatic region $t \ll -t^*$ if, for all values of the control parameter $a = ct$ inside this region, we have $|a| \gg a_f$ [cf. (12.1.27)]. Since the maximum value of $|a| = a^* = c^{1/2}$ is reached at the end of this region, i.e. at $t = -t^*$, we obtain the condition $a^* \gg a_f$. It turns out [12.10] that the same condition is sufficient to justify linearization in the nonadiabatic region $|t| \ll t^*$. Substitution of the explicit expressions for a^* and a_f into $a^* \gg a_f$ yields the following condition, which allows us to use the linearized equation (12.2.1) up to $t \sim t^*$:

$$c \gg \left(\frac{bS}{D^{d/2}}\right)^{4/(4-d)} . \qquad (12.2.19)$$

Note that for $t > 0$ the symmetric state $\eta = 0$ in (12.2.1) is unstable and fluctuations of η begin to grow with time. However, at the early stage (while $t \ll t^*$) such deterministic growth of fluctuations is small compared to their steady production by noise $f(r, t)$ in (12.2.1). For $|t| \ll t^*$ even the linear term $ct\eta$ in (12.2.1) can be neglected.

When the system exits the nonadiabatic region at $t \approx t^*$, the linear term in (12.2.1) becomes overwhelming and fast exponential deterministic growth of fluctuations sets in. This burst of fluctuations is eventually saturated due to the nonlinear cubic term in the Ginzburg-Landau equation (12.1.15) and the medium becomes transformed into the mixture of domains of two opposite phases. The characteristic spatial extent of the domains produced is determined by the correlation radius $r_c(t^*) \approx D^{1/2}c^{-1/4}$ at the time $t = t^*$. After this abrupt transformation, inside each domain occupied by a definite phase, the fluctuations are again small and the system is close to equilibrium. At $t \gg t^*$ it adiabatically follows the increase of the control parameter $a = ct$.

Consequently, in the case of fast sweeping [when condition (12.2.19) holds] the actual formation of domains with new phases is *delayed*[4] until the control parameter a reaches a value of about $a^* = c^{1/2}$. Only at that moment do the fluctuations burst out on a kind of explosion accompanied by the production of the mixture of phase domains.

If the sweeping is not fast enough and condition (12.2.19) is violated, there would also be some nonadiabatic region around the point $a = 0$ inside which the fluctuations are damped. However, the system then arrives at the boundary of this nonadiabatic region with fluctuations that are already large, so that their nonlinear interactions cannot be neglected. Furthermore, when this system exists the nonadiabatic region at $a \approx a_f$, it initially enters the adiabatic region with large-amplitude equilibrium fluctuations. Only later (at $a \approx a_f$) are the distinct domains of two definite phases formed.

12.3 The Biased Transition

Let us now return to the original model of two "hostile" species, used in Sect. 12.1 to derive the Ginzburg-Landau equation (12.1.15). Suppose that one of these species (for instance, A) has a slightly higher reproduction rate, so that the system is described by

$$\dot{n}_A = g_A n_A - \beta n_A^2 - \kappa n_A n_B + D \Delta n_A + \mu(n_B - n_A) \, ,$$
$$\dot{n}_B = g_B n_B - \beta n_B^2 - \kappa n_A n_B + D \Delta n_B + \mu(n_A - n_B) \, , \tag{12.3.1}$$

where $g_A > g_B$ and $(g_A - g_B) \ll g_A$. Repeating the steps taken in Sect. 12.1, we obtain the equation

$$\dot{\eta} = a\eta - b\eta^3 + D \Delta \eta + h \, , \tag{12.3.2}$$

where the additional term

$$h \approx \frac{1}{4\beta} \left(g_A^2 - g_B^2 \right) \tag{12.3.3}$$

specifies the *bias* towards one particular sign of the order parameter η. Although we derived (12.3.2) for the special case of a reproductive system, it is quite general. In the statistical physics of phase transitions, it is used to describe a second-order phase transition in the presence of an "external field"[5].

---

[4] The delay effects for bifurcations in deterministic systems under conditions of steady or periodic sweeping are discussed in [12.7–9] (see also a review [12.10]).

[5] If a transition from the paramagnetic to the ferromagnetic state of a crystal is considered, η represents the local magnetic moment of the crystal and h is the external magnetic field.

The stationary uniform solutions to (12.3.2) are the roots of the cubic equation

$$a\eta - b\eta^3 + h = 0 .$$

(12.3.4)

Their dependence on the control parameter a is shown in Fig. 12.3. We see that application of the bias h changes the type of bifurcation. When the control parameter is increased, the attractive fixed point $\eta \approx 0$ moves along the upper branch in Fig. 12.3. However, when the control parameter a reaches a_c (where $a_c = 2^{-2/3}b^{1/3}h^{2/3}$), two new fixed points appear. Initially (at $a = a_c$) both of them have $\eta = -(h/2b)^{1/3}$ and they then become separated, giving rise to the lower (stable) and middle (unstable) branches. The minimum distance between the upper and lower stable branches is realized at $a = a_c$, when it is equal to $\delta\eta_c = 3(h/2b)^{1/3}$. Thus, for $a > a_c$ two uniform steady states are possible: one with a positive order parameter η, favored by the bias, and the other with a negative order parameter. The system is therefore bistable in this region.

Suppose we have a flat interface separating two regions occupied by opposite phases. Using the methods described in Chap. 2 of the first volume, it can be shown that the interface will move into the region occupied by the phase with the negative order parameter and not favored by the bias. The velocity of this motion can be estimated as $v_i \sim hD^{1/2}b^{1/2}/a$. Hence, the favored phase expands with time, eliminating the domains of the opposite phase.

The lower branch in Fig. 12.3 is *metastable*: if a large nucleus of another phase (corresponding to the upper branch in Fig. 12.3) is created, it grows indefinitely and thus the medium adopts a new uniform phase. The shape and size of the critical nucleus can be calculated in the same way as in Sect. 2.3 of the first volume.

We will now discuss fluctuation behavior in the presence of a bias described by the stochastic differential equation

$$\dot{\eta} = a\eta - b\eta^3 + D\Delta\eta + h + f(r, t) .$$

(12.3.5)

As before, it is assumed that the Gaussian random force $f(r, t)$ has a vanishingly short memory and very small correlation radius [cf. (12.1.16)].

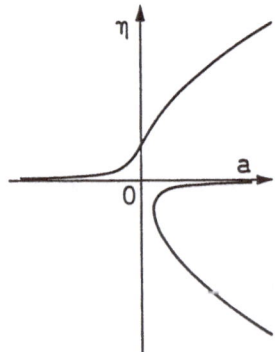

Fig. 12.3. Dependence of the order parameter η on the control parameter a in the presence of a positive bias h

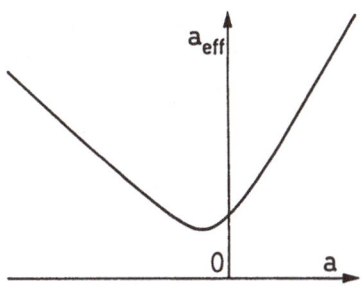

Fig. 12.4. Dependence of a_{eff} on a for a strong bias

Let $\eta(h)$ denote the uniform stationary solution to (12.3.4) which corresponds to the upper branch in Fig. 12.3, and let the deviation $\delta\eta$ be defined as $\delta\eta = \eta - \eta(h)$. This deviation obeys the following exact nonlinear equation

$$\frac{\partial \delta\eta}{\partial t} = -a_{\text{eff}}(a)\delta\eta - 3b\eta(h)\delta\eta^2 - 3b\delta\eta^3 + D\Delta\delta\eta + f(\mathbf{r}, t) , \qquad (12.3.6)$$

where

$$a_{\text{eff}}(a) = -a + 3b\eta(h)^2 . \qquad (12.3.7)$$

The dependence of a_{eff} on a is shown in Fig. 12.4. We see that a_{eff} is always positive. Its minimum value can be estimated as $\min\{a_{\text{eff}}\} \sim h^{2/3}b^{1/3}$. The same order-of-magnitude estimate holds for the value of $a_{\text{eff}}(a)$ at $a \approx a_{\text{c}}$, i.e. when the lower metastable solution first appears (Fig. 12.3).

If a_{eff} is large, fluctuations are suppressed and their characteristic amplitude is small. We can then neglect the nonlinear terms in (12.3.13). The resulting equation for $\delta\eta$ has the same form as in the absence of bias, the only exception being that the coefficient a is now replaced by $a_{\text{eff}}(a)$. In the same way as in Sect. 12.1, we can estimate the mean-square fluctuation averaged over the correlation volume and further use this estimate to determine the condition for validity of the linearized equation. This condition reads $\min\{a_{\text{eff}}\} \gg a_{\text{f}}$. Substituting the explicit expressions for $\min\{a_{\text{eff}}\}$ and a_{f}, we find that fluctuations are heavily suppressed if the bias h is much stronger than

$$h_{\text{c}} = b^{-1/2}\left(\frac{bS}{D^{d/2}}\right)^{3/(4-d)} . \qquad (12.3.8)$$

The condition $h \gg h_{\text{c}}$ is known [12.11] as the criterion of *strong bias* (or strong external fields). When it is satisfied, the characteristic amplitude of fluctuations always remains small compared to the order parameter $\eta(h)$ induced by the bias. It implies that the probability of spontaneous creation of a large domain of a metastable phase (with a size greater than the correlation radius) inside a region occupied by the stable phase, favored by the bias, is exponentially small. Under the condition of strong bias, the noise $f(\mathbf{r}, t)$ in the Ginzburg-Landau equation (12.3.5) does not produce large fluctuations of the order parameter and can be neglected.

In the opposite case of *weak bias* ($h \ll h_c$), fluctuations dominate over the bias near the critical point $a = 0$. This point is then surrounded by a region with strong fluctuations. The fluctuation behavior for weak biases is practically the same as in their absence.

Suppose further that the control parameter is swept with time at a constant rate c through the critical region, i.e. $a = ct$. If the bias is weak, the fluctuation behavior would be the same as in the case without bias studied in the previous section. For strong biases, the effects are fundamentally different. As shown above, for strong biases at equilibrium the fluctuations are small and can be neglected. In the context of a sweeping experiment this means that, by changing the control parameter *very* slowly (so that the equilibrium distribution is maintained), we can cross the critical region without any significant enhancement of fluctuations and move the system into a homogeneous asymmetric phase, singled out by the bias.

The situation is different if we go *rapidly* through the critical region. We show below that then, on leaving the nonadiabatic region, the mixture of domains of two opposite phases is abruptly produced in the same way as if the bias were weak.

Consider first the deterministic dynamical equation for the order parameter in the presence of a bias, i.e.

$$\dot{\eta} = ct\eta - b\eta^3 + h . \tag{12.3.9}$$

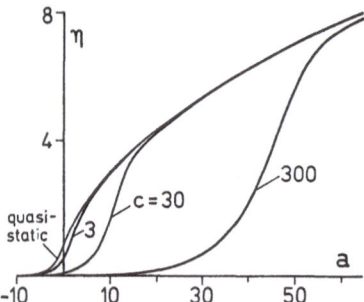

Fig. 12.5. Dependence of the order parameter η on $a = ct$ for different sweeping rates c. Numerical integration of (12.3.9) for $b = 1$ and $h = 1$

The results of numerical integration of (12.3.9) for various sweeping rates c are plotted in Fig. 12.5. When the sweeping rates are high, the instantaneous value $\eta(t)$ of the order parameter inside the critical region is much smaller than its respective value at equilibrium with the same control parameter a.

This result is easily obtained analytically: Provided the order parameter η remains sufficiently small, we can linearize (12.3.9) and simply write

$$\dot{\eta} = ct\eta + h . \tag{12.3.10}$$

The solution to (12.3.10) is

$$\eta(t) = h \exp\left(\frac{ct^2}{2}\right) \int_{-\infty}^{t} \exp\left(\frac{-ct_1^2}{2}\right) dt_1 . \tag{12.3.11}$$

Investigation of (12.3.10) and (12.3.11) reveals that in the left-hand adiabatic region $t \ll -t^*$ (where $t^* = c^{-1/2}$), we have $\eta(t) = h/c|t|$ which coincides with the equilibrium result. In the nonadiabatic region (where $|t| \ll t^*$) we find the linear dependence $\eta = \eta(0) + ht$ with $\eta(0) = (\pi/2c)^{1/2}h$.

By the time $t \approx t^*$ of departure from the nonadiabatic region, the order parameter is about $\eta(t^*) \sim h/c^{1/2}$. On the other hand, the equilibrium value for the corresponding instantaneous value $a^* = ct^*$ of the control parameter is $\eta_{eq} \sim (h/b)^{1/3}$. Hence, a marked reduction of the order parameter in comparison to its equilibrium value is observed if the sweeping rate is sufficiently large, i.e., if

$$c \gg b^{2/3}h^{4/3} . \tag{12.3.12}$$

It can be directly checked that the same condition justifies the neglect of the nonlinear cubic term in (12.3.9).

Within the nonadiabatic region, the linear term $ct\eta$ in (12.3.9) is much smaller than the bias h. However, when the system leaves this region at $t \approx t^*$, this term becomes dominant and leads to a fast exponential increase of η. This growth later becomes suppressed by the cubic term, and the order parameter approaches its equilibrium value. This sharp increase of η at $t \approx t^*$ is clearly seen in numerical integrations (Fig. 12.5). Hence, we can say that under the condition (12.3.12) of fast sweeping, the actual transition to the asymmetric state is delayed until the moment $t \approx t^*$, even in the absence of any fluctuations (see [12.9, 10]).

Next we discuss fluctuation effects. Let us consider the deviation $\delta\eta$ of the order parameter from its time-dependent deterministic value $\eta_{det}(t)$, which is a solution to (12.3.9). Such a deviation again obeys the stochastic differential equation (12.3.6), the only exception being that $\eta(h)$ is now replaced by $\eta_{det}(t)$. In particular, $a_{eff}(a)$ given by (12.2.7) is replaced by

$$a_{eff}(t) = -ct + 3b\eta_{det}^2(t) . \tag{12.3.13}$$

In the adiabatic region, the instantaneous damping rate $a_{eff}(t)$ of fluctuations coincides with its respective value at equilibrium, but inside the nonadiabatic region $|t| \ll t^*$ it might be much smaller (and can even become negative). Indeed, we have seen above that, for sufficiently fast sweeping, $\eta_{det}(t)$ is much lower than in equilibrium conditions. Obviously, the profound decrease in the damping rate a_{eff} may strongly enhance fluctuations. This effect was investigated analytically by *Grossmann* and *Mikhailov* [12.6] who showed that for sufficiently high sweeping rates, when

$$c \gg \left(\frac{h^2 D^{d/2}}{S}\right)^{4/(d+2)} , \tag{12.3.14}$$

the system arrives at the boundary of the nonadiabatic region with large fluctuations.

If (12.3.14) holds, it is almost equally probable to find both the negative and positive values of the order parameter η at the time of exit from the nonadiabatic region. At a latter stage, the linear term $ct\eta$ in (12.3.9) becomes dominant, which leads to a deterministic exponential enhancement of the *absolute* value of the order parameter, but preserves its sign in a given fluctuation.

Suppose that, at the exit time t^*, the order parameter was negative (but small) in some spatial region. Then in this region it will grow with time, reaching large negative values. Eventually, this exponential enhancement will be saturated by the cubic nonlinear term in (12.3.5) and, in this region, the system will approach a metastable state where the sign of the order parameter is opposite to that favored by the bias. We see that, if (12.3.14) is satisfied, the system transforms into a mixture of domains of the two opposite phases soon after the moment $t^* \approx c^{-1/2}$.

Hence, even for strong biases, the state of the system immediately after passing the critical region can represent a complex pattern of phase domains, provided that sweeping is fast enough. The characteristic spatial size of the domains produced is determined by the correlation radius $r_c(t^*) \sim D^{1/2}/c^{1/4}$ at the moment of exit from the nonadiabatic region [12.6].

The pattern produced is not stationary at $t > t^*$. Domains of the stable phase grow at the expense of the metastable domains with a negative order parameter and merge together, so that the uniform state ultimately occupies the entire medium.

Until now we have considered the phenomena in very large systems. If the volume of a system is finite, the fluctuation behavior may be different. Suppose, for instance, that the linear size of a volume element is small in comparison with the characteristic correlation radius $r_c(t^*)$. There is then a very high probability that, at time t^*, the entire volume will be occupied by a single domain of a certain phase. Since there are no interfaces, this phase will persist indefinitely in such a volume, even if it is metastable, i.e. if it has a negative order parameter.

The problem of sweeping through the critical region in the presence of a bias was investigated by *Nicolis* and *Prigogine* [12.13] in the context of prebiological evolution. Certain important chemical compounds, such as sugars and amino acids, are found in all living organisms, but only in a particular chiral form (right enantiomers for sugars and left enantiomers for amino acids), whereas in inorganic matter molecules of both chiral forms are detected in equal proportions. The competition between replication processes relying on two alternative chiralities, can be described by the model of "hostile" species (12.1.6). Spontaneous breaking of the chiral symmetry is then interpreted as a second-order phase transition at an early stage of prebiological evolution [12.1, 12–14].

Prebiological evolution is a very slow process, with a characteristic time-scale of millions of years. On such time-scales, turbulent flows in the ocean are able to maintain a uniform distribution of reagents. Therefore, in this case it is natural to consider the problem of sweeping in a non-distributed system (i.e. formally with the dimensionality $d = 0$). This problem was investigated in [12.15, 16].

Another motivation for studies of time-dependent phase transitions came from the field of nonlinear optics. Influence of noise on delayed bifurcations in optical systems was analyzed in [12.10, 17, 18].

12.4 Medium-Populating Transitions

In this section we consider a special kind of transition found in birth–death systems. Below the explosion threshold, the population in such systems dies out with time and, hence, only the trivial attractive steady state $n = 0$ is present. By varying the system's parameters, the explosion threshold can be crossed. This leads to a rapid explosive growth of the population and its subsequent saturation due to the nonlinear effects. As a result, a populated steady state with $n > 0$ is established while the trivial steady state $n = 0$ becomes unstable. In mathematical terms, the transition from the unpopulated to the populated state of the medium represents a certain bifurcation. Below we study fluctuations in the populated state in the vicinity of such a transition.

The simplest mathematical model of this phenomenon is given by the *logistic equation*

$$\dot{n} = -k_{\text{death}}n + k_{\text{birth}}n - \beta n^2 + D\nabla^2 n , \qquad (12.4.1)$$

where k_{birth} and k_{death} are birth and death rates, β is the coefficient controlling nonlinear saturation, and D is the diffusion constant of the reproducing species.

When the birth rate exceeds the death rate, this system undergoes a transition to the populated state. Below the transition point (at $k_{\text{birth}} < k_{\text{death}}$) any initial population dies out with time, whereas above this point even small initial populations eventually spread over the entire medium and give rise to a steady state with a constant population density, i.e., we have

$$n = \begin{cases} 0, & \text{for } k_{\text{birth}} < k_{\text{death}} \\ \dfrac{1}{\beta}(k_{\text{birth}} - k_{\text{death}}), & \text{for } k_{\text{birth}} > k_{\text{death}} . \end{cases} \qquad (12.4.2)$$

The characteristic relaxation time required to reach the steady state (12.4.2) diverges as

$$t_{\text{rel}} = |k_{\text{birth}} - k_{\text{death}}|^{-1} \qquad (12.4.3)$$

near the transition point. We can also introduce the correlation length r_c which is defined as the characteristic length of the region over which the influence of a local perturbation (a source or a sink) extends. This correlation length,

$$r_c = \left(\frac{D}{k_{\text{birth}} - k_{\text{death}}}\right)^{1/2} , \qquad (12.4.4)$$

also diverges near the transition point.

In fluctuating media, the birth and death rates in the logistic equation are random functions of time and space coordinates. The effects of such fluctuations at the initial stage of the population explosion, described by the linearized version of equation (12.4.1), have been discussed in Chap. 10. We have seen that fluctuations shift the explosion threshold and lead to the formation of intermittent spatiotemporal distributions. Below, we focus our attention on the statistical properties of the populated steady state that is established in a fluctuating medium above the transition point.

We begin our analysis with the simplest case of vanishing diffusion. Then the processes at different points in the medium are independent and evolution of the population density at a given point is described by the stochastic differential equation:

$$\dot{n} = \alpha n - \beta n^2 + f(t)n , \tag{12.4.5}$$

where α is the mean difference of birth and death rates and $f(t)$ is the fluctuating component of the same difference. The rate fluctuations are Gaussian with a very short correlation time, so that $f(t)$ is modeled by a white noise,

$$\langle f(t)f(t')\rangle = 2s\delta(t - t') . \tag{12.4.6}$$

Since we consider the limiting case of a random process $f(t)$ with a very short, but still finite, correlation time, the Stratonovich interpretation of the stochastic differential equation (12.4.5) should be used.

The probability distribution for the random variable n satisfies the Fokker–Planck equation:

$$\frac{\partial p}{\partial t} = -\frac{\partial}{\partial n}\left[(\alpha n - \beta n^2 + sn)p\right] + \frac{\partial^2}{\partial n^2}(sn^2 p) . \tag{12.4.7}$$

Its stationary solution, yielding the probability distribution in the populated steady state, is

$$p_s(n) = Z n^{(\alpha/s)-1} \exp\left(-\frac{\beta n}{s}\right) , \quad \alpha > 0 \tag{12.4.8}$$

Here Z is the normalization constant,

$$Z^{-1} = \left(\frac{\beta}{s}\right)^{-(\alpha/s)} \Gamma\left(\frac{\alpha}{s}\right) \tag{12.4.9}$$

and $\Gamma(x)$ is the gamma-function. When $\alpha < 0$, the stationary probability distribution reduces to

$$p_s(n) = \delta(n) . \tag{12.4.10}$$

Thus, in the nonlinear model (12.4.5), a steady state with a nonvanishing population is possible only for $\alpha > 0$. Surprisingly, the explosion threshold in the respective linear model (10.3.4) with $\beta = 0$ is reached, as shown in Sect. 10.3, already at $\alpha = -s$, for a *lower* value of the parameter α. What is the reason for this difference?

For the linear model, a strongly intermittent behavior is typical. Close to the explosion threshold, determined by the time dependence of the mean population density in the system, this density actually decreases with time almost everywhere in the medium, except in a small area occupied by growing population spikes.

However, rapid growth of the population in the spikes more than compensates for the decrease in other regions and, as a result, the mean population density $\langle n \rangle$ can increase with time.

Nonlinear damping prevents the unlimited growth of the population density in such rare spikes. In the absence of the increasing contribution from the spikes, the mean population density is determined by more typical realizations. Consequently, the self-supporting populated state becomes possible not at $\alpha = -s$ (when $\langle n \rangle$ starts to grow in the linear model) but only at $\alpha = 0$.

In the interval $-s < \alpha < 0$, a very interesting transient behavior is observed. Suppose that at the initial time moment the population density everywhere in the medium is so small that the linearization of equation (12.4.5) is justified. Then, according to the predictions of our study in Sect. 10.3, the mean population density must start to grow. This growth reflects the formation of an intermittent pattern with very rare but strong spikes. The development of such a pattern continues while the level of the population density in the spikes is still significantly below the saturation level. If the coefficent β characterizing the strength of nonlinear damping is small, this intermittent stage can have a long duration.

Eventually, however, the spikes will reach saturation and cease to grow. Afterwards, the mean density $\langle n \rangle$ will start to slowly *decrease*, because of the general decrease of the population density in typical realizations and finite lifetimes of the spikes. We see that the final extinction of the population is preceded in this case by a temporary growth of the mean population density. The smaller the nonlinear damping coefficient β, the longer the transient with the increasing population density.

Since the stationary probability distribution in the steady populated state at $\alpha > 0$ is known, the statistical properties of this state can be analyzed in more detail. Examination of equation (12.4.8) reveals that the probability distribution undergoes a further qualitative change at the positive parameter value $\alpha = s$ (see Fig. 12.6). If $0 < \alpha < s$ the probability density p_s decreases monotonically with n and diverges at $n = 0$ (but still remains normalizable). On the other hand, when $\alpha > s$ this probability density vanishes at $n = 0$ and has a maximum at a certain positive value of the variable n. This qualitative change of the probability distribution was first found (in the Ito interpretation) by *Horsthemke* and *Malek-Mansour* [12.19] who considered it as an example of a noise-induced transition (see also the next section).

From the viewpoint of our previous discussion in Sect. 10.3, it can be said that the fluctuation behavior in the nonlinear model (12.4.5) preserves, inside the region $0 < \alpha < s$, some remnants of the intermittency characteristic for the linear model (10.3.4).

Using equation (12.4.8), the statistical moments of the random variable n in the steady state are determined as

$$\langle n^k \rangle = \left(\frac{s}{\beta} \right)^k \frac{\Gamma(k + \alpha/s)}{\Gamma(\alpha/s)} \, , \qquad k = 1, 2, \ldots \tag{12.4.11}$$

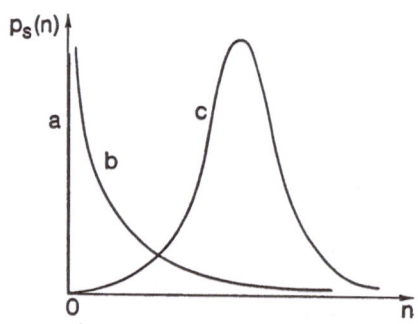

Fig. 12.6. Stationary probability distribution $p_s(n)$ for the stochastic equation of logistic growth: (a) $\alpha < 0$, (b) $0 < \alpha < s$, (c) $\alpha > s$

To estimate the magnitude of these moments, we can compare them with with the respective powers of the mean population density as

$$\langle n \rangle = \frac{\alpha}{\beta} \ . \tag{12.4.12}$$

in the same steady state. We find that

$$\frac{\langle n^k \rangle}{\langle n \rangle^k} = \left(\frac{s}{\alpha}\right)^k \frac{\Gamma(k + \alpha/s)}{\Gamma(\alpha/s)}$$

$$= \left(\frac{s}{\alpha}\right)^k \left(k + \frac{\alpha}{s}\right) \left(k - 1 + \frac{\alpha}{s}\right) \ldots \left(1 + \frac{\alpha}{s}\right) \ . \tag{12.4.13}$$

Therefore, within the interval $0 < \alpha < s$ this ratio increases with k approximately as $(s/\alpha)^k k!$ for the moments of high orders k. This indicates persisting weak intermittency of the stationary distribution. On the other hand, for $\alpha \gg s$ we have $\langle n^k \rangle \approx \langle n^k \rangle$ up to $k \sim \alpha/s$ and hence no intermittency is found for statistical moments in this region.

Obviously, equation (12.4.5) is only an idealization, since it involves a continuous population density n. In reality, at very small population densities the discrete nature of birth–death processes should be taken into account. Each time the system visits the region of very small values of n, it has a large probability of extinction. Since for $s > \alpha > 0$ the stationary probability distribution (12.4.8) for the idealized model is concentrated in the vicinity of $n = 0$, stochastic extinction is highly probable here.

Until now we have neglected diffusion of reproducing particles. When this is taken into account, the stochastic equation of logistic growth reads

$$\dot{n} = \alpha n - \beta n^2 + D\nabla^2 n + f(r, t)n \ . \tag{12.4.14}$$

We consider the case when $f(r, t)$ is a Gaussian random field with vanishingly short correlation time and corrrelation radius, i.e.,

$$\langle f(r, t) \rangle = 0 \ , \quad \langle f(r, t)f(r', t') \rangle = 2s\delta(r - r')\delta(t' - t) \ . \tag{12.4.15}$$

Note that (12.4.14) resembles the Ginzburg-Landau equation (12.1.15) for second-order phase transitions, but there are a few essential differences. Firstly, the order parameter η in the Ginzburg-Landau equation can take both positive and negative values, while the "order parameter" of the logistic equation (i.e. the population density n) is strictly nonnegative. Therefore, to provide nonlinear damping (12.1.15) must include the *cubic* nonlinear term, whereas in (12.4.14) the same role may be played by the *quadratic* term. Secondly, the Ginzburg-Landau equation (12.1.15) involves an *additive* white noise $f(r, t)$, whereas (12.4.14) has only a *multiplicative* noise term. It can easily be seen that Gaussian noise can never enter in an additive manner into the equation for a nonnegative order parameter, since it would be able to produce negative fluctuations.

The particular form of the noise term in the logistic equation (12.4.14), i.e. linear in the order parameter n, is typical for cases where the fluctuations have an *external* origin. If the environment of a system fluctuates, this leads to random variations in the values of the parameters that control local breeding and decay of the particles. Assuming that the right-hand side of the kinetic equation for the local population density n of these particles is analytic in terms of n, we can expand it near the transition threshold in powers of n, keeping only the most significant linear (exciting) and quadratic (damping) terms. By taking into account random, externally induced variations in the coefficient of the linear term, we immediately arrive at (12.4.14).

Consider deviations $\Delta n = n - n_0$ from the uniform population distribution $n_0 = \alpha/\beta$ in the absence of fluctuations above the threshold. When these deviations are sufficiently small, their behavior is described by the linearized equation

$$\delta\dot{n} = -\alpha\delta n + D\Delta\delta n + f(r, t)n_0 \; . \tag{12.4.16}$$

This coincides with (12.1.17), but the random field $f(r, t)$ is now multiplied by the factor n_0.

In close analogy to the analysis of second-order phase transitions in Sect. 12.1, we can introduce the deviation $(\delta n)_V$ averaged over a volume element V and estimate the mean square of this quantity. It is given by [cf. (12.1.24)]

$$\langle(\delta n)_V^2\rangle \approx \frac{sn_0^2}{\alpha V} \; . \tag{12.4.17}$$

An estimate of the characteristic intensity of fluctuations can be obtained if we put here $V \sim r_c^d$, where $r_c \sim (D/\alpha)^{1/2}$ is the correlation radius for fluctuations of the population density n. This yields

$$\langle(\delta n)_{V_c}^2\rangle \approx \frac{sn_0^2}{D^{d/2}}\alpha^{(d-2)/2} \; . \tag{12.4.18}$$

We should compare the estimate obtained with the square of the mean order parameter n_0: the linearized equation (12.4.16) is valid as long as (12.4.18) remains much smaller than n_0^2. Explicitly, this condition reads

$$\frac{s}{D^{d/2}}\alpha^{(d-2)/2} \ll 1 \; . \tag{12.4.19}$$

Hence, we again find that the fluctuation behavior is very sensitive to the dimensionality of the medium. However, in contrast to the second-order phase transition studied in Sect. 12.1, the critical dimensionality of the medium is now equal to 2 (instead of 4).

If $d = 3$, the linear approximation becomes increasingly better as the critical point $\alpha = 0$ is approached. On the other hand, in the one-dimensional medium ($d = 1$) the characteristic intensity of fluctuations grows as a approaches zero. The fluctuations remain so weak that nonlinear interactions between them can be neglected only while $\alpha \gg s^2/\mathcal{D}$. In the marginal case of the two-dimensional medium, (12.4.19) predicts that the fluctuations remain weak right down to the critical point $\alpha = 0$ provided the condition $s \ll D$ is satisfied. However, the logarithmic corrections, which were not taken into account in the above crude estimates can still lead, in this case, to the enhancement of fluctuations in the immediate vicinity of the critical point.

Mikhailov and *Uporov* [12.20, 21] (see also the reviews [12.22, 23] have investigated the fluctuation behavior in the vicinity of the population settling-down transition in the more general situation in which the random field $f(r, t)$ in the stochastic logistic equation has finite correlation time τ_0 and finite correlation radius r_0. They found that the critical dimensionality then remains equal to 2, but that the estimates depend explicitly on r_0 and τ_0. The fluctuation behavior in the case of weak random *breeding centers* (Sect. 10.3) was analyzed in [12.21–25].

Above we have neglected the *internal* noise in the logistic equation, which results from the stochastic discrete nature of birth, death, and diffusion processes. This noise was considered in Sect. 9.2, where it was modeled by adding special fluctuating forces to the equation for the population density of particles [see (9.2.43)]. Simple estimates show that the internal noise of diffusion does not significantly influence the density fluctuations with large characteristic lengths, such as the correlation length r_c near the transition point. When only internal noises due to reproduction and death processes are taken account, the stochastic logistic equation is

$$\dot{n} = \alpha n - \beta n^2 + D\nabla^2 n + (k_{\text{birth}} n)^{1/2} f_{\text{birth}}(r, t) + (k_{\text{death}} n)^{1/2} f_{\text{death}}(r, t) \, , (12.4.20)$$

where $f_{\text{birth}}(r, t)$ and $f_{\text{death}}(r, t)$ are two independent white noises of unit intensity, i.e.,

$$\langle f_{\text{birth(death)}}(r, t) f_{\text{birth(death)}}(r', t') \rangle = \delta(r - r')\delta(t - t') \tag{12.4.21}$$

and the Ito interpretation is employed.

For equation (12.4.20) we can repeat the steps taken above when we investigated the fluctuation behavior for the stochastic logistic equation (12.4.14). Considering the linearized equation for deviations δn, we find that in this case the mean square fluctuation, averaged over the correlation volume $V_c \sim r_c^d$, is

$$\langle (\delta n)_{V_c}^2 \rangle \approx \frac{1}{2} (k_{\text{birth}} + k_{\text{death}}) \frac{n_0}{D^{d/2}} \alpha^{(d-2)/2} \, . \tag{12.4.22}$$

The fluctuations are weak when this mean square fluctuation is much less than $n_0^2 = \alpha/\beta$, i.e., if

$$\alpha^{(4-d)/2} \gg \left(k_{\text{birth}} + k_{\text{death}}\right) \frac{\beta}{D^{d/2}} \, . \tag{12.4.23}$$

Hence, in media with the dimensionality less than $d = 4$, the fluctuations caused by the internal noise always become strong sufficiently close to the transition threshold and the mean-field approximation is not valid in this region, similar to second-order phase transitions discussed in Sect. 12.1. In other words, the critical dimensionality for the fluctuations caused by internal noises is $d = 4$. This result has been derived by *Janssen* [12.26, 27] using the diagrammatic perturbation technique. Note that, as shown above, the critical dimensionality for fluctuations caused by external noise is $d = 2$.

When both external and internal noises are simultaneously taken into account in the logistic equation, the effects of the internal noise always dominate in sufficiently close proximity to the critical point. Indeed, this multiplicative noise enters into the equation with the factor $n^{1/2}$, whereas the external noise has the factor n [compare equations (12.4.20) and (12.4.14)]. Therefore, at very small mean population densities, a crossover to the fluctuation behavior typical for the second-order phase transitions is expected.

12.5 Noise-Induced Phase Transitions: Competition and Coexistence in the Fluctuating Environment

The usual effect of noise is that it brings about fluctuations. If the noise is strong enough, however, it can also produce significant qualitative changes in the behavior of a system. These phenomena are called *noise-induced phase transitions*.

A simple example of such a transition has already been considered in the previous section. We noted, following *Horsthemke* and *Malek-Mansour* [12.19], that the stationary probability distribution for a system described by the stochastic logistic equation (12.4.5) undergoes an important change when the noise intensity reaches a critical value. While for all smaller noise intensities this distribution decreases monotonically with the variable n, above the critical point it has a maximum at $n \neq 0$. This resembles the behavior found when a dynamical system passes through a bifurcation that makes the fixed point $n = 0$ unstable and creates a new attractive fixed point in its vicinity.

In another stochastic model [12.28], the stationary probability distribution changes its shape in such a way that, at the threshold, a local probability maximum splits and gives rise to two new maxima that diverge from one another as the noise intensity is further increased. This can be interpreted as a transition from a monostable to an effectively bistable system, caused by noise.

A detailed discussion of noise-induced transitions is given in the book by *Horsthemke* and *Lefever* [12.29]. Below we consider an example of a noise-induced transition in a distributed active system with diffusion that has been studied by *Mikhailov* [12.30, 31].

Suppose that we have two biological species with population densities N and n, which compete for the same renewable resource – namely, for the food m that grows at a rate Q. If we assume that the birth rates of both species are proportional to the amount of food m available, this ecological system will obey the equations

$$\dot{N} = (Bm - A)N ,$$
$$\dot{n} = (\beta m - \alpha)n ,$$
$$\dot{m} = Q - \gamma m - UN - un .$$

(12.5.1)

The third equation in (12.5.1) describes the food dynamics. Individuals belonging to the two species consume amounts U and u of food per unit time, respectively. The food itself grows at a constant rate Q and, even in the absence of the two species, its maximum density is limited by some decay mechanism represented by the term $-\gamma m$ in the above equation.

When the inequalities

$$\frac{A}{B} < \frac{\alpha}{\beta} , \qquad \frac{A}{B} < \frac{Q}{\gamma}$$

(12.5.2)

are satisfied, the species with population density N is strong and that with population density n is weak. In the long time limit, the strong species survives and the weak species becomes extinct, so that as $t \to \infty$ a steady state

$$N = N_1 , \qquad m = m_1 , \qquad n = 0$$

(12.5.3)

is reached, where (assuming that $Q > \gamma A/B$)

$$m_1 = \frac{A}{B} , \qquad N_1 = \frac{Q - \gamma m_1}{U} .$$

(12.5.4)

This is a classical result, which frequently goes by the name of the "ecological theorem": two biological species that totally rely on the same resource cannot coexist in a stationary state.

Now suppose that the food growth rate Q fluctuates in time, $Q \to Q + f(t)$, with $\langle f \rangle = 0$. Can this random variation change the result of the competition and, specifically, can it make coexistence possible? The answer is *no*. Indeed, the first two equations of (12.5.1) can be rewritten as

$$\frac{d}{dt} \ln N = Bm - A , \qquad \frac{d}{dt} \ln n = \beta m - \alpha .$$

(12.5.5)

By excluding m and integrating, we obtain from (12.5.5) the equation

$$\left\langle n^{B/\beta} \right\rangle = \text{const } \exp(-Bp_c t) \langle N \rangle ,$$

(12.5.6)

where

$$p_c = \frac{\alpha}{\beta} - \frac{A}{B} \; . \tag{12.5.7}$$

Note that conditions (12.5.2) imply $p_c > 0$. Hence, in a steady state (reached at $t \to \infty$) we find $\langle n^{B/\beta} \rangle = 0$ and, furthermore, $n = 0$. It can also be shown that, if individuals of the weak species are mobile (i.e. capable of diffusion) but the food growth rate is constant, this does not prevent extinction of the weak species.

The situation is radically different when the rate of growth of food fluctuates in both space and time, and the weak species is mobile. The basic result of *Mikhailov* [12.30, 31] (see also [12.23, 32]) is that, beginning with a certain critical noise intensity, it is possible for two competing species to coexist statistically in a stationary state, i.e. a new steady state is established in which $\langle n \rangle > 0$ and $\langle N \rangle > 0$. The transition to this state, which is accompanied by the appearance of a nonzero average density $\langle n \rangle$, is an example of a population settling-down transition.

Our conclusion is important from the standpoint of mathematical ecology. In a fluctuating environment, simple mobility turns out to be an essential factor ensuring evolutionary advantage and the possibility of coexistence with a stronger but immobile species. Mobile individuals belonging to the weaker species survive because they are capable of "eating up" the fluctuations!

This system is described by

$$\dot{N} = (Bm - A)N \; ,$$
$$\dot{n} = (\beta m - \alpha)n + D\Delta n \; , \tag{12.5.8}$$
$$\dot{m} = Q - \gamma m - UN - un + f(r,t) \; .$$

Here $f(r,t)$ is a Gaussian noise with the pair correlation function

$$\langle f(r,t)f(r',t') \rangle = 2\gamma\theta \, \exp(-k_0|r - r'|) \, \delta(t - t') \; . \tag{12.5.9}$$

Thus, $r_0 = 1/k_0$ is the typical spatial size of an individual fluctuation and θ characterizes the intensity of fluctuations. In the absence of both species (when $N = n = 0$) we have $\langle \delta m^2 \rangle_0 = \theta$.

It is now convenient to introduce the new variables

$$p = m - m_1 \; , \qquad q = \ln\left(\frac{N}{N_1}\right) \; , \tag{12.5.10}$$

which characterize deviations of the densities m and N from their stationary values given by (12.5.3) in the absence of fluctuations. In terms of these new variables, equations (12.5.1) become

$$\dot{q} = Bp \; ,$$
$$\dot{p} = -\gamma p - \frac{\Omega_0^2}{B}(e^q - 1) - un + f(r,t) \; , \tag{12.5.11}$$
$$\dot{n} = \beta(p - p_c)n + D\Delta n \; ,$$

where we have introduced the notation $\Omega_0^2 = BUN_1$. Note that p_c given by (12.5.7) characterizes the deficiency of food in the stationary state (12.5.4) for the persistence of individuals of the weaker species.

If the intensity θ of noise $f(r, t)$ is sufficiently small (see the discussion below), we have $\langle q^2 \rangle \ll 1$. Then, after linearization of (12.5.11) with respect to q, we obtain

$$\dot{q} = Bp$$

$$\dot{p} = -\gamma p - \frac{\Omega_0^2}{B} q + f(r, t) - un \qquad (12.5.12)$$

$$\dot{n} = \beta(p - p_c)n + D\Delta n .$$

These equations describe a set of identical damped oscillators of frequency Ω_0, located at each point of space and interacting through the field $n(r, t)$. When all the oscillators are in the unexcited state, this field decays exponentially to the value $n = 0$. Excitation of the oscillators by the random force $f(r, t)$ provides us, however, with the possibility of reproduction of this field. The spatial structure of the reproductive regions (in which $p > p_c$) varies randomly in space and time. When the increase in n in the reproductive regions begins to compensate for the reduction in the field n outside such regions, a medium-populating transition takes place for the field n.

Next we perform a separation of the fast and slow variables of the problem. The characteristic short time-scales in equations (12.5.12) are (βp_c^{-1}), Γ^{-1}, Ω_0^{-1} and $(Dk_0^2)^{-1}$ Slowly varying components are obtained by taking time averages of the fields n, p and q over time intervals, T, which are longer than all these short time-scales, i.e.,

$$\tilde{n}(r, t) = T^{-1} \int_0^T n(r, t + t_1) \, dt_1 \equiv \langle n(r, t) \rangle_T \qquad (12.5.13)$$

and $\tilde{q}(r, t) = \langle q(r, t) \rangle_T$, $\tilde{p}(r, t) = \langle p(r, t) \rangle_T$. Fast variables are then defined as $\delta n = n - \tilde{n}$, $\delta q = q - \tilde{q}$, and $\delta p = p - \tilde{p}$.

Our aim is to derive a closed equation for the slow component $\tilde{n}$. We start by applying the operation of time averaging to equations (12.5.12). This yields

$$\dot{\tilde{q}} = B\tilde{p} , \qquad (12.5.14)$$

$$\dot{\tilde{p}} = -\gamma\tilde{p} - \frac{\Omega_0^2}{B}\tilde{q} + \tilde{f}(r, t) - u\tilde{n} , \qquad (12.5.15)$$

$$\dot{\tilde{n}} = -\beta p_c\tilde{n} + \beta\tilde{p}\tilde{n} + D\nabla^2\tilde{n} + \beta\langle\delta p\delta n\rangle_T . \qquad (12.5.16)$$

Taking the time derivative of equation (12.5.15) and then using equation (12.5.14) to eliminate $\tilde{q}$ we obtain

$$\ddot{\tilde{p}} = -\gamma\dot{\tilde{p}} - \Omega_0^2\tilde{p} + \dot{\tilde{f}} - u\dot{\tilde{n}} . \qquad (12.5.17)$$

Since the characteristic time-scale for variation of the slow components is much longer than the relaxation time γ^{-1} in equation (12.5.17), we approximately have

$$\tilde{p}(r, t) \approx -\frac{u}{\Omega_0^2}\frac{\partial \tilde{n}}{\partial t} + \frac{1}{\Omega_0^2}\frac{\partial \tilde{f}}{\partial t} , \qquad (12.5.18)$$

where $\tilde{f} = \langle f \rangle_T$.

The equations for the fast components have the form

$$\delta\dot{q} = B\delta p , \qquad (12.5.19)$$

$$\delta\dot{p} = -\gamma\delta p - \frac{\Omega_0^2}{B}\delta q - u\delta n + \delta f(r, t) , \qquad (12.5.20)$$

$$\delta\dot{n} = -\beta(p_c - \tilde{p})\delta n + \beta\tilde{n}\delta p + D\nabla^2\delta n + \beta(\delta n\delta p - \langle\delta n\delta p\rangle_T) , \qquad (12.5.21)$$

where $\delta f = f - \tilde{f}$. The slow variables $\tilde{n}$ and $\tilde{p}$ in these equations should be treated as if they were constants. The probability distribution for fast variables adjusts adiabatically to their given values. The time averages $\langle\delta n^2\rangle_T$ and $\langle\delta n\delta p\rangle_T$ can therefore be replaced with the respective statistical averages $\langle\delta n^2\rangle$ and $\langle\delta n\delta p\rangle$ which are found using the stationary probability distribution for fast variables established under fixed values of the slow variables $\tilde{n}$ and $\tilde{p}$.

To determine $\langle\delta n^2\rangle$ and $\langle\delta n\delta p\rangle$, the mean-field approximation can be applied. If fluctuations are so weak so that the condition

$$(\langle\delta n^2\rangle)^{1/2} \ll \tilde{n} \qquad (12.5.22)$$

holds, the only nonlinear term $\beta(\delta n\delta p - \langle\delta n\delta p\rangle)$ in equation (12.5.21) is, on average, small in comparison with the term $\beta\tilde{n}\delta p$ in the same equation and can thus be neglected. Then evolution of fast variables is approximately described by a set of *linear* stochastic differential equations that can be exactly solved.

The solution is best performed by taking the time Fourier transform of the linearized equations (12.5.19)–(12.5.21), solving the resulting system of linear algebraic equations for the Fourier amplitudes, and then returning to the time-dependent variables. Following these steps, we find for the three-dimensional medium:

$$\langle\delta n\delta p\rangle = \frac{1}{(2\pi)^3}\int \frac{\beta\theta\tilde{n}g(k)h(k)dk}{g(k)[g(k) + \gamma] + \Omega_0^2 + \beta u\tilde{n}[1 + g(k)/\gamma]} . \qquad (12.5.23)$$

Here $g(k) = \beta(p_c - \tilde{p}) + Dk^2$, and $h(k) = 8\pi k_0(k^2 + k_0^2)^{-2}$.

Substitution of (12.5.18) and (12.5.23) into (12.5.16) results in a closed equation for the slow component of the population density $\tilde{n}$. This equation can be further simplified if we take into account that, near the transition point, the population density of the weak species is small. Expanding the right-hand side of this equation in powers of $\tilde{n}$ and keeping only the terms up to $\tilde{n}^2$, we derive an effective evolution equation

$$\dot{\tilde{n}} = \beta p_c \left(\frac{\theta}{\theta_c} - 1\right)\tilde{n} - \varepsilon\tilde{n}^2 + D\nabla^2\tilde{n} + \Phi(r, t)\tilde{n} , \qquad (12.5.24)$$

where the coefficients are given by

$$\theta_c = \frac{p_c}{\beta} D k_0^2 , \tag{12.5.25}$$

$$\varepsilon = \frac{3\beta p_c u}{2^{3/2}\gamma(\Omega_0 D k_0^2)^{1/2}} , \tag{12.5.26}$$

provided that the conditions

$$D k_0^2 \gg \Omega_0 \gg \gamma \gg \beta p_c \tag{12.5.27}$$

are satisfied.

The random Gaussian force $\Phi(r, t)$ in equation (12.5.24) is defined as

$$\Phi(r, t) = \frac{1}{\Omega_0^2} \frac{\partial \tilde{f}}{\partial t} . \tag{12.5.28}$$

Hence, its Fourier transform $\Phi(k, \omega)$ has correlations

$$\langle \Phi(k, \omega)\Phi(k', \omega')\rangle = S(k, \omega)\delta(k + k')\delta(\omega + \omega') , \tag{12.5.29}$$

where

$$S(k, \omega) = \begin{cases} 0, & |\omega| > \dfrac{2\pi}{T} , \\ 2\gamma\theta\left(\dfrac{\omega}{\Omega_0}\right)^2 h(k), & |\omega| < \dfrac{2\pi}{T} . \end{cases} \tag{12.5.30}$$

The intensity of this force approaches zero as the averaging interval T is increased.

The evolution equation (12.5.24) has the same form as the stochastic equation of logistic growth (12.4.14). A transition to the self-maintained populated state of the medium takes place when the coefficient α, i.e., the mean difference of birth and death rates, changes its sign and becomes positive. As we see from equation (12.5.24), for the present ecological model this coefficient is

$$\alpha = \beta p_c \left(\frac{\theta}{\theta_c} - 1\right) . \tag{12.5.31}$$

Therefore, the effective mean birth rate of the weak species in this model is *controlled by the noise intensity* θ. When θ exceeds a threshold θ_c, the weak species populates the medium. Above the transition point, the mean population density of the weak species is

$$n = \frac{\beta p_c}{\varepsilon} \left(\frac{\theta}{\theta_c} - 1\right) , \quad \theta \geq \theta_c . \tag{12.5.32}$$

Hence, for sufficiently strong fluctuations in the rate of food growth, coexistence of a stronger species with the weaker but mobile species becomes established.

This is a remarkable result. Above we have noted that in the deterministic ecological model, where two species are using the same food resource, coexistence was never possible. Now we see that the situation is radically changed in the presence of fluctuations in the system's environment. In a sense, the fluctuations can be viewed as an additional resource that is more effectively utilized by the mobile species. It is better to be clever, but sometimes it suffices to be mobile... .

According to equation (12.5.25), the critical noise intensity is proportional to the diffusion constant D, i.e., it increases with the mobility of the individuals of the weak species. This property has a simple explanation: If their mobility is large, weak individuals traverse the effective breeding regions (where $p > p_c$) too rapidly, and therefore cannot efficiently use the excess of food available inside these regions. Note that equation (12.5.25) holds only when (12.5.27) is satisfied and the diffusion coefficient D is sufficiently large.

The above analysis was based on the assumption that fluctuations in the population density n of the weak species were sufficiently small, so that the condition (12.5.22) was satisfied. Using linearized equations (12.5.19)–(12.5.21), the mean square fluctuation $\langle \delta n^2 \rangle$ in the steady populated state can be determined. If the conditions (12.5.27) are realized, we find for a three-dimensional medium (see [12.31]) that

$$\frac{(\langle \delta n^2 \rangle)^{1/2}}{\langle n \rangle} = 2^{1/4} \left(\frac{\beta p_c}{\Omega_0} \right)^{1/2} \left(\frac{\Omega_0}{D k_0^2} \right)^{1/4} \ll 1 \tag{12.5.33}$$

and hence the fluctuations are indeed negligible. This agrees with the results of Sect. 12.4 where medium-populating transitions in the presence of external noises were considered.

References

Chapter 1

1.1 E. Lorenz: J. Atm. Sci. **20**, 130–141 (1963)
1.2 Ya.G. Sinai: Dokl. Akad. Nauk SSSR **153**, 1261–1264 (1963)
1.3 B.A. Huberman, J.P. Crutchfield, N.H. Packard: Appl. Phys. Lett. **37**, 750–752 (1980)
1.4 D. Ruelle, F. Takens: Commun. Math. Phys. **20**, 167–192 (1971)
1.5 A.J. Lichtenberg, M.A. Lieberman: *Regular and Stochastic Motion* (Springer, Berlin, Heidelberg 1983)
1.6 H.G. Schuster: *Deterministic Chaos* (Physik-Verlag, Weinheim, 1984)
1.7 G.M. Zaslavskii: *Stochasticity of Dynamical Systems* (Nauka, Moscow 1984)
1.8 K. Kaneko: Physica D **34**, 1–41 (1989)
1.9 Y. Kuramoto: *Chemical Oscillations, Waves, and Turbulence* (Springer, Berlin, Heidelberg 1984)
1.10 P. Coullet, L. Gil, J. Lega: Physica D **37**, 91–103 (1989)
1.11 W. Feller: *An Introduction to Probability Theory and its Applications*, Vols. 1 and 2 (Wiley, New York 1968 and 1971)
1.12 L. Arnold: *Stochastic Differential Equations: Theory and Applications* (Wiley, New York 1974)
1.13 Z. Schuss: *Theory and Applications of Stochastic Differential Equations* (Wiley, New York 1980)
1.14 N.G. van Kampen: *Stochastic Processes in Physics and Chemistry* (North-Holland, Amsterdam 1983)
1.15 R.L. Stratonovich: *Topics in the Theory of Random Noise*, Vols. I and II (Gordon and Breach, New York 1963 and 1967)
1.16 C.W. Gardiner: *Handbook of Stochastic Methods* (Springer, Berlin, Heidelberg 1985)
1.17 H. Risken: *The Fokker-Planck Equation*, 2nd ed. (Springer, Berlin, Heidelberg 1989)
1.18 W. Horsthemke, R.L. Lefever: *Noise-Induced Transitions* (Springer, Berlin, Heidelberg 1984)
1.19 R. Thom: *Structural Stability and Morphogenesis* (Benjamin, Reading 1975)
1.20 V.I. Arnold: *Catastrophe Theory* (Springer, Berlin, Heidelberg 1986)
1.21 H.A. Ceccatto, B.A. Huberman: Physica Scripta **37**, 145–150 (1988)

Chapter 2

2.1 H. Goldstein: *Classical Mechanics* (Addison-Wesley Press, Cambridge 1951)
2.2 E.T. Whittaker: *Treatise on the Analytical Dynamics of Particles and Rigid Bodies* (Cambridge University Press, Cambridge 1961)
2.3 P. Appell: *Traité de Mécanique Rationnelle*, Vols. 1 and 2 (Gauthier-Villars, Paris 1953)
2.4 V.I. Arnold: *Mathematical Methods of Classical Mechanics* (Springer, Berlin, Heidelberg 1989)

2.5 M. Henon, C. Heiles: Astron. J. **69**, 73–79 (1964)
2.6 A.J. Lichtenberg, M.A. Lieberman: *Regular and Stochastic Motion* (Springer, Berlin, Heidelberg 1983)
2.7 P. Holmes: Physica D **5**, 335–347 (1982)
2.8 R.Z. Sagdeev, D.A. Usikov, G.M. Zaslavsky: *Nonlinear Physics* (Harwood, New York 1988)
2.9 G.M. Zaslavsky, R.Z. Sagdeev: *Introduction to Nonlinear Physics* (Nauka, Moscow 1988, in Russian)
2.10 A.N. Kolmogorov: Dokl. Akad. Nauk SSSR **98**, 527–530 (1954)
2.11 V.I. Arnold: Russian Math. Surveys **18**, 85 (1963)
2.12 J.K. Moser: Nachr. Acad. Wiss. Göttingen, Math. Phys. Kl **11a**, 1–20 (1962)
2.13 V.I. Arnold, A. Avez: *Ergodic Problems of Classical Mechanics* (Benjamin, New York 1968)
2.14 J.K. Moser: Math. Ann. **169**, 136–176 (1967)
2.15 V.I. Arnold: Sov. Math.–Dokl. **5**, 581 (1964)
2.16 B.V. Chirikov: Phys. Rep. **52**, 263–379 (1979)
2.17 R.S. MacKay, J.D. Meiss, I.C. Percival: *Transport in Nonlinear Systems* (Queen Mary College, London 1983)
2.18 P.J. Holmes, J.E. Marsden: J. Math. Phys. **23**, 669–675 (1982)
2.19 A.N. Kolmogorov, S.V. Fomin: *Elements of Function Theory and of Functional Analysis* (Nauka, Moscow 1989)
2.20 R. Balescu: *Equilibrium and Nonequilibrium Statistical Mechanics* (Wiley, New York 1975)
2.21 Ya.G. Sinai: Sov. Math.–Dokl. **4**, 1818 (1963)
2.22 Ya.G. Sinai: Russian Math. Surveys **25**, 137 (1970)

Chapter 3

3.1 A.J. Lichtenberg, M.A. Lieberman: *Regular and Stochastic Motion* (Springer, Berlin, Heidelberg 1983)
3.2 O.E. Lanford: "Strange attractors and turbulence", in *Hydrodynamic Instabilities and Transition to Turbulence*, ed. by H.L. Swinney, J.P. Gollub (Springer, Berlin, Heidelberg 1981) pp. 7–31
3.3 J.P. Eckmann: Rev. Mod. Phys. **53**, 643–654 (1981)
3.4 J. Guckenheimer, P. Holmes: *Nonlinear Oscillations, Dynamical Systems and Bifurcation of Vector Fields* (Springer, Berlin, Heidelberg 1983)
3.5 V.I. Arnold, V.S. Afraimovich, Yu.S. Ilyashenko, L.P. Shilnikov: "Bifurcation theory", in *Dynamical Systems*, Vol. 5, ed. by V.I. Arnold (Springer, Berlin, Heidelberg 1989)
3.6 V.S. Afraimovich: "Attractors", in *Nonlinear Waves 1. Dynamics and Evolution*, ed. by J. Engelbrecht, A.V. Gaponov-Grekhov, M.I. Rabinovich (Springer, Berlin, Heidelberg 1989) pp. 14–28
3.7 J.M.T. Thompson, H.B. Stewart: *Nonlinear Dynamics and Chaos* (Wiley, New York 1986)
3.8 Ya.G. Sinai: "Stochasticity of dynamical Systems", in *Nonlinear Waves*, ed. by A.V. Gaponov-Grekhov, M.I. Rabinovich (Nauka, Moscow 1979) pp. 192–212
3.9 J. Milnor: Commun. Math. Phys. **99**, 177–196 (1985)
3.10 S. Newhouse: Ann. N.Y. Acad. Sci. **357**, 292–299 (1980)
3.11 D. Ruelle, F. Takens: Commun. Math. Phys. **20**, 167–192 (1971)
3.12 R. Shaw: Z. Naturforsch. A **36**, 80–112 (1981)
3.13 V.S. Afraimovich, L.P. Shilnikov: "On strange attractors and quasiattractors", in *Nonlinear Dynamics and Turbulence* (Pitman, Boston 1983) pp. 1–34
3.14 E.N. Lorenz: J. Atmos. Sci. **20**, 130–141 (1963)

3.15 C. Sparrow: *The Lorenz Equations: Bifurcations, Chaos and Strange Attractors* (Springer, Berlin, Heidelberg 1982)

3.16 L.A. Bunimovich, Ya.G. Sinai: "Stochasticity of the attractor in the Lorenz model", in *Nonlinear Waves*, ed. by A.V. Gaponov-Grekhov, M.I. Rabinovich (Nauka, Moscow 1979) pp. 212–226

3.17 V.S. Afraimovich, V.V. Bykov, L.P. Shilnikov: Sov. Phys.–Dokl. **22**, 563 (1977)

3.18 J.A. Yorke, E.D. Yorke: J. Stat. Phys. **21**, 263–277 (1979)

3.19 O.E. Lanford: "Computer pictures of the Lorenz attractor", in *Lecture Notes in Mathematics*, Vol. 615 (Springer, Berlin, Heidelberg 1977) pp. 113–116

3.20 S. Smale: Bull. Amer. Math. Soc. **73**, 747–817 (1967)

3.21 R.F. Williams: Global Analysis. Amer. Math. Soc. **14**, 361–393 (1970)

3.22 A.N. Kolmogorov: Dokl. Akad. Nauk SSSR **119**, 861–864 (1958)

3.23 A.N. Kolmogorov: Dokl. Akad. Nauk SSSR **124**, 754–755 (1959)

3.24 Ya.G. Sinai: Dokl. Akad. Nauk SSSR **124**, 768–771 (1959)

3.25 Yu.L. Klimontovich: *Statistical Physics* (Harwood, New York 1986)

3.26 G. Benettin, L. Galgani, J.M. Strelcyn: Phys. Rev. A **14**, 2338–2345 (1976)

3.27 G. Benettin, C. Froeschle, J.P. Scheidecker: Phys. Rev. A **19**, 2454–2460 (1979)

3.28 P. Hartman: *Ordinary Differential Equations* (Wiley, New York 1964)

3.29 I. Shimada, T. Nagascima: Progr. Theor. Phys. **61**, 1605–1616 (1979)

3.30 S.D. Feit: Commun. Math. Phys. **61**, 249–260 (1978)

3.31 G. Benettin, L. Galgani, A. Giogilli, J.M. Strelcyn: Meccanica **15**, 9–31 (1980)

3.32 J.S. Turner, J.-C. Roux, W.D. McCormick, H.L. Swinney: Phys. Lett. A **85**, 9–12 (1981)

Chapter 4

4.1 B. Mandelbrot: *Fractals: Form, Chance and Dimension* (Freeman, San Francisco 1977)

4.2 B. Mandelbrot: *The Fractal Geometry of Nature* (Freeman, San Francisco 1982)

4.3 J. Feder: *Fractals* (Plenum, New York 1988)

4.4 T. Vicsek: J. Phys. A **16**, L647–L652 (1983)

4.5 J. Feder, L. Hinrichsen, K.J. Maloy, T. Jossany: Physica D **38**, 104–111 (1989)

4.6 J. Hittman, G. Daccord, H.E. Stanley: "When viscous 'fingers' have fractal dimension?", in *Fractals in Physics*, Proc. 6th Trieste Int. Symp. on Fractals in Physics, ITCP, Trieste, July 9–12, 1985, ed. by L. Pietronero, E. Tosatti (Elsevier, Amsterdam 1986)

4.7 T.A. Witten, L.M. Sander: Phys. Rev. B **27**, 5686 (1983)

4.8 M. Henon: Commun. Math. Phys. **50**, 69–77 (1976)

4.9 J.D. Farmer, E. Ott, J.A. Yorke: Physica D **7**, 153–180 (1983)

4.10 A.N. Kolmogorov, V.M. Tikhomirov: Usp. Mat. Nauk. **14**, 3–86 (1959)

4.11 P. Grassberger: Phys. Lett. A **97**, 227–230 (1983)

4.12 H.G.E. Hentschel, I. Procaccia: Physica D **8**, 435–444 (1983)

4.13 C. Halsey, M.H. Jensen, L. Kadanoff, I. Procaccia, B.I. Shraiman: Phys. Rev. A **33**, 1141 (1986)

4.14 M. Sano, S. Sato, Y. Sawada: Progr. Theor. Phys. **76**, 945 (1986)

4.15 S. Sato, M. Sano, Y. Sawada: Progr. Theor. Phys. **77**, 1 (1987)

4.16 D.K. Umberger, G. Mayer-Kress, E. Jen: Hausdorf dimension for sets with broken scaling symmetry, in *Dimensions and Entropies in Chaotic Systems*, ed. G. Mayer-Kress (Springer, Berlin, Heidelberg 1986) pp. 42–53

4.17 B.M. Smirnov: *Physics of Fractal Aggregates* (Nauka, Moscow 1991)

4.18 G. Paladin, A. Vulpiani: Phys. Rep. **156**, 147 (1987)

4.19 T.S. Akhromeyeva, S.P. Kurdyumov, G.G. Malinetskii, A.A. Samarskii: Phys. Rep. **176**, 189 (1989)

4.20 A.F. Turbin, N.V. Pracevity: *Fractal Sets, Functions, Distributions* (Naukova Dumka, Kiev 1992)
4.21 D. Ruelle, F. Takens: Commun. Math. Phys. **20**, 167–192 (1971)
4.22 P. Grassberger, I. Procaccia: Physica D **9**, 189–208 (1983)
4.23 D.A. Russel, J.D. Hanson, E. Ott: Phys. Rev. Lett. **45**, 1175–1178 (1980)
4.24 J.L. Kaplan, J.A. Yorke: *Chaotic Behavior of Multidimensional Difference Equations,* Lecture Notes in Mathematics, Vol. 730 (Springer, Berlin, Heidelberg 1979) pp. 204–227
4.25 J.D. Farmer: Physica D **4**, 366–393 (1982)
4.26 L.-S. Young: Ergod. Theory and Dyn. Systems **2**, 109–124 (1982)
4.27 Ya.G. Sinai (ed.): *Dynamical Systems II* (Springer, Berlin, Heidelberg 1988)
4.28 P. Constantin, C. Foias: Commun. Pure App. Math. **38**, 1–27 (1985)
4.29 F. Ledrappier: Commun. Math. Phys. **81**, 229–238 (1981)
4.30 G. Julia: J. Math. Pures et Appl. **8**, 47 (1918)
4.31 P. Fatou: Bull. Soc. Math. Fr. **47**, 161 (1919/1920).
4.32 H.-O. Peitgen, P.H. Richter: *The Beauty of Fractals. Images of Complex Dynamical Systems* (Springer, Berlin, Heidelberg 1988)
4.33 C. Grebogi, E. Ott, J.A. Yorke: Phys. Rev. Lett. **50**, 935 (1983)
4.34 F.C. Moon, G.-X. Li: Phys. Rev. Lett. **55**, 1439 (1985)

Chapter 5

5.1 M. Henon: Commun. Math. Phys. **50**, 69–77 (1976)
5.2 N.V. Butenin, Yu.I. Neimark, N.A. Fufayev: *Introduction to the Theory of Nonlinear Oscillations* (Nauka, Moscow 1987, in Russian)
5.3 P. Collet, J.P. Eckmann: *Iterated Maps of the Interval as Dynamical Systems* (Birkhäuser, Boston 1980)
5.4 A.N. Sharkovskii: Ukr. Mat. Zh. **1**, 61–71 (1964)
5.5 A.N. Sharkovskii, Yu.A. Maistrenko, E.Yu. Romanenko: *Difference Equations and Their Applications* (Naukova Dumka, Kiev 1986, in Russian)
5.6 T.Y. Li, J.A. Yorke: Amer. Math. Mon. **82**, 985–992 (1975)
5.7 M.J. Feigenbaum: J. Stat. Phys. **19**, 25–52 (1978)
5.8 M.J. Feigenbaum: J. Stat. Phys. **21**, 669–706 (1979)
5.9 M.J. Feigenbaum: Los Alamos Science **1**, 4–27 (1980); Physica D **7**, 16–39 (1983)
5.10 S. Grossmann, S. Thomae: Z. Naturforsch. **32a**, 1353–1363 (1977)
5.11 B. Hu: "Functional renormalization-group equation approach to the transition to chaos", in *Chaos and Statistical Methods*, ed. by Y. Kuramoto (Springer, Berlin, Heidelberg 1984)
5.12 O.E. Lanford: Bull. Amer. Math. Soc. **6**, 427–434 (1982)
5.13 P. Collet, J.P. Eckmann, O.E. Lanford: Commun. Math. Phys. **76**, 211–254 (1980)
5.14 H.G. Schuster: *Deterministic Chaos* (Physik-Verlag, Weinheim 1984)
5.15 E.B. Vul, Ya.G. Sinai, K.M. Khanin: Russian Math. Surveys **39:3** (1984)
5.16 J.A. Yorke, C. Grebogi, E. Ott, L. Tedeschini-Lalli: Phys. Rev. Lett. **54**, 1093 (1985)
5.17 M. Bucher: Phys. Rev. **33A**, 3544–3546 (1986)

Chapter 6

6.1 A.A. Andronov, S. Khaikin: *Theory of Oscillations* (ONTI, Moscow 1939) [English transl.: Princeton University Press, Princeton 1949]
6.2 E. Hopf: Phys. Sachs. Acad. Wiss. Leipzig **94**, 1–22 (1942)
6.3 V.I. Arnold (ed.): *Dynamical Systems V* (Springer, Berlin, Heidelberg 1990)

6.4   B.D. Hassard, N.D. Kazarinoff, Y.-H. Wan: *Theory and Applications of Hopf Bifurcation* (Cambridge University Press, Cambridge 1981)

6.5   J.E. Marsden, M. McCracken: *The Hopf Bifurcation and Its Applications* (Springer, Berlin, Heidelberg 1976)

6.6   G. Ioss, D.D. Joseph: *Elementary Stability and Bifurcation Theory* (Springer, Berlin, Heidelberg 1980)

6.7   S.N. Chow, J.K. Hale: *Methods of Bifurcation Theory* (Springer, Berlin, Heidelberg 1982)

6.8   L.D. Landau: Dokl. Acad. Nauk. SSSR **44**, 339–342 (1944)

6.9   L.D. Landau, E.M. Lifshitz: *Mechanics of Continuous Media* (Gostekhizdat, Moscow 1954)

6.10 E. Hopf: Comm. Pure Appl. Math. **1**, 303–322 (1948)

6.11 E. Hopf: In Proc. Conf. Diff. Eqs., Univ. of Maryland, 1956, 49–57

6.12 J.-P. Eckmann: Rev. Mod. Phys. **53**, No. 4, 643–654 (1981)

6.13 D. Ruelle, F. Takens: Commun. Math. Phys. **20**, 167–192 (1971)

6.14 S. Newhouse, D. Ruelle, F. Takens: Commun. Math. Phys. **64**, 35–40 (1978)

6.15 A. Andronov, L. Pontryagin: Dokl. Akad. Nauk SSSR **14**, 247–250 (1937)

6.16 C. Grebogi, E. Ott, J.A. Yorke: Phys. Rev. Lett. **51**, 339–342 (1983)

6.17 R.W. Walden, P. Kolodner, A. Ressner, C.M. Surko: Phys. Rev. Lett. **53**, 242–245 (1984)

6.18 C. Grebogi, E. Ott, J.A. Yorke: Physica D **15**, 354–373 (1985)

6.19 M.J. Feigenbaum, L.P. Kadanoff, S.J. Shenker: Physica D **5**, 370–386 (1982)

6.20 S. Ostlund, D. Rand, J. Sethna, E. Siggia: Physica D **8**, 303–342 (1983)

6.21 S.J. Shenker: Physica D **5**, 405–411 (1982)

6.22 B. Hu: Phys. Lett. A **98**, 79–82 (1983)

6.23 H.G. Schuster: *Deterministic Chaos* (Physik-Verlag, Weinheim 1984)

6.24 J.P. Gollub, H.L. Swinney: Phys. Rev. Lett. **35**, 927–930 (1975)

6.25 H.L. Swinney, J.P. Gollub: Physics Today **31**, No. 8, 41–49 (1978)

6.26 S. Fauve, A. Libchaber: "Rayleigh-Benard experiment in a low Prandtl number fluid mercury", in *Chaos and Order in Nature*, ed. by H. Haken (Springer, Berlin, Heidelberg 1981) pp. 25–35

6.27 H. Haken: *Advanced Synergetics* (Springer, Berlin, Heidelberg 1983)

6.28 O.E. Rossler: Phys. Lett. A **57**, 397–398 (1976)

6.29 D. Gurel, O. Gurel: *Oscillations in Chemical Reactions* (Springer, Berlin, Heidelberg 1983)

6.30 J.P. Crutchfield, J.D. Farmer, N. Packard, R. Shaw, G. Jones, R.J. Donnely: Phys. Lett. A **76**, 1–4 (1980)

6.31 B.A. Huberman, J.P. Crutchfield, N.H. Packard: Appl. Phys. Lett. **37**, 750–752 (1980)

6.32 B.A. Huberman, J.P. Crutchfield: Phys. Rev. Lett. **43**, 1743–1747 (1979)

6.33 J.P. Crutchfield, J.D. Farmer, B.A. Huberman: Phys. Rep. **92**, 45–82 (1982)

6.34 B. Shraiman, C.E. Wayne, P.C. Martin: Phys. Rev. Lett. **46**, 935–939 (1981)

6.35 Y. Pomeau, P. Manneville: Commun. Math. Phys. **74**, 189–197 (1980)

6.36 P. Manneville, Y. Pomeau: Physica D **1**, 219–226 (1980)

6.37 V.S. Afraimovich, L.P. Shilnikov: Sov. Math.–Dokl. **15**, 1761 (1974)

6.38 V.I. Luk'yanov, L.P. Shilnikov: Sov. Math.–Dokl. **19**, 1314 (1978)

6.39 B. Hu: "Functional renormalization group equations approach to the transition to chaos", in *Chaos and Statistical Methods*, ed. by Y. Kuramoto (Springer, Berlin, Heidelberg 1984) pp. 72–82

6.40 B. Hu, J. Rudnick: Phys. Rev. Lett. **48**, 1645–1648 (1982)

6.41 B. Hu, J. Rudnick: Phys. Rev. A **26**, 3035–3036 (1982)

6.42 P. Berge, M. Dubois, P. Manneville, Y. Pomeau: J. Physique Lett. (Paris) **41**, L341–L345 (1980)

6.43 M. Dubois, M.A. Rubio, P. Berge: Phys. Rev. Lett. **51**, 1446–1449 (1983)

6.44 C. Vidal: "Dynamical instabilities observed in the Belousov-Zhabotinsky system", in *Chaos and Order in Nature*, ed. by H. Haken (Springer, Berlin, Heidelberg 1981) pp. 68–82

6.45 Y. Pomeau, J.C. Roux, A. Rossi, C. Vidal: J. Physique Lett. (Paris) **42**, L271–L272 (1981)

6.46 W.L. Ditto, S. Rauseo, R. Cawley et al.: Phys. Rev. Lett. **63**, 923–926 (1989)

6.47 R.K. Tavakol, A.J. Tworkowski: Phys. Lett. A **126**, 318–324 (1988)

6.48 J.-Y. Huang, J.J. Kim: Phys. Rev. A **36**, 1495–1497 (1987)

6.49 D.J. Biswas, Vas Dev, U.K. Chatterjee: Phys. Rev. A **35**, 456–458 (1987)

6.50 K. Kaneko: Progr. Theor. Phys. **72**, 202–215 (1984)

6.51 M.R. Bassett, J.L. Hudson: Physica D **35**, 289–298 (1989)

6.52 K. Kaneko: *Collapse of Tori and Genesis of Chaos in Dissipative Systems* (World Scientific, Singapore 1986)

6.53 V.S. Anishchenko, T.E. Letchford, M.A. Safonova: Radiophys. and Quantum Electron. **27**, 381–390 (1984)

6.54 A.I. Goldberg, Ya.G. Sinai, K.M. Khanin: Russian Math. Surveys **38:1**, 187 (1983)

6.55 B. Hu, I.I. Satija: Phys. Lett. A **98**, 143–146 (1983)

6.56 H. Dadio: Progr. Theor. Phys. **70**, 879–882 (1983)

6.57 M. Sano, Y. Sawada: "Chaotic attractors in Rayleigh-Benard Systems", in *Chaos and Statistical Methods*, ed. by Y. Kuramoto (Springer, Berlin, Heidelberg 1984) pp. 226–231

6.58 E. Ott, C. Grebogi, J.A. Yorke: Phys Rev. Lett. **64**, 1196 (1990)

6.59 T. Shinbrot, C. Grebogi, E. Ott, J.A. Yorke: Nature **363**, 411 (1993)

6.60 D. Auerbach, E. Ott, C. Grebogi, J.A. Yorke: Phys. Rev. Lett. **69**, 3479 (1992)

6.61 T. Shinbrot, E. Ott, C. Grebogi, J.A. Yorke: Phys. Rev. A **45**, 4165 (1992)

6.62 T. Shinbrot, C. Grebogi, E. Ott, A. Yorke: Phys. Lett. A **169**, 349 (1992)

6.63 B.A. Huberman, E. Lumer: IEEE Trans. Circ. Systs. **37**, 547 (1990)

6.64 A. Hübler, R. Georgii, M. Kuckler, W. Stelyl, E. Lüscher: Helv. Phys. Acta **61**, 897 (1988)

6.65 E.A. Jackson, A. Hübler: Physica D **44**, 407 (1990)

6.66 V.V. Alexeev, A. Yu. Loskutov: Sov. Phys.–Dokl. **32**, 270 (1987)

6.67 A. Yu. Loskutov, A.I. Shishmarev: Chaos **4**, 391 (1994)

6.68 A. Yu. Loskutov: J. Phys. A **26**, 4581 (1993)

6.69 N.L. Komarova, A. Yu. Loskutov: Matem. Modelir. **7**, 133 (1995)

6.70 I.M. Starobinets, A.S. Pikovsky: Phys. Lett. A **181**, 149 (1993)

6.71 R. Lima, M. Pettini: Phys. Rev. A **41**, 726 (1990)

6.72 Yu. S. Kivshar, F. Rödelsperger, H. Benner: Phys. Rev. E **49**, 319 (1994)

6.73 Ph. V. Bayly, L.N. Virgin: Phys. Rev. E **50**, 604 (1994)

6.74 K. Pyragas: Phys. Lett. A **170**, 421 (1992)

6.75 K. Pyragas: Z. Naturforsch. A **48**, 629 (1993)

6.76 D. Vassiliadis: Physica D **71**, 319 (1994)

6.77 S. Bielawski, D. Derozier, P. Glorieux: Phys. Rev. E **49**, 971 (1994)

Chapter 7

7.1 A.V. Babin, M.I. Vishik: Russian Math. Surveys **38**, 151 (1983)

7.2 D. Ruelle: Commun. Math. Phys. **93**, 285–300 (1984)

7.3 O.A. Ladyzhenskaya: Russian Math. Surveys **42**, 27 (1987)

7.4 P.R. Gromov, A.B. Zobnin, M.I. Rabinovich, A.M. Reiman, M.M. Suschik: Sov. Phys.–Dokl. **32**, 9 (1987)

7.5 P. Constantin, C. Foias, B. Nikolaenko, R. Temam: *Integral Manifolds for Dissipative Partial Differential Equations* (Springer, Berlin, Heidelberg 1989)

7.6 P. Constantin, C. Foias, G.D. Gibbon: Nonlinearity **2**, 241 (1989)

7.7 R. Temam: "Attractors for Navier–Stokes equations", in *Nonlinear Differential equations and Their Applications* (Pitman, London 1985) pp. 272–292

7.8 D.V. Lyubimov, G.F. Putin, V.I. Chernatynskii: Sov. Phys.–Dokl. **22**, 360 (1977)

7.9 M.I. Rabinovich: Sov. Phys.–Usp. **21**, 443 (1978)

7.10 H. Haken: *Synergetics* (Springer, Berlin, Heidelberg 1978)

7.11 N.H. Packard, J.P. Crutchfield, J.D. Farmer, R.S. Shaw: Phys. Rev. Lett. **45**, 712–716 (1980)

7.12 F. Takens: "Detecting strange attractors in turbulence", in *Dynamical Systems and Turbulence*, Lecture Notes in Mathematics, Vol. 898 (Springer, Berlin, Heidelberg 1981) pp. 366–381

7.13 M.V. Hirsh: *Differential Topology* (Springer, Berlin, Heidelberg 1989)

7.14 J.D. Farmer: Physica D **4**, 366–393 (1982)

7.15 R. Mane: "On the dimension of the compact invariant sets of certain nonlinear maps", in *Dynamical Systems and Turbulence*, Lecture Notes in Mathematics, Vol. 898 (Springer, Berlin, Heidelberg 1981) pp. 230–242

7.16 S.N. Lukashchuk, A.A. Predtechenskii, G.E. Falkovich, A.I. Chernykh: "Calculation of attractor dimensions from experimental data", preprint 280, Inst. Automatics and Electrometry, Novosibirsk 1985

7.17 D.S. Broomhead, G.P. King: Physica D **20**, 217–236 (1986)

7.18 A.I. Smirnov: Sov. Phys.–JETP **67**, 969 (1988)

7.19 P. Grassberger, I. Procaccia: Phys. Rev. Lett. **50**, 346–349 (1983)

7.20 P. Grassberger, I. Procaccia: Physica D **9**, 189–208 (1983)

7.21 B. Malraison, P. Atten, P. Berge, M. Dubois: J. Physique Lett. (Paris) **44**, L897–L902 (1983)

7.22 A. Babloyantz, J.N. Salazar, G. Nicolis: Phys. Lett. A **111**, 152–156 (1985)

7.23 A.M. Fraser, H.L. Swinney, Phys. Rev. A **33**, 1134–1140 (1986)

7.24 S. Sato, M. Sano, Y. Sawada: Progr. Theor. Phys. **77**, 1–5 (1987)

7.25 G. Mayer-Kress (ed.): *Dimensions and Entropies in Chaotic Systems* (Springer, Berlin, Heidelberg 1986)

7.26 N.B. Abraham, A.M. Albano, B. Das et al.: Phys. Lett. A **114**, 217–221 (1986)

7.27 J. Theiler: Phys. Rev. A **36**, 4456–4462 (1987)

7.28 P. Grassberger, T. Schreiber, C. Schaffrath: Int. J. Bifurcation and Chaos **1**, 521–543 (1991)

7.29 J. Dvorak, J. Klaschka: Phys. Lett. A **145**, 225–231 (1990)

7.30 I. Aranson, A.M. Reiman, V.G. Shekhov: "Measurement methods for correlation dimensions", in *Nonlinear Waves 2. Dynamics and Evolution*, ed. by J. Engeibrecht, A.V. Gaponov-Grekhov, M.I. Rabinovich (Springer, Berlin, Heidelberg 1989) pp. 29–33

7.31 K. Pawelzik, H.G. Schuster: Phys. Rev. A **35**, 975–977 (1984)

7.32 A. Wolf, J. Swift, H.L. Swinney, J. Vastano: Physica D **16**, 285–317 (1985)

7.33 A. Ben-Mizrachi, I. Procaccia, P. Grassberger: Phys. Rev. A **29**, 975–977 (1984)

7.34 R. Cerf: "Attractor-ruled dynamics in neurobiology: Does it exist? Can it be measured?" in *Interdisciplinary Approaches to Nonlinear Complex Systems*, ed. by H. Haken, A. Mikhailov (Springer, Berlin, Heidelberg 1993) pp. 201–214

7.35 H. Haken: "Spatial and temporal patterns formed by systems far from equilibrium", in *Nonequilibrium Dynamics in Chemical Systems*, ed. by C. Vidal, A. Pacault (Springer, Berlin, Heidelberg 1984) pp. 7–21

7.36 A.V. Gaponov-Grekhov, M.I. Rabinovich: "Dynamic chaos in ensembles of structures and spatial development of turbulence in unbounded systems", in *Self-Organization by Nonlinear Irreversible Processes*, ed. by W. Ebeling, H. Ulbricht (Springer, Berlin, Heidelberg 1984) pp. 37–46

7.37 Q. Quyang, H.L. Swinney: Chaos **1**, 411 (1991)

7.38 Q. Quyang, H.L. Swinney: "Onset and beyond Turing pattern formation", in *Chemical Waves and Patterns*, ed. by R. Kapral, K. Showalter (Kluwer, Amsterdam 1995) pp. 269–298

7.39 A.S. Mikhailov: *Foundations of Synergetics I. Distributed Active Systems*, 2nd revised edition (Springer, Berlin, Heidelberg 1994)

7.40 J.J. Perraud, K. Agladze, E. Dulos, P. De Kepper: Physica A **188**, 1 (1992)

7.41 J. Boissonade, E. Dulos, P. De Kepper: "Turing patterns: From myth to reality", in *Chemical Waves and Patterns*, ed. by R. Kapral, K. Showalter (Kluwer, Amsterdam 1995) pp. 221–268

7.42 J.E. Pearson: Science **261**, 189 (1993)

7.43 K.-J. Lee, W.D. McCormick, J.E. Pearson, H.L. Swinney: Nature **369**, 215 (1994)

7.44 K. Krischer, A. Mikhailov: Phys. Rev. Lett. **73**, 3165–3168 (1994)

7.45 J.R. Mines: Trans. Roy. Soc. Can. **4**, 43–53 (1914)

7.46 N. Wiener, R. Rosenblueth: Arch. Inst. Cardiol. Mex. **16**, 205 (1946)

7.47 G.K. Moe, W.C. Rheinboidt, J.A. Abidskov: Am. Heart J. **67**, 338 (1964)

7.48 V.I. Krinsky,. In *Problemy Kibernetiki* (Problems of Cybernetics), Vol. 20 (Nauka, Moscow 1968) p. 59

7.49 V.I. Krinsky, A.S. Mikhailov: *Avtovolny* (Autowaves) (Znanie, Moscow 1984)

7.50 K.I. Agladze, V.I. Krinsky, A.M. Pertsov: Nature **308**, 834–835 (1984)

7.51 J. Maselko, K. Showalter: Physica D **49**, 21 (1991)

7.52 A.S. Mikhailov, V.I. Zykov, Physica D **52**, 379 (1991)

7.53 A.V. Panfilov, A. Holden: Phys. Lett. A **151**, 23 (1990)

7.54 A.V. Panfilov, P. Hogeweg: Phys. Lett. A **156**, 295 (1993)

7.55 M. Courtemanche, A.T. Winfree: Int. J. Bifurcation and Chaos **1**, 219 (1991)

7.56 A. Karma: Phys. Rev. Lett. **71**, 1103 (1993)

7.57 M. Bär, M. Eiswirth: Phys. Rev. E **48**, R1635 (1993)

7.58 O.A. Druzhinin, A. S. Mikhailov: Phys. Lett. A **148**, 429–433 (1990)

7.59 K. Kaneko: Physica D **34**, 1–41 (1989)

7.60 L.A. Bunimovich, Ya. G. Sinai: Nonlinearity **1**, 491–516 (1988)

7.61 W. Eckhaus: *Studies in Nonlinear Stability Theory* (Springer, Berlin, Heidelberg 1965)

7.62 B. Janiaud, A. Pumir, D. Bensimon, V. Croquette, H. Richter, L. Kramer: Physica D **55**, 269–286 (1992)

7.63 K. Nozaki, N. Bekki: J. Phys. Soc. Jap. **53**, 1581–1582 (1984)

7.64 N. Bekki, K. Nozaki: Phys. Lett. A **110**, 133–135 (1985)

7.65 A.C. Newel: In Lectures in Applied Mathematics, Vol. 15 (Springer, Berlin, Heidelberg 1974) p.157

7.66 H. Sakaguchi: Prog. Theor. Phys. **85**, 417 (1991)

7.67 H. Chate, P. Manneville: Phys. Lett. A **171**, 183–188 (1992)

7.68 S. Popp, O. Stiller, I. Aranson, L. Kramer: Physica D **84**, 398 (1995)

7.69 M. Bazhenov, M.I. Rabinovich: Phys. Lett. A **179**, 191–197 (1993)

7.70 P.C. Hagan: SIAM J. Appl. Math. **2**, 400 (1981)

7.71 P. Ortoleva, J. Ross: J. Phys. Chem. **58**, 5673–5680 (1973)

7.72 Y. Kuramoto: Prog. Theor. Phys. Suppl. **64**, 346–362 (1978)

7.73 Y. Kuramoto: *Chemical Oscillations, Waves, and Turbulence* (Springer, Berlin, Heidelberg 1984)

7.74 P. Coullet, L. Gil, J. Lega: Physica D **37**, 91–103 (1989)

7.75 Y. Kuramoto, T. Tsuzuki: Prog. Theor. Phys. **54**, 687–699 (1975); **55**, 356–369 (1976)

7.76 G. Sivashinsky: Acta Astronautica **4**, 1177–1206 (1977)

7.77 J.M. Hyman, B. Nicoiaenko, S. Zaleski: Physica D **23**, 265 (1986)

7.78 V. Yakhot: Phys. Rev. A **24**, 642–644 (1981)

7.79 S. Zaleski: Physica D **34**, 427–438 (1989)

7.80 K. Sneppen, J. Krug, M.H. Jensen, C. Jayaprakash, T. Bohr: Phys. Rev. A **46**, 7351–7354 (1992)

7.81 F. Hayot, C. Jayaprakash, Ch. Josserand: Phys. Rev. E **47**, 911 (1993)

7.82 M. Kardar, G. Parisi, Y. Zhang: Phys. Rev. Lett. **56** 889–892 (1986)

7.83 D. Forster, D. Nelson, M.J. Stephen: Phys. Rev. A **16**, 732 (1977)

7.84 J.M. Kosterlitz, D.J. Thouless: Prog. Low Temp. Phys. **78**, 371 (1978)

7.85 A.S. Mikhailov, I.V. Uporov: Dokl. Akad. Nauk SSSR **249**, 733–736 (1979)

7.86 Ya.B. Zeldovich, B.A. Malomed: Sov. Phys.–Dokl. **25**, 721–723 (1980)

7.87 P. Coullet, L. Gil, J. Lega: Phys. Rev. Lett. **62**, 1619–1622 (1989)

7.88 B.I. Shraiman, A. Pumir, W. van Saarlos, P.C. Hohenberg, H. Chate, M. Holen: Physica D **57**, 241–248 (1992)

7.89 M.V. Bazhenov, M.I. Rabinovich, A.L. Fabrikant: Phys. Lett. A **163**, 87–94 (1992)

7.90 H. Chate: Nonlinearity **7**, 185–204 (1994)

7.91 D.A. Egolf, H.S. Greenside: Phys. Rev. Lett. **74**, 1751–754 (1995)

7.92 F. Mertens, R. Imbihl, A. Mikhailov: J. Chem. Phys. **101**, 9903–9908 (1994)

7.93 Y. Pomeau: Physica D **23**, 3 (1986)

7.94 M. Bazhenov, M. Rabinovich: Physica D **73**, 318 (1994)

7.95 D.S. Cohen, J.C. Neu, R.R. Rosales: SIAM J. Appl. Math. **35**, 536 (1978)

7.96 T. Erneux, M. Herchkowitz-Kaufman: Bull. Math. Bio. **41**, 767 (1979)

7.97 J.M. Greenberg: SIAM J. Appl. Math. **39**, 301 (1980)

7.98 J.M. Greenberg: Adv. Appl. Math. **2**, 450 (1981)

7.99 Ya.B. Zeldovich, B.A. Malomed: Sov. Phys.–Dokl. **25**, 721 (1980)

7.100 G.T. Gurija, M.A. Lifshits: Phys. Lett. A **97**, 175 (1983)

7.101 P. Coullet, L. Gil, J. Lega: Phys. Rev. Lett. **62**, 1619 (1989)

7.102 L. Gil, J. Lega, J.L. Meunier: Phys. Rev. A **41**, 1138 (1990)

7.103 P. L. Ramazza, S. Residori, G. Giacomelli, F.T. Arecchi: Europhys. Lett. **19**, 475 (1992)

7.104 M. Hildebrand, M. Bär, M. Eiswirth: Phys. Rev. Lett. **75**, 1503 (1995)

7.105 M. Hildebrand: Diploma Thesis, Free University of Berlin, 1995

7.106 J.P. Crutchfield, K. Kaneko: Phys. Rev. Lett. **60**, 2715 (1988)

7.107 A. Politi, R. Livi, G.-L. Opp, R. Kapral: Europhys. Lett. **22**, 571 (1993)

7.108 B.I. Shraiman: Phys. Rev. Lett. **57**, 325 (1986)

7.109 A. Wacker, S. Bose, E. Schöll: Europhys. Lett. **31**, 257 (1995)

7.110 S. Wolfram: Rev. Mod. Phys. **55**, 601 (1983)

Chapter 8

8.1 S. Wolfram: Physica D **22**, 385–399 (1986)

8.2 S.A. Kauffman: Physica D **10**, 145–156 (1984)

8.3 G. Grinstein, C. Jayaprakash, Yu. He: Phys. Rev. Lett. **55**, 2527–2530 (1985)

8.4 E. Domany, W. Kinzel: Phys. Rev. Lett. – 1984, V. 53, p. 311–314

8.5 W. Kinzel: Z. Phys. B **58**, 229–244 (1985)

8.6 G.Y. Vichniac: "Cellular automata models of disorder and organization", in *Disordered Systems and Biological Organization*, ed. by E. Bienenstock et al. (Springer, Berlin, Heidelberg 1986) pp. 3–20

8.7 O.A. Druzhinin, A.S. Mikhailov: Izv. VUZ. Radiofizika **32**, 444–450 (1989)

8.8 A.T. Bharucha-Reid: *Elements of the Theory of Markov Processes and Their Applications* (McGraw-Hill, New York 1960)

8.9 W. Feller: *An Introduction to Probability Theory and its Applications*, Vols. 1 and 2 (Wiley, New York 1968 and 1971)

8.10 M. Loeve: *Probability Theory*, Vols. 1 and 2 (Springer, Berlin, Heidelberg 1977 and 1978)

8.11 R. von Mises: *Mathematical Theory of Probability and Statistics* (Academic, New York 1964)
8.12 Yu.V. Prokhorov, Yu.A. Rozanov: *Probability Theory* (Springer, Berlin, Heidelberg 1968)
8.13 C.W. Gardiner: *Handbook of Stochastic Methods* (Springer, Berlin, Heidelberg 1985)
8.14 H. Haken: *Synergetics, An Introduction*, 3rd ed. (Springer, Berlin, Heidelberg 1983)
8.15 J.L. Doob: *Stochastic Processes* (Wiley, New York 1953)
8.16 R.L. Stratonovich: *Topics in the Theory of Random Noise*, Vols. I and II (Gordon and Breach, New York 1963 and 1967)
8.17 N.G. van Kampen: *Stochastic Processes in Physics and Chemistry* (North-Holland, Amsterdam 1983)
8.18 H. Risken: *The Fokker-Planck Equation*, 2nd ed. (Springer, Berlin, Heidelberg 1989)
8.19 W. Horsthemke, R.L. Lefever: *Noise-Induced Transitions* (Springer, Berlin, Heidelberg 1984)
8.20 F. Moss, P.V.E. McClintock (eds.): *Noise in Nonlinear Dynamical Systems; Theory, Experiment, Simulations* (3 volumes) (Cambridge University Press, Cambridge 1989)
8.21 R.F. Pawula: Phys. Rev. **162**, 186 (1967)

Chapter 9

9.1 G.E. Uhlenbeck, L.S. Ornstein: Phys. Rev. **36**, 823 (1930)
9.2 S. Chandrasekhar: Rev. Mod. Phys. **15**, 1 (1943)
9.3 T. Hida: *Brownian Motion* (Springer, Berlin, Heidelberg 1980)
9.4 R.L. Stratonovich: *Topics in the Theory of Random Noise*, Vols. I and II (Gordon and Breach, New York 1963 and 1967)
9.5 L. Arnold: *Stochastic Differential Equations: Theory and Applications* (Wiley, New York 1974)
9.6 Z. Schuss: *Theory and Applications of Stochastic Differential Equations* (Wiley, New York 1980)
9.7 N.G. van Kampen: *Stochastic Processes in Physics and Chemistry* (North-Holland, Amsterdam 1983)
9.8 C.W. Gardiner: *Handbook of Stochastic Methods* (Springer, Berlin, Heidelberg 1985)
9.9 M. San Miguel, J.M. Sancho: "Langevin equations with coloured noise", in *Noise in Nonlinear Dynamical Systems; Theory, Experiment, Simulations*, Vol. 1, ed. by F. Moss, P.V.E. McClintock (Cambridge University Press, Cambridge 1989) pp. 110–160
9.10 R.L. Stratonovich: SIAM J. Control **4**, 362 (1966)
9.11 K. Ito: Nagoya Math. J. **1**, 35 (1950)
9.12 E. Wong, M. Zakai: Ann. Math. Stat. **36**, 1560 (1965)
9.13 W. Horsthemke, R.L. Lefever: *Noise-Induced Transitions* (Springer, Berlin, Heidelberg 1984)
9.14 O.A. Druzhinin, A.S. Mikhailov: Izv. VUZ. Radiofizika **32**, 444–450 (1989)
9.15 M. Doi: J. Phys. A **9**, 1465–1477 (1976)
9.16 Ya.B. Zeldovich, A.A. Ovchinnikov: Sov. Phys. JETP **47**, 829 (1978)
9.17 A.S. Mikhailov: Phys. Lett. A **85**, 214 (1981); A **85**, 427 (1981)
9.18 A.M. Gutin, A.S. Mikhailov, V.V. Yashin: Sov. Phys. JETP **65**, 535 (1987)
9.19 A.S. Mikhailov: Phys. Rep. **184**, 308–374 (1989)
9.20 P.S. Martin, E.D. Siggia, H.A. Rose: Phys. Rev. A **8**, 423 (1973)
9.21 R. Graham: In Lecture Notes in Physics, Vol. 84 (Springer, Berlin, Heidelberg 1978) p. 82
9.22 H.K. Janssen: Z. Phys. B **23**, 377 (1976)
9.23 R. Phytian: J. Phys. A **10**, 777 (1977)

9.24 P. Hänggi: "Noise in continuous dynamical systems: A functional calculus approach", in *Noise in Nonlinear Dynamical Systems; Theory, Experiment, Simulations*, Vol. 1, ed. by F. Moss, P.V.E. McClintock (Cambridge University Press, Cambridge 1989) pp. 307–328

9.25 L. Pesquera, H.A. Rodrigues, E. Santos: Phys. Lett. A **94**, 287 (1983)

9.26 A. Förster, A.S. Mikhailov: "Application of path integrals to stochastic reaction-diffusion equations", in *Selforganization by Nonlinear Irreversible Processes*, ed. by W. Ebeling, H. Ulbricht (Springer, Berlin, Heidelberg 1986) pp. 89–94

9.27 M.I. Dykman: Phys. Rev. A **42**, 2020 (1990)

9.28 A. Förster, A.S. Mikhailov: Phys. Lett. A **126**, 459 (1988)

9.29 R. Landauer: "Noise-activated escape from metastable states: A historical review", in *Noise in Nonlinear Dynamical Systems; Theory, Experiment, Simulations*, Vol. 1, ed. by F. Moss, P.V.E. McClintock (Cambridge University Press, Cambridge 1989) pp. 1–15

9.30 P. Hänggi, P. Talkner, M. Borkovec: Rev. Mod. Phys. **62**, 251 (1990)

Chapter 10

10.1 M. Kardar, G. Parisi, Y.C. Zhang: Phys. Rev. Lett. **56**, 889 (1986)

10.2 N.S. Goel, N. Richter-Dyn: *Stochastic Models in Biology* (Academic, New York 1974)

10.3 G. Nicolis, F. Baras (eds.): *Chemical Instabilities* (Reidel, Dordrecht 1984)

10.4 B. Zeldovich, A. S. Mikhailov: Sov. Phys. Usp. **30**, 977 (1988)

10.5 A.N. Kolmogorov: Izv. Akad. Nauk SSSR, Mat. No. 3, 355 (1937)

10.6 Ya.B. Zeidovich, S.A. Molchanov, A.A. Ruzmaikin, D.D. Sokolov: Sov. Phys. Usp. **30**, 353 (1987)

10.7 D.D. Sokolov, T.S. Shumkina: Vestn. Mosk. Univ. Fiz. Astron. **29**, 23 (1988)

10.8 L.D. Landau, E.M. Lifshitz: *Quantum Mechanics. Nonrelativistic Theory* (Pergamon, Oxford 1977)

10.9 F.A. Berezin, G.P. Pokhil, V.M. Filkenberg: Vestn. Mosk. Univ. **1**, 21 (1964); see also L.D. Faddeev, in *Solitons*, ed. by R.K. Bullough, P.J. Cauldrey (Springer, Berlin, Heidelberg 1980) p. 339

10.10 A.S. Mikhailov: J. Phys. A **24**, L757 (1991)

10.11 A.S. Mikhailov: Physica A **188**, 367 (1992)

10.12 V.E. Zakharov, A.B. Shabat: Sov. Phys. JETP **61**, 118 (1971)

10.13 A.S. Mikhailov, I.V. Uporov: Sov. Phys. Usp. **27**, 695 (1984)

10.14 A.S. Mikhailov: Phys. Rep. **184**, 307 (1989)

10.15 H.W. Wyld: Ann. Phys. **14**, 143 (1961)

10.16 A.-H. Nayfeh: *Perturbation Methods* (Wiley, New York 1973)

10.17 A.S. Mikhailov, I.V. Uporov: Sov. Phys. JETP **57**, 863 (1983)

10.18 D. Stauffer: *Introduction to Percolation Theory* (Taylor and Francis, London 1985)

10.19 S.P. Obukhov: Physica A **101**, 145 (1980)

10.20 T. Sanada: Phys. Rev. A **44**, 6480 (1991)

10.21 D. Zanette, A.S. Mikhailov: Phys. Rev. E **50**, 1638 (1994)

Chapter 11

11.1 B. Felderhof, I. Deutch: J. Chem. Phys. **64**, 4551–4558 (1976)

11.2 J.R. Lebenhaft, R. Kapral: J. Stat. Phys. B **20**, 125 (1979)

11.3 M. Bixon, R. Zwanzig: J. Chem. Phys. **75**, 2354 (1981)

11.4 M. Muthukumar, R.J. Cukier: J. Stat. Phys. **26**, 453 (1981)

11.5 B. Felderhof, I. Deutch, U. Titulauer: J. Chem. Phys. **76**, 4178–4184 (1982)

11.6 T.R. Kirkpatrick: J. Chem. Phys. **76**, 4255–4259 (1982)

11.7 M. Tokuyama, R.I. Cukier: J. Chem. Phys. **76**, 6202 (1982)

11.8 B.Ya. Balagurov, V.G. Vaks: Sov. Phys. JETP **38**, 968 (1974)

11.9 P. Grassberger, I. Procaccia: J. Chem. Phys. **77**, 6281–6284 (1982)

11.10 I.M. Lifshitz, S.A. Gredeskul, L.A. Pastur: *Introduction to the Theory of Disordered Systems* (Nauka, Moscow 1982)

11.11 I.M. Lifshitz: Sov. Phys. Usp. **7**, 549 (1965)

11.12 I.M. Lifshitz: Sov. Phys. JETP **53**, 462 (1967)

11.13 A.S. Mikhailov, I.V. Uporov: Sov. Phys. Usp. **27**, 695 (1984)

11.14 R.F. Kayser, J.B. Hubbard: Phys. Rev. Lett. **51**, 79–82 (1983)

11.15 R.F. Kayser, J.B. Hubbard: J. Chem. Phys. **80**, 1127–1130 (1984)

11.16 A. Blumen, J. Klafter, G. Zumofen: Phys. Rev. B **28**, 6112–6115 (1983)

11.17 J. Klafter, A. Blumen: J. Chem. Phys. **80**, 875–877 (1984)

11.18 T.R. Waite: Phys. Rev. **107**, 463–470 (1957)

11.19 S.A. Rice: *Diffusion-Limited Reactions* (Elsevier, Amsterdam 1985)

11.20 Ya.B. Zeldovich: Sov. Electrochem. **13**, 581 (1977)

11.21 A.A. Ovchinnikov, Ya.B. Zeldovich: Chem. Phys. **28**, 215 (1978)

11.22 D. Toussaint, F. Wilczec: J. Chem. Phys. **78**, 2642–2647 (1983)

11.23 S.F. Burlatskii, A.A. Ovchinnikov: Sov. Phys. JETP **65**, 908 (1987)

11.24 Ya.B. Zeldovich, A.S. Mikhailov: Sov. Phys. Usp. **30**, 977–992 (1987)

11.25 Ya.B. Zeldovich: Zh. Tekh. Fiz. **19**, 1199 (1949)

11.26 I.M. Sokolov: JETP Lett. **44**, 67 (1986)

11.27 Ya.B. Zeldovich, A.S. Mikhailov: "Chirality wave propagation and the first evolutionary catastrophe", in *Thermodynamics and Pattern Formation in Biology*, ed. by I. Lamprecht, A.I. Zotin (de Gruyter, Berlin 1988) pp. 57–68

11.28 A.S. Mikhailov: Phys. Rep. **184**, 308–374 (1989)

11.29 Ya.B. Zeldovich, A.A. Ovchinnikov: Sov. Phys. JETP **65**, 535 (1978)

11.30 Ya.B. Zeldovich, A.A. Ovchinnikov: JETP Lett. **26**, 440 (1977)

11.31 S.F. Burlatskii: *Fluctuational Kinetics of Diffusional Processes*, Dr. Sc. Thesis (Institute for Chemical Physics, Moscow 1988)

11.32 S.F. Burlatskii, A.A. Ovchinnikov, G.S. Oshanin: Sov. Phys. JETP **68**, 1153 (1989)

11.33 G.S. Oshanin, S.F. Burlatskii, A.A. Ovchinnikov: J. Phys. A **22**, 977 (1989)

11.34 A.A. Ovchinnikov, S.F. Burlatskii: Sov. Phys. JETP Lett. **43**, 638 (1986)

11.35 A.M. Gutin, A.S. Mikhailov, V.V. Yashin: Sov. Phys. JETP **65**, 535 (1987)

11.36 Y.C. Zhang: Phys. Rev. Lett. **59**, 1726–1728 (1987)

11.37 V.L. Vinetskii, A.V. Kondrachuk: Ukr. Fiz. Zh. **27**, 383–387 (1982)

11.38 Yu.H. Kalnins, Yu.Yu. Krikis: Izv. Akad. Nauk Latv. SSR. Ser. Fiz.-Tekh. Nauk. No. 1, 104–106 (1983)

11.39 F.V. Pirogov, E.I. Palagishvili: Izv. Akad. Nauk Latv. SSR. Ser. Fiz.-Tekh. Nauk. No. 4, 46–53 (1984)

11.40 L.W. Anacker, R. Kopelman: Phys. Rev. Lett. **58**, 289–291 (1987)

Chapter 12

12.1 Ya.B. Zeldovich, A.S. Mikhailov: Sov. Phys. Usp. **30**, 977–992 (1987)

12.2 H. Haken: Rev. Mod. Phys. **47**, 67 (1975)

12.3 A. Nitzan, P. Ortoleva, J. Deutch, J. Ross: J. Chem. Phys. **61**, 1056 (1974)

12.4 D. Walgraef, G. Dewel, P. Borckmans: Adv. Chem. Phys. **49**, 311 (1982)

12.5 S. K. Ma: *Modern Theory of Critical Phenomena* (Benjamin, Reading 1976)

12.6 S. Grossmann, A.S. Mikhailov: Z. Phys. B **78**, 1–10 (1990)

12.7 M.A. Shishkova: Sov. Math.–Dokl. **14**, 483 (1973)

12.8 N.R. Lebovitz, R.J. Schaar: Stud. Appl. Math. **54**, 229–260 (1975)

12.9 R. Haberman: SIAM J. Appl. Math. **37**, 69–105 (1979)

12.10 P. Mandel, H. Zeghlache, T. Erneux: "Time-dependent phase transitions", in *Far from Equilibrium Phase Transitions*, ed. by L. Garrido (Springer, Berlin, Heidelberg 1988) pp. 217–235

12.11 L.D. Landau, I.M. Lifshitz: *Statistical Physics* (Pergamon, Oxford 1980)

12.12 G. Nicolis, I. Prigogine: Proc. Natl. Acad. Sci. USA **78**, 659 (1981)

12.13 D. Kondepudi, I. Prigogine: Physica A **107**, 1 (1987)

12.14 L.L. Morozov, V.I. Goldanskii: "Violation of symmetry and self-organization in prebiological evolution", in *Self-Organization. Autowaves and Structures Far from Equilibrium*, ed. by V.I. Krinsky (Springer, Berlin, Heidelberg 1984) pp. 224–231

12.15 D. Kondepudi, G.W. Nelson: Phys. Rev. Lett. **50**, 1013 (1983)

12.16 F. Moss, D. Kondepudi, P.V.F. McClintock: Phys. Lett. A **112**, 293 (1985)

12.17 G. Broggi, A. Colombo, L.A. Lugiato, P. Mandel: Phys. Rev. A **33**, 3635–3637 (1986)

12.18 H. Zeghlache, P. Mandel, C. van den Broeck: Phys. Rev. A **40**, 286–294 (1989)

12.19 W. Horsthemke, M. Malek-Mansour: Z. Phys. B **24**, 307 (1976)

12.20 A.S. Mikhailov, I.V. Uporov: Sov. Phys. JETP **52**, 989 (1980)

12.21 A.S. Mikhailov, I.V. Uporov: Sov. Phys. JETP **57**, 863 (1983)

12.22 A.S. Mikhailov, I.V. Uporov: Sov. Phys. Usp. **27**, 695 (1984)

12.23 A.S. Mikhailov: Phys. Rep. **184**, 307 (1989)

12.24 V.R. Chechetkin, V.S. Lutovinov: Nuovo Cimento B **99**, 103–116 (1987)

12.25 V.R. Chechetkin, V.S. Lutovinov: Z. Phys. B **69**, 129–135 (1987)

12.26 H.K. Janssen: Z. Phys. B **42**, 151 (1981)

12.27 H.K. Janssen: Z. Phys. B **58**, 311 (1985)

12.28 L. Arnold, W. Horsthemke, R. Lefever: Z. Phys. B **29**, 867 (1978)

12.29 W. Horsthemke, R. Lefever: *Noise-Induced Transitions* (Springer, Berlin, Heidelberg 1984)

12.30 A.S. Mikhailov: Phys. Lett. A **73**, 143 (1979)

12.31 A.S. Mikhailov: Z. Phys. B **41**, 277 (1981)

12.32 A.S. Mikhailov: "Fluctuations and critical phenomena in reaction-diffusion systems", in *Far from Equilibrium Phase Transitions*, ed. by L. Garrido (Springer, Berlin, Heidelberg 1988) p. 139

Subject Index